Occurrence, Distribution and Toxic Effects of Emerging Contaminants

I0050857

Editors

Uma Shanker

Department of Chemistry
Dr B R Ambedkar National Institute of Technology Jalandhar
Punjab, India

Manviri Rani

Department of Chemistry
Malaviya National Institute of Technology Jaipur
Rajasthan, India

CRC Press
Taylor & Francis Group
Boca Raton London New York

CRC Press is an imprint of the
Taylor & Francis Group, an **informa** business

A SCIENCE PUBLISHERS BOOK

Cover credit: Thank you to Dr. Manviri Rani and Mrs. Jyoti Yadav for designing the cover image.

First edition published 2024
by CRC Press
2385 NW Executive Center Drive, Suite 320, Boca Raton FL 33431

and by CRC Press
4 Park Square, Milton Park, Abingdon, Oxon, OX14 4RN

© 2024 Uma Shanker and Manviri Rani

CRC Press is an imprint of Taylor & Francis Group, LLC

Reasonable efforts have been made to publish reliable data and information, but the author and publisher cannot assume responsibility for the validity of all materials or the consequences of their use. The authors and publishers have attempted to trace the copyright holders of all material reproduced in this publication and apologize to copyright holders if permission to publish in this form has not been obtained. If any copyright material has not been acknowledged please write and let us know so we may rectify in any future reprint.

Except as permitted under U.S. Copyright Law, no part of this book may be reprinted, reproduced, transmitted, or utilized in any form by any electronic, mechanical, or other means, now known or hereafter invented, including photocopying, microfilming, and recording, or in any information storage or retrieval system, without written permission from the publishers.

For permission to photocopy or use material electronically from this work, access www. copyright.com or contact the Copyright Clearance Center, Inc. (CCC), 222 Rosewood Drive, Danvers, MA 01923, 978-750-8400. For works that are not available on CCC please contact mpkbookspermissions@tandf.co.uk

Trademark notice: Product or corporate names may be trademarks or registered trademarks and are used only for identification and explanation without intent to infringe.

Library of Congress Cataloging-in-Publication Data (applied for)

ISBN: 978-1-032-37187-0 (hbk)
ISBN: 978-1-032-37188-7 (pbk)
ISBN: 978-1-003-33575-7 (ebk)

DOI: 10.1201/9781003335757

Typeset in Times New Roman
by Prime Publishing Services

Preface

Emerging contaminants are compounds characterized by a perceived or veridical threat to the environment or human health, lacking published health criteria. An "emerging" contaminant could also be identified from an unknown source, a new exposure to humans, or a novel detection approach or technology. Emerging contaminants include an extensive array of synthetic chemicals in global use, such as plastic additives, PAHs, perfluorinated compounds, water disinfection by-products, gasoline additives, pharmaceuticals, manufactured nanomaterials, and UV-filters, which are significant for the development of modern society. Micro and nano plastic contamination has become an environmental concern, as they have been found in various sources, including water, soil, and air. These particles can be generated from the fragmentation of large plastic objects, such as bottles and packaging, or directly originate from the composition of certain products, such as pharmaceutical and personal hygiene products.

Due to their rapidly increasing use in industry, transport, agriculture, and urbanization, these chemicals are entering the environment at increasing levels as hazardous wastes and non-biodegradable substances. Furthermore, adequate and robust epidemic information on their behaviour and fate in the global environment, as well as on human exposure, serum, and tissue concentrations, and threats to ecological and human health, have yet to be well documented. Therefore, the present book emphasizes comprehensive information on emerging contaminants overview, distribution, environmental occurrence, analysis, risk assessment, and toxicity assessment. Some chapters provide information on micro and nano plastics estimation, occurrence, challenges, and strategies for their mitigation. Information on environmental toxicology has also been presented, focusing specifically on the fate and risk assessment of emerging contaminants. It provides valuable insights into the behaviour, distribution, and potential risks associated with these contaminants, enabling researchers, practitioners, and policymakers to develop effective strategies for monitoring, managing, and mitigating their environmental impact. This book also provides environmental, legal, and health concerns of the ECs. An updated status from an industrial point of view has also been given in this book.

Contents

CHAPTER 1

Emerging Contaminants in the Environment

Introduction

Himanshu Gupta[1],* and *Soniya Dhiman*[2]

1. Introduction

The technological development of resources at the revolutionary level has generated many chemicals in the environment. Consequently, the number of chemicals which possess a threat to the living beings has also increased. The continuous enhancement in the quantity of unregulated contaminants leads to the potential environmental hazard as well as to mankind (Malule et al., 2020). As per the US Geological Survey "Emerging contaminants (ECs) are synthetic or natural compounds and microorganisms produced and used by humans that tend to have ecological and human health effects once they reach the environment" (U.S Geological Survey, 2017). Emerging contaminants also include microbes such as antibiotic resistance genes/bacteria (Glassmeyer et al., 2017). The matter related to emerging contaminants has now gained concerns of various non-governmental as well as government organizations for more research on this burning topic. However, pollution or environmental damage due to ECs has been found to exist for several years. The advances of technologies and industries are directly proportional to the enhancement of ECs. Environmental contamination has been extended to an unmanageable extent due to warfares and world wars releasing biological, nuclear, and chemical contamination to the environment (Sauve and Desrosiers, 2014).

The origin of contaminants can be through intentional or unintentional applications. Any contaminant can be referred as emerging, if it is found to reach

[1] Department of Chemistry, School of Sciences, IFTM University, Lodhipur Rajput, Moradabad-244102, India.
[2] Department of Biochemical Engineering and Biotechnology, Indian institute of Technology Delhi, India.
* Corresponding author: hims.research@gmail.com

living beings through any new pathway or new source. It could appear through newly developed treatment technology or through new detection methods (Kassinos and Michael, 2013). Moreover, as only scarce data is present about the behaviour and toxic effects of ECs on environment and human health, the understanding of environmental contaminants poses a real challenge to researchers around the world. In a period of 25 yr, the analytical methods such as mass spectrometry are rapidly developing , which has provided methods to identify emerging contaminants even at a trace amount (Philip et al., 2018). Moreover, in comparison to developed countries, emerging contaminant issues lack enough information in developing countries (Rehman et al., 2015). India is a developing country, where emerging contaminants are produced as well as consumed at a higher rate day after day. Furthermore, at the global level, India appears at number 05 on the list of emerging pharmaceutical markets (Kalotra, 2014; CCI, 2015). In terms of revenues, 70% of the market share of the Indian pharmaceutical sector is from the largest segment of generic drugs. In terms of volume, 20% of the worldwide generic exports is from the largest global drug supplier country, India. In the coming years, the supply capacity of India is further expected to expand (IBEF, 2017). The lack of patent protection, cheap production, educated personnel, high-quality research and innovative technology has become an attractive option for foreign multinational companies in order to outsource their drug production (Kurunthachalam, 2012). Presently, the Indian pharmaceutical industry is producing various kinds of medicines including complex cardiac medicines, sophisticated antibiotics as well as simple headache pills (Bansal, 2011; Mukherjee, 2006). The water treatment plants and untreated water consists of at least 25 and 121 different kinds of unregulated compounds as well as microbes (Agunbiade and Moodley, 2014). As per the 2013 statistics of EUROSTAT, during 2002 to 2011, environmentally harmful compounds are more than half of the total chemical production and greater than 70% significantly harms the environment. Among them, the ecotoxicological effects related to various ECs have raised concerns about their production and utilization (Eurostat Statistics Explained, 2023). Across the country, correct methods for the EC treatment has not been developed or implemented (Parida et al., 2022).

2. Distribution of Emerging Contaminants

Inspite of sufficient information about various emerging contaminants still being unavailable, a number of synthetic as well as natural chemicals are included in the list of emerging contaminants. These may be classified on the basis of different parameters, generally these are classified on the basis of their application (Flores et al., 2017). Therefore, ECs include compounds such as plasticizers, personal care products, pharmaceuticals, Poly cyclic Aromatic Hydrocarbons (PAHs), flame retardants, hormones, surfactants, biocides, and nanomaterials. Figure 1.1 provides an overview of different groups of contaminants which are referred as ECs.

A brief discussion of various kinds of emerging contaminants is presented as follows:

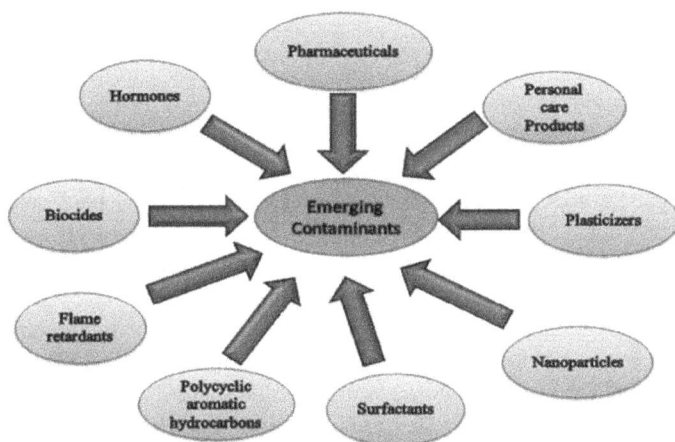

Figure 1.1 Emerging contaminants.

2.1 Pharmaceuticals

Pharmaceuticals could enter the environment as a result of human or veterinary use, industrial discharge and improper disposal pharmaceuticals can be classified by their therapeutic use: analgesics, antibiotics, antihistamines, antidepressants, beta-blockers, cytostatics, lipids regulators, stimulating and X-ray contrast media. They have been detected in water bodies throughout the world and could pose a risk to humans and the environment because of their continuous release (Tiwari et al., 2017; Gojkovic et al., 2019). Pharmaceuticals are given in the list of ECs, as these are still unregulated or at some places these are still under the regularization process, but the legal framework or directives are not clear. Many countries have reported the contamination caused by pharmaceuticals in ground as well as surface waters as a matter of concern for the environment. A few of pharmaceuticals reported in water are antibiotics (chloramphenicol, fluoroquinolones, sulfonamides, quinolones, penicillins, tetracyclines, etc.), antiulcer drugs and antihistamines (famotidine, ranitidine), β-blockers (metoprolol, propranolol, atenolol), antiepileptics (carbamazepine), antidepressant ((benzodiazepines), analgesics and anti-inflammatories (diclofenac, ibuprofen, acetylsalicylic acid, paracetamol). It is also reported that pharmaceuticals absorb radiation and these may be photodegraded, as they consist of functional groups, hetero atoms and aromatic rings which allows photosensitizing reactions inducing photon-assisted decay in different matrices. Generally, using UV irradiation the photodegradation of pharmaceuticals is < 30%, which was found to be enhanced to 100% on the addition of activated carbon/TiO_2, TiO_2, H_2O_2. The risk studies related to the pharmaceuticals need to be reported for various pharmacologically active chemicals as well as their intermediates (Rivera-Utrilla et al., 2013).

2.2 Plasticizers

Plasticizers are synthetic chemicals that are commonly used in polyvinyl chloride (PVC) based products, food packaging, children's toys, medical devices, and adhesives (Cook and Halden, 2022). There are about 30000 chemicals which can potentially be used as plasticizers (Godwin, 2017). Phthalate plasticizers are a commonly used compound, comprising up to 85% of the total plasticizers in the market (Qadeer et al., 2022). It has been reported that some plasticizers are toxic and exhibit endocrine-disrupting properties. Phthalate plasticizers have been regarded as hazardous compounds due to numerous reports based on its toxicological effects, including bioaccumulation potential, endocrine disruption, carcinogenicity, and developmental defects (Jurewicz and Hanke, 2011; Mankidy et al., 2013). Plasticizers are also reported for their toxicity as well as are suspected to alter the endocrine and lipid metabolism on exposure to animals and human beings. Plasticizers could reach the feotus by crossing the human placenta and may cause embryonic or fetal disruptions. Preterm delivery is also linked to plasticizers (Huang et al., 2019). The urine of a pregnant woman has been reported to contain 8-isoprostane, which serves as oxidative-stress biomarker (Ferguson et al., 2017). It is also reported that the urinary levels of bisphenol A can be correlated with the levels of oxo-2' –deoxyguanosine, a biomarker for oxidative stress (Liu et al., 2019; Pérez-Albaladejo et al., 2020).

2.3 Personal Care Products

PCPs are compounds that can be found in consumer products marketed for use on the human body. These active ingredients are present in cosmetics, skin care, dental and hair care products, sunscreen agents, soaps, antibacterial products, fragrances, and insect repellents (Ebele et al., 2017). As there are not prescribed as pharmaceuticals, they may be over used and can be directly introduced into the environment through regular use, such as showering, spraying or disposal. Thus due to the great use and the conditions released, they can be classified as pseudo persistent (Kagle et al., 2009). Earlier studies have reported the presence of triclosan with its metabolite, monoethoxylates, alkylphenols and synthetic musk in the tissue of fishes (Rudel et al., 2006; Wenzel et al., 2004; Boehmer et al., 2004; Subedi et al., 2012).

2.4 Polycyclic Aromatic Hydrocarbon

Polycyclic Aromatic Hydrocarbons (PAHs) comprise a large group of compounds with two or more fused benzene rings, widely distributed in the environment (Abdel-Shafy et al., 2016). PAHs are recognized as potent carcinogens or mutagens and are listed as priority pollutants by the United States Environmental Protection Agency (USEPA) and European commission (EC) (Keith, 2015; Gupta and Dhiman, 2023). PAHs are introduced into the aquatic and terrestrial systems as a result of both, natural activities, namely forest fires, volcanic eruptions, and anthropogenic activities like incomplete combustion of fuels, cigarette smoke, vehicular traffic, fumes, and wastewater releases from various industries. PAHs could also contribute by oil spillage and tire abrasion (Gonzalez et al., 2006; Ravindra et al., 2007).

PAHs contributed by industrial and domestic waste play a significant role in the deterioration of the urban aquatic and terrestrial bodies. After emission from various sources and due to persistent characteristics, PAHs can disperse into the environment through atmospheric transport and deposit on the soil surface (Abhilash et al., 2013; Gupta and Gupta, 2015). PAHs generally differ from persistent organic pollutants due to their emerging concerns and these are also produced as fossil fuel by-products unintentionally (Balmer et al., 2019). As the PAHs are persistent, these could cause atmospheric transfer and get deposited on surfaces at longer or shorter distances. The exposure of PAHs to living beings directly through soil, water or air is highly toxic. These could also be generated through various industrial plants. Various researchers have found these pollutants in wastewater or water (Gonzalez et al., 2012; Olenycz et al., 2015; WlodarczykMakula, 2005), air (Liu et al., 2014; Masih et al., 2010), sediment (Christensen and Bzdusek, 2005; Nikolaou et al., 2009; Soclo et al., 2000) and soil (Xing-Hong et al., 2006; Yang et al., 2002; Skrbic et al., 2005; Malawaska and Wilkomirski, 2001).

2.5 Surfactants

Surfactants possess both hydrophilic (a polar charged or uncharged head group) and hydrophobic (a non-polar hydrocarbon tail) and thus are regarded as amphipathic molecules (Mungray and Kumar, 2009). Surfactants are used for various domestic and industrial applications due to their unique physicochemical properties (Collivignarelli et al., 2019). However, the persistence of the transformed products in an environment is of great concern to environmental sustainability and healthy ecosystems (Liu et al., 2021; Yao et al., 2021; Hao et al., 2021).

Mainly due to the extensive applications of surfactants, their huge concentration from mostly urban or industrial and domestic wastewater can end up in Municipal Wastewater Treatment Plants (MWTPs) or directly being discharged into the environment (Baustista-Toledo et al., 2014; Camacho-Munoz et al., 2014). Several years ago, it was estimated that surfactants in domestic and industrial wastewater are between 1 to 10 mg/L and 300 mg/L, respectively (Zhang et al., 1999). There are chances that surfactants can pass into drinkable water by MWTPs, which poses a significant risk to aquatic organisms, animal, and human health (Baustista-Toledo et al., 2014). Moreover, the electron acceptors, environmental systems (anaerobic or aerobic), chemical structure and fate of surfactants generally contributes to ecological effects, degradability, and persistence of surfactants in the environment (Zhu et al., 2018).

2.6 Biocides

Biocides are chemical compounds of biological or chemical characteristic controls harmful for organisms or deter, destroy, or remain harmless. For disinfection purposes, biocides are used largely in healthcare, industrial and domestic environments (Maillard, 2005). Along with the application as antibiotics, biocides are used in controlling infections for several years. The physician Ignaz Semmelweis used chlorinated handwash as biocide in the 19th Century (Fraise et al., 2008;

Cookson, 2005; Fraise, 2002). Presently, biocides comprise of peroxygens, chlorine-releasing agents, biguanides, quaternary ammonium compounds, topical agents and disinfectants (Wesgate et al., 2016). In order to prevent infection, the advances in science have suggested the application of biocidal compounds to different items including socks, soaps, mouthwashes and surgical scrubs (Gilbert et al., 2002). Biocides are divided into various types including acaricides, rodenticides, algaecides, piscicides, molluscicides, avicides, preservatives, disinfectants, and insecticides (Henry et al., 2012). Among insecticides, due to the prohibition of organophosphate and organochlorine chemicals, neonicotinoids and pyrethroids are developed as new categories. Due to the higher toxicity towards non-targeted organisms, toxicologists are particularly concerned about these categories of compounds (Maund et al., 2012).

2.7 Flame Retardants

Flame Retardants (FRs) are substances which could prevent or delay the process of combustion. Four categories of FRs are available which includes nitrogen, phosphorus, halogenated organics, and other inorganic compound-based FRs. Phosphorus-based flame retardants (PFRs) are largely applied in electronics, building materials, textiles, furniture and other chemical processes and are considered as widely applied flame retardant (Reemtsma et al., 2008). PFRs are generally applied as epoxy resins, engineering thermoplastics, coatings, and plasticizers in floor polishes (Du et al., 2015). PFRs are unique in the environment with regards to the various physiological and physical characteristics such as the bioconcentration factor, vapour pressure, log K_{ow} value and solubility. The impact of organic PFRs on the organism and environmental behaviour could be assessed by the application of these properties (Greaves and Letcher, 2014). In comparison to heavier and larger PFRs, volatile PFRs including tri(2-chloroethyl)phosphate (TCEP), triethyl phosphate (TEP) and tributylphosphat (TBP) tend to discharge readily in air and deposit on dust (Veen and Boer, 2012; Cristale et al., 2016), whereas, alkyl and aryl PFRs having greater molecular weights are highly hydrophobic and consist of greater soil and sediment affinity and a similar bioconcentration factor. Aquatic animals are continuously threatened by chlorinated PFRs due to their higher solubility in water (Wei et al., 2015).

Due to its effective and less expensive nature Brominated Flame Retardants (BFRs) are widely used . However, due to toxicity, bioaccumulation evidence and higher environmental persistence, few BFRs like polybrominated diphenyl ethers (PBDEs) are ruled out from the world market of flame retardants. Some of the FRs are banned globally including octabromodiphenyl ethers (OctaBDE) and pentabromodiphenyl ether (PentaBDE) (Hou et al., 2016).

2.8 Hormones

Hormones are synthetic or natural compounds which make a large impact on animal and human health. In 1994, the endocrine-distrupting characteristics of hormones were first reported which are responsible for fish hermaphroditism due to their presence in water (Purdom et al., 1994). In recent years, livestock farming and application of hormone therapy has been found to increase. Additionally, due to the natural occurrence of estrogenic chemicals, it is expected that such compounds are present in environment and wastewaters (Adeel et al., 2017).

2.9 Nanomaterials

During the last decade, the rapid growth in industrial and inter disciplinary technological research area is of nantechnology (Gupta et al., 2011). The nanomaterials can be referred to as the manufactured, incidental, or natural substances, containing at least one dimension ranging between 1 to 100 nm (Khan and Khan, 2019). Nanoparticles are used in catalytic, energy, cosmetics, pharamaceuticals, biomedical and electronic applications. The smaller size and higher applications in various sectors make them substances which could contaminate the environment (Khan and Khan, 2019). Nanomaterials present in accidental spillage, direct discharges, wastewater effluents and solid wastes can undergo transportation into water systems through run off or through the wind (Klaine et al., 2008). Due to the slower gravity settlement and smaller size, nanomaterials could remain suspended in water or air for long periods. In order to find the effect of nanomaterials on wildlife species, studies are still being conducted (Federici et al., 2007). The biodegradability of nanomaterials is also a concern, as C_{70} and C_{60} fullerenes are reported to degrade after many months. Zero valent iron is applicable in various remediation processes, but the small size could cause increased mobility and migration from the contaminated area to drinking water and other matrices. Investigation on such processes is also required (EPA, 2010).

3. Physio-chemical Characteristics of Emerging Contaminants

The different categories of emerging contaminants do not show similar characteristics, analytical or environmental behaviour. The physio-chemical characteristics of the different compounds are presented in Table 1.1.

The diverse sources of water, i.e., ground water, reused water, and fresh water, serve as major pathways in order to introduce these compounds in water. Therefore, solubility in water acts as major factor for the presence of contaminants in the environment. However, pK_a and $logK_{ow}$ could be used as parameters from the analytical point of view. The non-polar compound has a higher Kow value. Therefore, the extraction of various compounds from different matrices depends on the Kow value of the compound such as PAHs with the high Kow value (Table 1.1). On the other hand, acid dissociation constant (pKa) can be used as a chemical property. The extraction of any compound is also dependent on the pH of the solution.

4. Main pathways of Emerging Contaminants into the Environment

In the environment ECs can enter by different pathways. Figure 1.2 shows the major pathways of emerging contaminants in the environment.

Table 1.1 Physico-chemical properties of some representative ECs.

S. No	Compounds	Class	LogK$_{ow}$	Molecular weight (g/moL)	pKa	Water solubility (mg/L) at 25°C	References
1.	Aspirin	Pharmaceuticals	1.2	180	3.5	0.003	Samal et al., 2022
2.	Diclofenac	Pharmaceuticals	4.51	296.2	4.91	0.0	Samal et al., 2022
3.	Ibuprofen	Pharmaceuticals	4.5	206.3	4.15	< 1 mg/mL	Samal et al., 2022
4.	Paracetamol	Pharmaceuticals	0.46	151.2	9.38	0.13	Samal et al., 2022
5.	Naproxen	Pharmaceuticals	3.18	230.3	4.15	15.9	Samal et al., 2022
6.	Sulphanethoxazole	Pharmaceuticals	0.89	253.27	5.6	< 1 mg/mL	Samal et al., 2022
7.	Erythromycin	Pharmaceuticals	2.48	733.93	8.88	0.05	Samal et al., 2022
8.	Trimethoprim	Pharmaceuticals	0.73	290.32	7.12	0.0004	Samal et al., 2022
9.	Acetyl tributyl citrate	Plasticizers	4.92	402.5		0.02 g/L	Qadeer et al., 2022
10.	Tri phenyl phosphate	Plasticizers	4.59	326.3		1.9 mg/L	Qadeer et al., 2022
11.	Glyceryl tri acetate	Plasticizers	0.25	218.20		0.6	Qadeer et al., 2022
12.	Diisononyl cyclohexane-1,2dicarboxylate	Plasticizers	10	424.7		< 0.02 mg/L	Qadeer et al., 2022
13.	Tris-2-ethyltrimellitate	Plasticizers	5.94	546.78		< 1 mg/mL	Qadeer et al., 2022
14.	Methyl paraben	PCPs	1.96	152.15	8.17	0.0025	Lincho et al., 2021
15.	Propyl paraben	PCPs	3.04	180.2	8.35	0.5	Lincho et al., 2021
16.	Methyl triclosan	PCPs	4.8	303.6		10	Lincho et al., 2021
17.	Triclosan	PCPs		289.54		0.002	Lincho et al., 2021
18.	Naphthalene	PAHs	3.29	128.18		31.6	Maagd et al., 1998
19.	Acenaphthylene	PAHs	3.98	154.21		3.8	Maagd et al., 1998
20.	Acenaphthene	PAHs	4.07	152.20		16.1	Maagd et al., 1998
21.	Fluorene	PAHs	4.18	166.22		1.9	Maagd et al., 1998
22.	Phenanthrene	PAHs	4.45	178.23		1.1	Maagd et al., 1998
23.	Antharene	PAHs	4.45	178.23		0.045	Maagd et al., 1998
24.	Fluoranthene	PAHs	4.90	202.26		0.26	Maagd et al., 1998

No.	Name	Category				Reference
25.	Pyrene	PAHs	4.88	202.26	0.132	Maagd et al., 1998
26.	Benzo[a]anthracene	PAHs	5.61	228.29	0.011	Maagd et al., 1998
27.	Chrysene	PAHs	5.90	228.29	0.0015	Maagd et al., 1998
28.	Benzo[b]fluoranthene	PAHs	6.04	252.32	0.0015	Maagd et al., 1998
29.	Benzo[k]fluoranthene	PAHs	6.06	252.32	0.0008	Maagd et al., 1998
30.	Benzo[a]pyrene	PAHs	6.06	252.32	0.0038	Maagd et al., 1998
31.	Indeno[1,2,3-c,d]pyrene	PAHs	6.58	276.34	0.062	Maagd et al., 1998
32.	Benzo[g,h,i]perylene	PAHs	6.50	276.34	0.00026	Maagd et al., 1998
33.	Dibenzo[a,h]anthracene	PAHs	6.84	276.34	0.0005	Maagd et al., 1998
34.	Perfluorobutanoic acid	Surfactants	4.81	214.04	2290 mg/L	Bhhatarai and Gramatica, 2011
36.	Cetyltrimethyl ammonium bromide	Surfactants		364	3 g/L	Bielawska et al., 2013
37.	Octyl phenol diethoxylate	Surfactants	4.0	624	13.2 mg/L	Lintelmann et al., 2003
39.	Diuron	Biocides	2.87	233.09	0.042	Kahrilas et al., 2015
40.	Isoproturon	Biocides	2.48	203.28	0.07	Kahrilas et al., 2015
41.	Cybutryne	Biocides	3.95	253.37	0.007	Kahrilas et al., 2015
42.	Terbutryn	Biocides	3.74	241.36	0.025	Kahrilas et al., 2015
43.	Estrone	Hormones	3.1	270.36	0.0	Lintelmann et al., 2003
44.	Hexestrol	Hormones	4.8	270.37	0.0	Lintelmann et al., 2003
45	Diethylstilbestrol	Hormones	5.07	268.35	0.0	Lintelmann et al., 2003

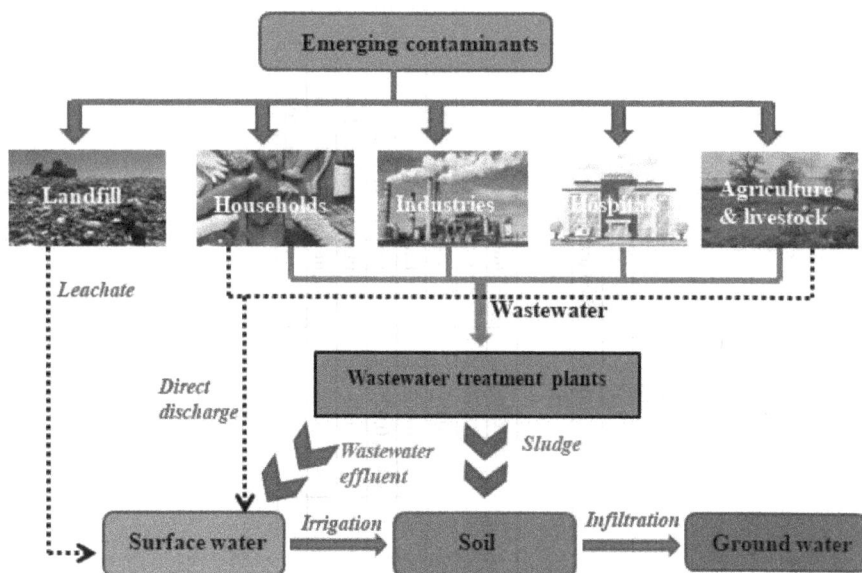

Figure 1.2 Pathways of emerging contaminants into the environment.

1) Application of ECs through human beings and urban pets released to the sewage.
2) Sewage systems due to release of hospital waste
3) Septic tank leaching
4) Wastewater treatment plant effluent for irrigation
5) Wastewater treatment plant biosolids for soil amendment
6) Drift runoff and leaching: Release from agriculture
7) Swimming, bathing, and washing: Direct release to open wasters
8) Industrial waste discharge
9) Landfills Disposal through hazardous wastes, medical wastes, and domestic refuse

As can be seen in Fig. 1.2, the best-case scenario will be that the contaminant water reaches a wastewater treatment plant (WWTP), but unfortunately, these days most of the ECs are not completely degraded during depurations processes. Consequently, they are discharged in treated effluents resulting in the contamination of freshwater, ground water and drinking water from farms are used in agricultural soils to increase their content in organic matter and nutrients, but ECs could be also introduced into agricultural soils. Finally, inappropriate disposal of these compounds is another source that cannot be ignored. Nonetheless, the input source is well known, but the fate and behaviour of these ECs in the environment and their possible introduction into the food chain need further study.

In spite of the sizeable amounts of ECs released and detected in the environment, environmental risk assessment and regulation are largely missing. Thus, reliable, and robust multiresidue methods are needed to assess their presence into different environmental matrices.

Conclusion

In the present case, emerging contaminants are released into the environment due to the technological enhancement in the generation of goods and services in order to meet the requirements of increasing populations. A vital consideration is required to the present environmental issue of emerging contaminants. Generally, using UV irradiation the photodegradation of pharmaceuticals is < 30%, which was found to be enhanced to 100% on addition of activated carbon/TiO_2, TiO_2, H_2O_2. The risk studies related to the pharmaceuticals should be reported for various pharmacologically active chemicals as well as their intermediates. The urine of a pregnant woman has been reported to contain 8-isoprostane, which serve as oxidative stress biomarker. Earlier studies reported the presence of triclosan with its metabolite, monoethoxylates, alkylphenols and synthetic musk in the tissue of fishes. PAHs generally differ from persistent organic pollutants due to their emerging concerns and these are also produced as fossil fuel by-products unintentionally. Among insecticides, due to the prohibition of organophosphate and organochlorine chemicals, neonicotinoids and pyrethroids have been developed as new categories. For disinfection purposes, biocides are largely used in healthcare, industrial, and domestic environments. In comparison to heavier and larger PFRs, volatile PFRs including tri(2-chloroethyl) phosphate (TCEP), triethyl phosphate (TEP) and tributylphosphat (TBP) tend to discharge readily in air and deposit on dust. In recent years, livestock farming and the application of hormone therapy is found to have increased. Due to slower gravity settlement and the smaller size, nanomaterials may be suspended in water or air for long periods . Zero valent iron is applicable in various remediation processes, but the small size could cause increased mobility and migration from the contaminated area to drinking water and other matrices. The extraction of various compounds from different matrices depends on the Kow value of the compound. It is required to develop newer techniques of production which uses starting materials which are less hazardous and eco-friendly to the environment for preventing ecosystem or aquatic biota from contamination.

References

Abdel-Shafy, H.I., Mona, S.M. and Mansour, A. 2016. Review on polycyclic aromatic hydrocarbons: Source, environmental impact, effect on human health and remediation. Egypt. J. Pet. 25: 107–123.

Abhilash, P.C., Dubey, R.K., Tripathi, V., Srivastava, P., Verma, J.P. and Singh, H.B. 2013. Remediation and management of POPs-contaminated soils in a warming climate: challenges and perspectives. Environ. Sci. Pollut. Res. 20(8): 5879–5885.

Adeel, M., Song, X., Wang, Y., Francis, D. and Yang, Y. 2017. Environmental impact of estrogens on human, animal and plant life: A critical review. Environ. Int. 99: 107–119.

Agunbiade, F.O. and Moodley, B. 2014. Pharmaceuticals as emerging organic contaminants in Umgeni River water system, KwaZulu-Natal, South Africa. Environ. Monit. Assess. 1: 7273–7291.

Balmer, J.E., Hung, H., Yu, Y., Letcher, R.J. and Muir, D.C.G. 2019. Sources and environmental fate of pyrogenic polycyclic aromatic hydrocarbons (PAHs) in the Arctic. Emerging Contaminants 5: 128–142.

Bansal, M. 2011. Financial analysis of selected pharmaceutical industries. Annamalai. Int. J. Bus. Stu. Res. 3: 89–102.

Bautista-Toledo, M.I., Rivera-Utrilla, J., Méndez-Díaz, J.D., Sánchez-Polo, M. and Carrasco-Marín, F. 2014. Removal of the surfactant sodium dodecylbenzenesulfonate from water by processes based on adsorption/bioadsorption and biodegradation. J. Colloid. Interface. Sci. 418: 113–119.

Bhhatarai, B. and Gramatica, P. 2011. Prediction of aqueous solubility, vapor pressure and critical micelle concentration for aquatic partitioning of perfluorinated chemicals. Environ. Sci. Technol. 45(19): 8120–8128.

Bielawska, M., Chodzinnska, A., Janczuk, B. and Zdziennicka, A. 2013. Determination of CTAB CMC in mixed water + short-chain alcohol solvent by surface tension, conductivity, density and viscosity measurements. Colloids. Surf. A 424: 81−88.

Boehmer, W., Ruedel, H., Wenzel, A. and Schroeter-Kermani, C. 2004. Retrospective monitoring of triclosan and methyl-triclosan in fish: Results from German Environmental Specimen Bank. Organohalogen Compd. 66: 1516−1521.

Camacho-Muñoz, D., Martín, J., Santos, J.L., Aparicio, I. and Alonso, E. 2014. Occurrence of surfactants in wastewater: hourly and seasonal variations in urban and industrial wastewaters from Seville (Southern Spain). Sci. Total Environ. 468–469: 977–984.

C.C.I. 2015. A Brief Report on Pharmaceutical Industry in India. Corporate Catalyst (India) Pvt. Ltd. pp. 1−11.

Christensen, E.R. and Bzdusek, P.A. 2005. PAHs in sediments of the Black River and the Ashtabula River, Ohio: source apportionment by factor analysis. Water Res. 39: 511–524.

Collivignarelli, M.C., Carnevale Miino, M., Baldi, M., Manzi, S., Abbà, A. and Bertanza, G. 2019. Removal of non-ionic and anionic surfactants from real laundry wastewater by means of a full-scale treatment system. Process. Saf. Environ. Prot. 132: 105–115.

Cook, C.R. and Halden, R.U. 2020. Plastic Waste and Recycling. Academic Press; Cambridge, MA, USA: Ecological and health issues of plastic waste.

Cookson, B. 2005. Clinical significance of emergence of bacterial antimicrobial resistance in the hospital environment. J. Appl. Microbiol. 99: 989–996.

Cristale, J., Hurtado, A., Gómez-Canela, C. and Lacorte, S. 2016. Occurrence and Sources of Brominated and Organophosphate Flame Retardants in Dust from Di_erent Indoor Environments in Barcelona, Spain. Environ. Res. 149: 66–76.

Diaz-Garduno, B., Pintado-Herrera, M.G., Biel-Maeso, M., Rueda-Marquez, J.J., Lara-Martín, P.A., Perales, J.A. et al. 2017. Environmental risk assessment of effluents as a whole emerging contaminant: Efficiency of alternative tertiary treatments for wastewater depuration. Water Research 119: 136−149.

Du, Z.K., Wang, G.W., Gao, S.X. and Wang, Z.Y. 2015. Aryl organophosphate flame retardants induced cardiotoxicity during zebra fish embryogenesis: By disturbing expression of the transcriptional regulators. Aquat. Toxicol. 161: 25–32.

Ebele, A.J., Abdallah, M.A.E. and Harrad, S. 2017. Pharmaceuticals and personal care products (PPCPs) in the freshwater aquatic environment. Emerg. Contam 3: 1–16.

EPA 2010. Fact Sheet: Emerging Contaminats – Nanomaterials. United States Environmental Protection Agency, EPA 505-F-10-008.

Eurostat Statistics Explained. 2023. Chemical production and consumption statistics.

Federici, G., Shaw, B.J. and Handy, R.R. 2007. Toxicity of Titanium Dioxide Nanoparticles to Rainbow Trout: Gill Injury, Oxidative Stress, and Other Physiological Effects. Aquativ Toxicology. Volume 84: 415–430.

Ferguson, K.K., Chen, Y.H., Vanderweele, T.J., McElrath, T.F., Meeker, J.D. and Mukherjee, B. 2017. Mediation of the relationship between maternal phthalate exposure and preterm birth by oxidative stress with repeated measurements across pregnancy. Environ. Health Perspect. 125: 488–494.

Flores, C., Francisco, G., Jimenez, D. and Jose, A. 2017. Emerging Pollutants: Origin, Structure and Properties, December DOI:10.1002/9783527691203.

Fraise, A.P. 2002. Biocide abuse and antimicrobial resistance—A cause for concern? J. Antimicrob. Chemother. 49: 11–12.

Fraise, A.P., Lambert, P.A. and Maillard, J.Y. 2008. Editors. Russell, Hugo & Ayliffe's Principles and Practice of Disinfection, Preservation and Sterilization. John Wiley & Sons; Hoboken, NJ, USA.

Glassmeyer, S.T., Furlong, E.T., Kolpin, D.W., Batt, A.L., Benson, R., Boone J.S. et al. 2017. Nationwide reconnaissance of contaminants of emerging concern in source and treated drinking waters of the United States. Sci. Total Environ. 581: 909–922.

Gilbert, P., McBain, A.J. and Bloomfield, S.F. 2002. Biocide abuse and antimicrobial resistance: Being clear about the issues. J. Antimicrob. Chemother. 50: 137–139.

Godwin, A.D. 2017. Plasticizers. *In*: Appl. Plast. Eng. Handbook 2nd Ed: 533–553.

Gojkovic Z., Lindberg, R.H., Tysklind, M. and Funk, C. 2019. Northern green algae have the capacity to remove active pharmaceutical ingredients. Ecotoxicol. Environ. Saf. 170: 644–656.

Gonzalez, J.J., Vinas, L., Franco, M.A., Fumega, J., Soriano, J.A., Grueiro, G. et al. 2006. Spatial and temporal distribution of dissolved/dispersed aromatic hydrocarbons in seawater in the area affected by the Prestige oil spill. Mar. Poll. Bull. 53: 250–259.

González, D., Ruiz, L.M., Garralón, G., Plaza, F., Arévalo, J. and Parada, J. 2012. Wastewater polycyclic aromatic hydrocarbons removal by membrane bioreactor. Desalin Water Treat 42: 94–99

Greaves, A.K. and Letcher, R.J. 2014. Body compartment distribution in female Great Lakes herring gulls and in ovo transfer of sixteen bioaccumulative organophosphate flame retardants. Environ. Sci. Technol. 48: 7942–7950.

Gupta, H. and Dhiman, S. 2023. Synthetic Nanoparticle-Based Remediation of Soils Contaminated with Polycyclic Aromatic Hydrocarbons, Springer Nature Switzerland AG 2023 U. Shanker et al. (eds.). Handbook of Green and Sustainable Nanotechnology, https://doi.org/10.1007/978-3-030-69023-6_102-1

Gupta, H. and Gupta, B. 2015. Photocatalytic degradation of polycyclic aromatic hydrocarbon benzo[a] pyrene by iron oxides and identification of degradation products. Chemosphere 138: 924–931.

Gupta, K., Bhattacharya, S., Chattopadhyay, D., Mukhopadhyay, A., Biswas H., Dutta J. et al. 2011. Ceria associated manganese oxide nanoparticles: synthesis, characterization and arsenic (V) sorption behavior. Chem. Eng. J. 172: 219–229.

Hao, M., Qiu, M., Yang, H., Hu, B. and Wang, X. 2021. Recent advances on preparation and environmental applications of MOF-derived carbons in catalysis. Sci. Total Environ. 760: 143333.

Henry, M., Requier, F., Rollin, O., Odoux, J.F., Aupinel, P., Aptel, J. et al. 2012. A common pesticide decreases foraging success and survival in honey bees: Science 336: 348–350.

Hou, R., Xu, Y.P. and Wang, Z.J. 2016. Review of OPFRs in animals and humans: Absorption, bioaccumulation, metabolism, and internal exposure research. Chemosphere 153: 78–90.

Huang, S., Li, J., Xu, S., Zhao, H., Li, Y., Zhou, Y. et al. 2019. Bisphenol A and bisphenol S exposures during pregnancy and gestational age—a longitudinal study in China. Chemosphere 237: 124426.

IBEF. 2017. Pharmaceuticals. India Brand Equity Foundation, Department of Commerce, Ministry of Commerce and Industry, Government of India, pp. 1–45.

Jurewicz, J. and Hanke, W. 2011. Exposure to Phthalates: Reproductive Outcome and Children Health. A Review of Epidemiological Studies. Int. J. Occup. Med. Environ. Health 24: 115–141.

Kagle, J., Porter, A.W., Murdoch, R.W., Cancel, G.R. and Hay, A.G. 2009. Biodegradation of pharmaceutical and personal care products. Adv Appl. Microbiol. 67: 65–108.

Kahrilas, G.A., Blotevogel, J., Stewart, P.S. and Borch, T. 2015. Biocides in hydraulic fracturing fluids: A critical review of their usage, mobility, degradation, and toxicity. Environ. Sci. Technol. 49: 16–32.

Kalotra, A. 2014. Marketing strategies of different pharmaceutical companies. J. Drug. Deliv. Ther. 4: 64–71.

Kassinos F.D. and Michael, C. 2013. Wastewater reuse applications and contaminants of emerging concern. Environ. Sci. Pollut. Res. 20: 3493–3495.

Keith, L.H. 2015. The Source of U.S. EPA's sixteen PAH priority pollutants. Polycycl. Aromat. Comp. 35: 147–160.

Khan, I., Saeed, K. and Khan, I. 2019. Nanoparticles: Properties, applications and toxicities. Arab. J. Chem. 12: 908–931.

Klaine, S.J., Alvarez, P.J.J., Batley, G.E., Fernandes, T.E., Handy, R.D., Lyon, D.Y. et al. 2008. Nanoparticles in the Environment: Behavior, Fate, Bioavailability and Effects. Environmental Toxicology and Chemistry. 27(9): 1825–1851.

Kurunthachalam, S.K. 2012. Pharmaceutical substances in India are a point of great concern? J Waste Water. Treat. Anal. 3: 1000–1103.

Lincho J., Martins, R.C. and Gones, J. 2021. Parabens Compounds-Part 1: An overview of their characteristics, detection and impacts. Appl. Sci. 11: 2307.

Lintelmann, J., Katayama, Kurihara, N., Shore, L. and Wenzel, A. 2003. Endocrine disruptors in the environment (IUPAC technical report). Pure. Appl. Chem. 75: 631–681.

Liu, D., Xu, Y., Chaemfa, C., Tian, C., Li, J., Luo, C. et al. 2014. Concentrations, seasonal variations, and outflow of atmospheric polycyclic aromatic hydrocarbons (PAHs) at Ningbo site, eastern China. Atmos. Pollut. Res. 5: 203–209

Liu, M., Jia, S., Dong, T., Han. Y., Xue, J. and Wanjaya, E.R, 2019. The occurrence of bisphenol plasticizers in paired dust and urine samples and its association with oxidative stress. Chemosphere 216: 472–478.

Liu, X., Ma, R., Zhuang, L., Hu, B., Chen, J., Liu, X. et al. 2021. Recent developments of doped g-C3N4 photocatalysts for the degradation of organic pollutants. Crit. Rev. Environ. Sci. Technol. 51: 751–790.

Maagd, P.J.D., Hulscher, D.D.T., Heuvel, V.D., Opperhuizen, A. and Sijm, D.T.H.M. 1998. Physicochemical properties of polycyclic aromatic hydrocarbons: aqueous solubilities, n-octanol/water partition coefficients, and henry's law constants. Environ. Toxicol. Chem. 17: 251–257.

Malule, H.R., Murillo, D.H.Q. and Duque, D.M. 2020. Emerging contaminants as global environmental hazards. A bibliometric analysis. Emerg. Contam. 6: 179–193.

Maillard, J.Y. 2005. Antimicrobial biocides in the healthcare environment: Efficacy, usage, policies, and perceived problems. Ther. Clin. Risk. Manag. 1: 307–320.

Malawaska, M. and Wilkomirski, B. 2001. An analysis of soil and plant (Taraxacum officinale) contamination with heavy metals and polycyclic aromatic hydrocarbons (PAHs) in the area of the railway junction Iława Glowna, Poland. Water Air. Soil. Poll. 127: 339–349.

Mankidy, R., Wiseman, S., Ma, H. and Giesy, J.P. 2013. Biological Impact of Phthalates. Toxicol. Lett. 217: 50–58.

Masih, J., Masih, A., Kulshrestha, A., Singhvi, R. and Taneja, A. 2010. Characteristics of polycyclic aromatic hydrocarbons in indoor and outdoor atmosphere in the North central part of India. J. Hazard. Mater. 177: 190–198.

Maund, S.J., Campbell, P.J., Giddilings, J.M., Hamer, M.J., Henry, K., Pilling, E.D. et al. 2012. Ecotoxicology of synthetic pyrethroids. pp. 137–165. *In*: N. Matsuo and T. Mori (eds.). Pyrethroids: From Chrysanthemum to Modern Industrial Insecticide: Topics in Current Chemistry, 314: Berlin, Springer-Verlag Berlin.

Mukherjee, R. 2006. Application and scope of IT in the Indian pharama sector. J. Med. Mark. 6: 146–150.

Mungray, A.K. and Kumar, P. 2009. Fate of linear alkylbenzene sulfonates in the environment: a review. Int. Biodeterior. Biodegrad. 63: 981–987.

Nikolaou, A., Kostopoulou, M., Lofrano, G. and Meric, S. 2009. Determination of PAHs in marine sediments: analytical methods and environmental concerns. Global. NEST J. 11(4): 391–405.

Olenycz, M., Sokolowski, A., Niewinska, A., Wolowicz, M., Namiesnik, J., Hummel, H. et al. 2015. Comparison of PCBs and PAHs levels in European coastal waters using mussels from the Mytilus edulis complex as biomonitors. Oceanologia 57: 196–211.

Parida, V.K., Sikarwar, D., Majumder, A. and Gupta, A.K. 2022. An assessment of hospital wastewater and biomedical waste generation, existing legislations, risk assessment, treatment processes, and scenario during COVID-19. J. Environ. Manage. 308: 114609.

Perez-Albaladejo, E., Sole, M. and Porte, C. 2020. Plastics and plastic additives as inducers of oxidative stress. Current Opinion in Toxicology 20–21: 69–76.

Philip, J.M., Aravind, U.K. and Aravindakumar, C.T. 2018. Emerging contaminants in Indian environmental matrices—A review. Chemosphere 190: 307–326.

Purdom, C.E., Hardiman, P.A., Bye, V.V.J., Eno, N.C., Tyler, C.R. and Sumpter, J.P. 1994. Estrogenic Effects of Effluents from Sewage Treatment Works: Chemistry and Ecology 8: 275–285.

Qadeer, A., Kirsten, K.L., Ajmal, Z., Jiang, X. and Zhao, X. 2022. Alternative Plasticizers As Emerging Global Environmental and Health Threat: Another Regrettable Substitution? Environ. Sci. Technol. 56: 1482–1488.

Qadeer, A., Kirsten, K.L., Ajmal, Z., Jiang, X. and Zhao, X. 2022. Alternative Plasticizers As Emerging Global Environmental and Health Threat: Another Regrettable Substitution? Environ. Sci. Technol. 56: 1482–1488.

Ravindra, K., Sokhi, R. and Grieken, R.V. 2007. Atmospheric polycyclic aromatic hydrocarbons: Source attribution, emission factors and regulation. Atmos. Environ. 42(13): 2895–2921.

Reemtsma, T., Quintana, J.B., Rodil, R., García-Lopez, M. and Rodríguez, I. 2008. Organophosphate flame retardants and plasticizers in water and air I. Occurrence and fate. Trends. Anal. Chem. 27: 727–737.

Rehman, M.S., Rashid, N., Ashfaq, M., Saif, A., Ahmad, N. and Han, J.I. 2015. Global risk of pharmaceutical contamination from highly populated developing countries. Chemosphere 138: 1045–1055.

Rivera-Utrilla, J., Sanchez-Polo, M., Ferro-Garcia, M.A., Prados-Joya, G. and Ocampo-Perez, R. 2013. Pharmaceuticals as emerging contaminants and their removal from water. A review. Chemosphere. DOI: http://dx.doi.org/10.1016/j.chemosphere.2013.07.059

Rüdel, H., Bohmer, W. and Schroter-Kermani, C. 2006. Retrospective monitoring of synthetic musk compounds in aquatic biota from German rivers and coastal areas. J. Environ. Monit. 8(8): 812–823. (19)

Samal, K., Mahapatra, S. and Ali, M.H. 2022. Pharmaceutical wastewater as Emerging Contaminants (EC): Treatment technologies, impact on environment and human health. Energy Nexus 6: 100076.

Sauve, S. and Desrosiers, M. 2014. A review of what is an emerging contaminant. Chem. Cent. J. 8: 1–7.

Shaver, D. 2011. Sources and Fate of Emerging Contaminants in Municipal Wastewater Treatment. M.TechThesis; The University of Guelph, Guelph, Ontario, Canada.

Skrbic, B., Cvejanov, J. and Durisic-Mladenovic, N. 2005. Polycyclic aromatic hydrocarbons in surfacesoils of Novi Sad and bank sediment of the Danube river. J. Environ. Sci. Health, Part A: Environ. Sci. Eng. 40: 29–42.

Soclo, H.H., Garrigues, P.H. and Ewald, M. 2000. Origin of polycyclic aromatic hydrocarbons (PAHs) in coastal marine sediments: case studies in Cotonou (Benin) and Aquitaine (France) areas. Mar. Pollut. Bull 40(5): 387–396.

Subedi, B., Du, B., Chambliss, C.K., Koschorreck, J., Rudel, H., Quack, M. et al. 2012. Occurrence of pharmaceuticals and personal care products in german fish tissue: A national study. Environ. Sci. Technol. 46: 9047–9054.

Thomaidis, N.S., Asimakopoulos, A.G. and Bletsou, A.A. 2012. Emerging contaminants: a tutorial mini-review. Global NEST J. 14(1): 72–79.

Tiwari, B., Sellamuthu, B., Ouarda, Y., Drogui, P., Tyagi, R.D. and Buelna, G. 2017. Review on fate and mechanism of removal of pharmaceutical pollutants from wastewater using biological approach. Bioresour. Technol. 224: 1–12.

U.S. Geological Survey. 2017. Contaminants of Emerging Concern in the Environment. Environmental Health - Toxic Substances Hydrology Program. U.S. Geological Survey.

Vasilachi, I.C., Asiminicesei, D.M., Fertu, D.I. and Gavrilescu, M. 2021. Occurrence and Fate of Emerging Pollutants in Water Environment and Options for Their Removal. Water 13: 181.

Veen I.V.D. and Boer, J.D. 2012. Phosphorus flame retardants: Properties, production, environmental occurrence, toxicity and analysis. Chemosphere 88: 1119–1153.

Wesgate, R., Grasha, P. and Maillard, J.Y. 2016. Use of a predictive protocol to measure the antimicrobial resistance risks associated with biocidal product usage. Am. J. Infect. Control. 44: 458–464.

Wei, G.L., Li, D.Q., Zhuo, M.N., Liao, Y.S., Xie, Z.Y., Guo, T.L. et al. 2015. Organophosphate flame retardants and plasticizers: Sources, occurrence, toxicity and human exposure. Environ. Pollut. 196: 29–46.

Wenzel, A., Bohmer, W., Muller, J., Rü del, H. and SchroeterKermani, C. 2004. Retrospective monitoring of alkylphenols and alkylphenol monoethoxylates in aquatic biota from 1985 to 2001: Results from the German Environmental Specimen Bank. Environ. Sci. Technol. 38(6): 1654–1661.

Wlodarczyk-Makula, M. 2005. The loads of PAHs in wastewater and sewage sludge of municipal treatment plant. Polycycl. Aromat. Compd. 25: 183–194.

Xing-Hong, L., Ling-ling, M.A., Xiu-fen, L., Shan, F., Hang-xin, C. and Xiao-bai, X. 2006. Polycyclic aromatic hydrocarbon in urban soil from Beijing, China. J. Environ. Sci. 18(5): 944–950.

Yang, Y., Zhang, X.X. and Korenaga, T. 2002. Distribution of polynuclear aromatic hydrocarbons (PAHs) in the soil of Tokushima, Japan. Water Air Soil. Poll. 138: 51–60.

Yao, L., Yang, H., Chen, Z., Qiu, M., Hu, B. and Wang, X. 2021. Bismuth oxychloride-based materials for the removal of organic pollutants in wastewater. Chemosphere. 2021(273): 128576.

Zhang, C., Valsaraj, K.T., Constant, W.D. and Roy, D. 1999. Aerobic biodegradation kinetics of four anionic and nonionic surfactants at sub- and supra-critical micelle concentrations (CMCs). Water Res. 33: 115–124.

Zhu, F.J., Ma, W.L., Xu, T.F., Ding, Y., Zhao, X., Li, W.L. et al. 2018. Removal characteristic of surfactants in typical industrial and domestic wastewater treatment plants in Northeast China. Ecotoxicol Environ. Saf. 153: 84–90.

CHAPTER 2

Recent Updates and Estimation of Emerging Contaminants in the Environment

*Nidhi Yadav and Md. Ahmaruzzaman**

1. Introduction

An "emerging contaminant" (EC) is a chemical that has been found in the environment for the first time and was previously undetected or even just found in extremely small concentrations. There are numerous types of chemicals that fall under the category of EC, such as endocrine disrupters, disinfection by-products, industrial waste chemicals, persisting organic pollutants, natural toxins, hormones, pesticides, flame retardants with bromine, nanomaterials and lifestyle constituents like personal care products, pharmaceuticals, artificial sweeteners, and caffeine. However, the examples of ECs are not over here. Illicit drugs, metabolites of ECs, altered genes, etc., are also considered in the class of ECs. Concluding, the examples of all ECs endocrine disrupters, pharmaceuticals, and products of personal care are mainly detected ECs. Since approximately 90% of oral medications end up in the environment after passing through the human system, ECs such as carbamazepine (a nanogram per liter) and acesulfame (a milligram per liter) are among the most prominent examples of ECs in the industry (Fatta-Kassinos et al., 2011; Yadav and Ahmaruzzaman, 2022a). The occurring source of ECs could be water, air, and soil in the environment, but wastewater is one of the major occurring sources of ECs. Households' wastewater, storm and commercial water, discrete and diffuse sources, and processes of water treatment are all major sources of ECs in municipal wastewater

Department of Chemistry National Institute of Technology Silchar, Assam -788010.
* Corresponding author: mda2002@gmail.com

(Yadav and Ahmaruzzaman, 2022b). It is possible for discrete or diffuse sources to lead to the contamination of groundwater, surface water, and soil resources. Discrete sources are associated with distinct points and can be accessed by a single discharge point. Examples of discrete sources include urban and industrial areas; hospitals; municipal sewage; accidental leaks; plants for water treatment; and garbage dumps (Jurado et al., 2012; Balderacchi et al., 2013). On the other hand, diffuse sources are defined as large areas that do not have any obvious standard sources of discharge. Some examples of diffuse sources include particles and gases emitted from vehicles, industrial contaminants, farmlands, vast breeding cattle zones, recycling sites of e-waste disposal, irrigation with untreated and treated wastewater, and fertilization with sludge from WWTPs (Bone et al., 2010; Stuart et al., 2012). A by-product of the treatment procedures of wastewater treatment plants (WWTPs) is sewage sludge and a broad range of organic and inorganic compounds with ECs, could be present in sewage sludge. The sludge usually focuses on organic chemicals with less biodegradability and heavy metals as well as possibly hazardous species (bacteria, viruses, etc.), present in wastes due to the biological, chemical, and physical methods engaged in the WWTPs (Martín-Pozo et al., 2019).

Soil is the second most significant sink of ECs. The most common methods by which pollutants get into the soil are called direct entry routes. These channels include deliberate discharge; leachates from landfills; irrigation with untreated or treated wastewater; sewage leaks; phosphogypsum additions; and widespread cow breeding (Mahmoud and Abd El-Kader, 2015). The dry and moisture discharge of contaminants directly onto soil, where they might pile up, is an example of one of the indirect pathways. Organic contaminants are more capable to collect in soils that have a high organic matter. In these soils, organic pollutants could remain for years or migrate for long or short distances as part of particles of solids and gases that travel through the atmosphere (Fabietti et al., 2010). The contaminants that are distributed on the surface of the soil will go through the soil profile and eventually reach groundwater.

Transportation, combustion and agriculture, cosmetic products, domestic use of fossil fuels, and power generation are the main factors that produce the air ECs. Combustion processes produce pollutants like furans and dioxins into the environment. Preindustrial dioxin deposition was aided by the burning of plant debris such as forests, range fibers, and brushes. Transportation and sources influence the spread of furans and dioxins in the environment. Soil deposition, combustion, and volatilized and dispersed particles that were buried and discharged into the atmosphere are all sources of furans and dioxins. Moreover, various foods containing furans and dioxins also include vital minerals like vitamins, calcium, proteins, iron, and fish. Agriculture products like pesticides are direct ECs present in the air.

The existence of ECs could impact human health, ecotoxicological effects, bioaccumulation, and degrading features will have an impact not only on aquatic life but also on the cost and performance of plants for wastewater treatment and cause environmental pollution. The majority of them have been shown to change the properties of natural assets, interfere with biochemical processes that occur in the environment, and negatively impact the health of living beings by imposing detrimental impacts like congenital disorders, endocrine disruption, cancer, mutagenesis, etc.

(Barroso et al., 2019). They induce fertility problems, long fecundation, poor fetal growth, fetal damage, and fetal anomalies. The majority of human and veterinary medications are found in low amounts in the environment and do not elicit toxic effects. Yet, certain species are exposed to low doses over extended periods, resulting in significant chronic toxic consequences. Pharmaceuticals have ecotoxicological impacts on certain species that share active locations with target species, like cells, tissues, organs, and active molecules (Ahmed and Ahmaruzzaman, 2015). Even though these pollutants have been present in the environment for a long time, the invention of even more sensitive analytical technology in the past few decades has made them more significant around the globe since they can be detected at levels as low as parts per trillion in water, air, and soil (Dimpe and Nomngongo, 2016).

2. Emerging Contaminants

2.1 Pharmaceuticals

Pharmaceuticals are mainly categorized into two types which are "human pharmaceuticals" as well as veterinary medicines. The majority of human and veterinary medications are found in low amounts in the environment, but do not trigger acute toxicity. On the other hand, few organisms are exposed to low dosages across extended periods, resulting in extraordinary chronic toxic consequences (Ahmed and Ahmaruzzaman, 2015). It is because of their unique mechanisms of action that they remain in the body for so long. Pharmaceuticals have a profound influence on land. Among them are antibiotics, illegal and legal pharmaceuticals, steroids and analgesics, beta-blockers, and more, the quantities ranging from ngL^{-1} to low μgL^{-1}, indicating widespread use in the aquatic environment (Gadore and Ahmaruzzaman, 2021). Transformation products and metabolites of pharmaceuticals are often present in the environment, which may have environmental impacts that are much more harmful than those of the source compounds themselves (Evgenidou et al., 2015). Natural, sludge, sediment waterways, and drinking water have all been shown to contain the contaminant. Soil microorganisms should acquire antibiotic resistance as a result of their use. More than 160 different medicines have been found in aquatic systems and aquatic life forms, yet little is known about their eco-toxicological effects. Aquatic creatures are one of the most important targets since they are constantly exposed to wastewater remains throughout their lives (Gogoi et al., 2018).

2.2 Hormones

Synthetic and biological sex hormones (progestogens and estrogen) coming in WWTPs from commercial and urban emissions go through a range of treatment procedures, with various degrees of success, before being discharged into surface waters in certain conditions (Desbrow et al., 1998; Ternes et al., 1999). The presence of synthetic and biological progestogens and estrogen in different kinds of water has been recorded in most cases to be in the range of low ng/L to tens of ng/L, with only a few exceptions where concentrations were greater, reaching µg/L levels.

Due to their enormous estrogenic efficacy, the synthetic estrogen ethynyl estradiol and the biological hormone estradiol are the most important estrogens to examine in environmental programs (Ternes et al., 1999; Shore et al., 1993). These two molecules, on the other hand, are typically the least often discovered in ambient fluids, with estrone and estriol, the primary estradiol metabolites, being the most common. According to (Mastrup et al., 2005), 13 to 92% of estrogens reaching a river valley end up in the bed-sediment compartment, with the majority of sorption occurring within the first 24 hr of exposure. Synthetic estrogens, such as ethynyl estradiol and mestranol, have also been proven to partition the sediment more compared to biological estrogens (Lai et al., 2000). However, sorption to bed sediments seems to be a very plausible mechanism, given these compounds have low polarity, which has octanol-water partition coefficients ranging from 103 to 106. These chemicals are likely to undergo modest biodegradation and photodecomposition under the dark, anaerobic conditions found in the stratigraphy of river sediments. Progestogens and estrogens could therefore accumulate in river sediments for an extended time period, be transferred to other regions, and finally be discharged back into the water stream (Lai et al., 2000).

2.3 Nanomaterials

Particles having around 1–100 nm in size are defined as manufactured nanomaterials, such as amorphous silicon dioxide (SiO_2), titanium dioxide (TiO_2), and carbon nanotubes (CNTs), which are all examples of manufactured nanomaterials (Morimoto et al., 2010). Skincare products, agriculture, healthcare, transportation, materials, energy, and telecommunications all contain manufactured nanomaterials to some extent (Morimoto et al., 2010; Yadav and Ahmaruzzaman, 2021). Toxic heavy metals (such as cobalt, mercury, lead, cadmium, and others), toxic gases (such as NO_2, SO_2, and others), as well as polycyclic aromatic hydrocarbons, microorganisms, pesticides, proteins, refractive indices, and nucleotides (biologically active substances) can all be adsorbents in nanomaterials due to the atomic interface covering 15 to 50% of the total surface area. Biological and chemical degradation procedures fail to degrade nanoparticles because of their catalytic nature, greater durability, and high strength (Morimoto et al., 2010). While adsorbing diverse organic and inorganic compounds on their surfaces, nanomaterials manufactured experience conversion processes, long-term migration, and complicated chemical interactions in the environment. Due to this, new contaminants have been spawned. Animal and *in vitro* testing has shown well-characterized carcinogenic potential, particularly for developmental and reproductive toxicity at larger concentrations of produced nanomaterials (Ema et al., 2010).

2.4 Disinfection By-products

Disinfectants used in drinking water and swimming pools play an important role in protecting people from water-borne infections (Richardson et al., 2007; Villanueva et al., 2015). These compounds, which are often oxidizing agents, have significant chemical activities that remove pathogenic pathogens as well as react with a wide range of deoxidizers. As a consequence, throughout the disinfection processes, unwanted by-products are produced. The broad and regular use of these chemicals results in Disinfection By-Products (DBPs), especially in purified water, and chlorinated DBPs (CDBPs), and almost all individuals in advanced countries are subjected to these chemicals via drinking water and swimming pools (G et al., 2007). Aldehydes, ketones, iodinated trihalomethanes (THMs), halomethanes, carboxylic acids, hydroxy acids, keto acids, alcohols, esters, and sometimes even nitrosamines (NDMA) have been identified, including about 600 DBPs (G et al., 2007). Trihalomethane (THM) is one of the most common forms of halogenated DBPs, contributing to more than 80% of the total (Villanueva et al., 2015).

2.5 Dioxins and Furans

Two kinds of dioxins and furans have been observed which are polybrominated dibenzo-p-dioxin and dibenzofuran (PBDD/Fs) as well as polychlorinated dibenzo-p-dioxins (dioxins) and polychlorinated dibenzofurans (furans). Nonferrous metallurgy activities are a major source of cancerous polybrominated dibenzo-p-dioxin and dibenzofuran (PBDD/Fs), which travel across the world (Yang et al., 2021). The presence of PBDD/Fs in urban air and particle matters has been verified, raising concerns about PBDD/Fs air intake concerns (Li et al., 2008). Furthermore, PBDD/Fs have been observed to have substantially greater toxic equivalent (TEQ) concentrations in soil, dust, and vegetation than PCDD/Fs, showing that PBDD/Fs in environmental compartments from particular places may represent significant concerns for biodiversity (Suzuki et al., 2010). PBDD/Fs were found in greater concentrations in Baltic Sea fish than their chlorinated equivalents, and PBDD/Fs were deemed a distinct category of marine pollutants (Peter Haglund et al., 2007). Diet, which includes dairy products, meat, fish, and shellfish, is responsible for approximately 90% of dioxin and furan consumption in the general population. Some sediments and soils have high quantities of dioxins and furans. Plants, water, and air all have very low amounts. Dioxins are not commercially manufactured in the United States. Instead, they occur as pollutants during the manufacture of chlorinated organic compounds. They are also created as a consequence of burning activities like fossil fuel combustion and waste incineration. Forest fires and volcanic activity could potentially produce dioxins and furans in the environment. When people are exposed to tobacco smoke or drink polluted water, they are exposed to dioxins and furans (Jain, 2021).

2.6 Others

Generally, organic compounds like personal care products, UV filters, flame retardants (halogen flame retardants), detergents, disinfectants, plasticizers, food preservatives, food additives, surfactants, and many other organic compounds produced by industries as well as human activities are found in wastewater, soil, and air. Beauty products, aromas, shampoos, sunscreens, synthetic hormones, steroids, and other PCPs are only a few examples. Among the most commonly encountered PCPs in aquatic environments are UV filters, which are known to have an estrogenic action (Jurado et al., 2014; Noguera-Oviedo and Aga, 2016). These organic compounds have been found in ng/L to µg/L concentrations in surface water, groundwater, soils, and WWTPs.

3. Occurring Sources of ECs

3.1 Sewage Sludge

A by-product obtained by the treatment of wastewater is termed as Sewage Sludge (SS) and could be used as a fertilizer in agriculture or for land restoration in forestry. It is a valuable nutrient source, but the vast variety of pollutants that they contain could remain during the process of treatment raising concerns about its reuse. According to research, antibiotics, antihypertensives, and anti-inflammatories are the most common substances, with up to 5 µg/g^{-1} found (ciprofloxacin). Seasonal changes in anti-inflammatory medicines and antibiotics were discovered, along with concentration disparities compared with other European nations (Riva et al., 2021). ECs enter the sewage system mostly as a result of their use in human activities, both industrially (i.e., dumping or leaks of water used in the process of manufacturing) and domestically (e.g., the use and inappropriate disposal of medications). There is proof that considerable quantities of ECs are maintained in SS, which is a complicated matrix (Fijalkowski et al., 2017). The physical and chemical characteristics of every component, which could have diverse modes of interaction, determine whether ECs are present in SS or not. As ECs are a diverse set of chemicals and therefore form a heterogenous group, the main interaction occurs with SS from hydrophobicity, but the formation of hydrogen bonds, electrostatic attraction, or exchange of ions are also sources of interaction with ECs (Rybacka and Andersson, 2016). As a result, dispersing SS on soil could be a key pathway for ECs to enter the environment, posing a risk to both the ecosystem and human health. SS contains ECs on soils that potentially enter the shallow aquifer and/or streams near shedding zones by land runoffs or leaching and then may reach the food chain via polluted drinking water or contaminated crops (Martín-Pozo et al., 2019).

3.2 Municipal Solid Waste

Although wastewater has been extensively studied as a significant source of ECs for the environment, landfill leachate from MSW has received much less attention as a possible source of these compounds. MSW is a kind of trash that includes items like plastic containers, food waste, and product packing materials. Due to the financial benefits, MSW is now mostly disposed of in landfills. Electrical items, batteries, pharmaceuticals, paints, and oils are all frequently disposed of in landfills. The principal route of human pharmaceutical compounds entering the waterways has been identified as the domestic pathway via municipal sewage treatment facilities. Landfill leachate is an aquatic by-product of landfilling that results from precipitation percolating via waste, biochemical reactions in waste cells, and the intrinsic water content of waste (Qi et al., 2018). The amount of landfill leachate produced at each landfill is determined by its surface area and topography, as well as regional meteorological factors such as rainfall and cover type. The kind and proportion of chemicals found in landfill leachate will be determined by the amount of trash existing in landfills and the amount of leachate created. It could contain a wide range of ECs derived from landslide MSW (Qi et al., 2018). For example, pharmaceutical concentrations in MSW from Xi'an City (the capital of Shaanxi Province in China) were 100.9 ± 141.81 mg/kg, 63.8 ± 37.7 mg/kg, and 47.9 ± 8.1 mg/kg, respectively for oxytetracycline, tetracycline, and sulfamethoxazole (Song et al., 2016). As a result, landfill leachate is seen as an ECs reservoir and an ECs source for the surrounding ecosystem. Toxicity tests using various testing species have established the hazardous effects of landfill leachates and the need to remediate them to fulfill discharge requirements in receiving waterways.

Leachate is seldom recovered in aged landfills or those without liners, and is prone to slowly trickling into the groundwater. The leachate is recovered and recycled on-site using membrane treatment, traditional treatment (biological and physical/chemical treatment), or transfer off-site to an STP for newly constructed landfills and those with liners. Most ECs, on the other hand, do not break down biologically. Instead, they may build up in biosolids or can be released into the effluent from WWTPs.

3.3 Wastewater

A large number of ECs are released into wastewater. Municipal wastewater is the primary source of ECs into the environment, along with point and nonpoint sources, wastewater from houses, industry, stormwater, and facilities for the treatment of water. ECs flow to MWWTP after being ingested by the community and stored in sewage systems, which are often not intended to eliminate them. Wu et al. (2016) assessed 10 antibiotics in four WWTPs in Shanghai (PRC), finding that they had been detected in most effluent samples. Antibiotic hotspots for receiving bodies of water were created by WWTPs. Dong et al. found consistent observations after studying the fate of 19 antibiotics in four WWTPs: one Designed Wetland (DW), one Activated Sludge (AS), one Stabilization Pond (SP), and one micropower biofilm (MP) (Yadav and Ahmaruzzaman, 2023). Despite the mean concentration of effluent of the targeted concentration of antibiotics in the effluent being typically lower than in the influent, their elimination was typically insufficient. The AS and DW systems outperformed the MP and SP systems, with the AS outperforming the DW. Summertime had better elimination efficiencies for both the AS and DW systems than winter, demonstrating that biodegradation could play a significant role in the degradation of antibiotics even if total elimination is not attained.

Personal care products (PCPs), pharmaceuticals, hormones, and EDCs are among the ECs introduced into hospital wastewater. These toxins have the potential to permeate natural habitats, both aquatic as well as terrestrial, posing a risk to aquatic life and human health. Traditional wastewater treatment facilities are not intended to handle all forms of chemical and biological pollutants, enabling toxins to be released into bodies of water, including streams, rivers, and ultimately groundwater. As a result, sufficient precautions and modern technology to avoid the release of ECs into aquatic habitats should be established (Khan et al., 2021). The wastewater disposed of by hospitals contains a majority of medications. Ampicillin values of 20 to 80 mg/L were discovered in the wastewater of a big German hospital. Ciprofloxacin was discovered in hospital effluent at quantities ranging from 0.7 to 124.5 mg/L. The concentrations of antibiotics in hospital effluents are on the same scale as the Minimal Inhibitory Concentration (MIC) for vulnerable pathogenic microorganisms. Since municipal wastewater also includes disinfectants from homes, antibiotic compounds, and veterinary sources, and to a lesser degree, livestock, hospital effluent dilution by municipal sewage would only modestly diminish antibiotic concentrations (Kümmerer et al., 2000). In terms of concentration and frequency, TCS (triclosan) and TCC (triclocarban) were amongst the top 10 most frequently identified organic chemicals in wastewater. It has been observed that the concentrations of TCS were higher than 10 mg/L in WWTP effluent (Brausch and Rand, 2011). Over the past 5 yr, TCC has been found at higher concentrations and more often in SW and WWTP effluent than TCS or M-TCS (Brausch and Rand, 2011). PBDEs enter WWTPs through wastewater from human activity, such as the washing of rugs, carpets, or gray water; water discharged from PBDE-containing industrial equipment; and leachate from PBDE-containing land-filled items. Designed NPs are discharged into the environment passively via the discharge of wastewater from items throughout their manufacturing and widely attractive use (cosmetics, pharmaceuticals, sunscreen formulas, paints, and so on).

3.4 *Emerging Contaminants in Air*

Air is the main resource of ECs obtained from transportation, combustion, agriculture products, combustion of fossil fuels, industrialization, etc. Fossil fuels are vital to the world's marketplace for a number of reasons, including transportation, energy generation, polymer and chemical production, heating, and many more applications. Nevertheless, fossil fuel processing and extraction, as well as their use , have significant environmental and natural resource implications, especially water. Large oil spills, such as the one caused by the Deepwater Horizon drilling rig, which spilled over 4.9 million barrels (780,000 cubic meters) of crude oil into the Gulf of Mexico, have heightened awareness of the potential for oil drilling catastrophes to pollute the environment. The increasing knowledge of the substantial hazards that natural gas fracking activities pose to groundwater and surface water quality brings additional concerns. Even the usual mining and processing of fossil fuels damages the environment (Allen et al., 2012). Through emissions of hazardous gases and particles as well as carbon dioxide (CO^2), a co-pollutant that is a significant cause of climate change, fossil fuel burning causes a number of substantial health and developmental effects. Each of the many toxins generated by fossil fuel combustion has the potential to have numerous and cumulative negative impacts on harmful air pollution and climate change, either indirectly or directly (McMichael et al., 2006).

Air plays a significant role as the transport medium for numerous ECs like pesticides, dioxins, furans, fire retardants, nanoparticles, and many more. For example, pesticides are released into the atmosphere via many mechanisms either during or after applications. Pesticides are dispersed effectively into the air through spraying pesticides, orchard mist blowers, and grounding spread sprays among other ways. Drifting is the transport of pesticide droplets from targeted to nontarget places via the air. Tiny pesticide droplets produced by high pressure and tiny nozzles are more easily drifted. Moreover, pesticides discharged near the ground are less likely than those launched from airplanes to disperse in the wind. Drifting is much less dangerous in calm weather than in stormy weather. For several BFRs (Brominated Fire Retardants), air is a significant mode of transportation to sites distant from the source. Outdoor and indoor air samples both include BFRs. BFR volatilization is dependent on the compound's vapor pressure. PBDEs (polybrominated diphenyl ethers) have been observed to be highly persistent in the environment in general (McMichael et al., 2006). Generally, nanoparticles are larger molecules than molecules of air (0.3 nm), yet smaller than the air quality monitoring standards' top limits. Spontaneous or inter-action-assisted aggregation of much smaller parent solid particles produces fractal-like suspended nanoparticles in gas. When nanoparticles are discharged into the environment or moved into the wet respiratory system, their shape, size, and density could alter.

3.5 Emerging Contaminants in Soil

Agrochemical residues are usually deposited in soil, but depending on their solubility, mobility, and decomposition processes, groundwater and surface water bodies could be the eventual receivers. The newly implemented regulation No. 1107/2009 and the old directive 91/414/EEC, which control the authorization of pesticide distribution on the market, address soil pollution by pesticides in the union of Europe. Pesticide occurrences and behavior in soil and sediment settings are mostly controlled by persistence, transportation, and transformation mechanisms, which are regulated by climate conditions, pesticide characteristics, method of entry, quality of soil, and plant or microbiological factors (Cheng, 1990).

Although soil has been suggested as a significant sink for several BFRs, there are few data available that assess the contamination level in the soil environment. BFRs have also been found in river sediments. Harrad and Hunter (2006) studied PBDE congenator concentrations in surface soil in the West Midlands of the United Kingdom. The primary components of the PBDEs discovered in the soils were the congenator BDE-47 and BDE-99. In these soils, the total of five congenator BDE-28, BDE -47, BDE -99, BDE -100, BDE -153, and BDE -154 ranged from 0.241 to 3.89 ng/g (Harrad and Hunter, 2006). A number of activities, including intended discharges through soil and water restoration methods, will expose soils to nanomaterials. chemicals used in agriculture, such as fertilizers, and inadvertent emissions through air, water, and sewage sludge sprayed on land. The migration of nanomaterials through all the soil profiles could pose a danger of groundwater pollution. The behavior, transport, and destiny of nanomaterials in soils are poorly understood. Nanomaterials might be taken up and destroyed by soil organisms, although there is little information on this. The information supplied is for nanomaterials used in polluted land cleanup. As efficient restoration demands particle mobility within the soil, most of this work has concentrated on nanomaterial transport through the soil.

Plasticizers have been found in a number of soil and sediment samples. As adsorption is a prominent demise process for several plasticizers, according to their carbon content, sediments may operate as a sink for dissolving plasticizers. The highest recorded amount of DnBP (di-n-butyl phthalate) in river sediments is 1,100 mg/kg, with observed quantities ranging from 100 to 1,100 mg/kg. Other amounts observed were in canal sediments of 0.2–150 mg kg1 DMP and in river sediments of 0.18–70 mg kg1 DEHP (HSDB 2001) (Kohli et al., 2017).

4. Adverse Effect of Contaminants

Environment pollution (water, soil, air) by ECs has become an aspect of primary concern because of the wide use of these compounds in human activities. This pollution directly or indirectly affects human health and other living things. From the source of contamination, these ECs enter directly or indirectly into the food chain system and affect the ecology. Water contains a high concentration of ECs in comparison to soil and air because water is the end-terminating source for ECs, so the major impact of ECs has been shown on aquatic life. Most ECs (such as DBPs, PFCs, UV filters, and personal care products) show carcinogenic behavior that causes various types of cancer like brain tumors, breast cancer, pancreatic cancer, and many more in human beings. Further issues such as infertility, fetal growth, sperm quality, and pregnancy problems have been caused by ECs. Pharmaceuticals, like psychotropic drugs, have harmful effects on aquatic life in trace amounts (i.e., ng/L to µg/L). Animals and humans could be infected by bacteria (multidrug-resistant) in the environment directly from the environment or from contaminated drinking water and food (Rousham et al., 2018).

4.1 Cancer

The carcinogenic behavior of ECs has been found by laboratory research on different animals and mouse models. Numerous research has validated THMs' (trihalomethanes) carcinogenic effects, showing that THMs in drinkable water are linked to colorectal cancers, breast cancer, and bladder tumors as a consequence of non-genetic toxicity. THM exposure through respiratory and oral routes, as well as cutaneous water absorption during swimming and bathing, were examined by Villanueva et al., Long-term THM consumption was associated with a twofold increase in the risk of bladder cancer, with an OR of 2.10 and a 95% confidence interval of 1.09 to 4.02 for THM concentrations of > 49 and 8 μg/liter, respectively. When compared to those who did not consume chlorinated water, those who consumed 35 μg of THM per day had an OR of 1.35 and a 95% confidence interval of 0.92 to 1.99. The OR for bath or shower time based on THM concentration was 1.83, with a 95% confidence interval of 1.17 to 2.87. Swimming pools were associated with an OR of 1.57 (95% CI: 1.18–2.09). Bladder cancer has been linked to long-term consumption of THMs from chlorinated water, which is common in developed regions. Chang et al. (2007) also discovered a link between the consumption of DBP and bladder cancer. The OR for rectal cancer was 0.91 (0.55–1.51), whereas, for 75 µg/L, the OR for colon cancer was 1.63 (1.07–2.48). DBP consumption has long been linked to an increased risk of colon cancer in men. Females who were exposed to DBP had no increased risk of colon cancer.

The most extensively used oxygenated bunker MTBE (methyl tert-butyl ether), is extensively applied as a new unleaded gasoline additive, especially in developing areas. MTBE may pollute groundwater and surface water, posing a major hazard to drinking water supplies. Due to the great half-life of MTBE in groundwater, its decomposition is difficult. The special properties and structure of MTBE make it more sustainable in groundwater and increase the half-life time. Inhalation exposures to MTBE absorb it quickly (McGregor, 2008). Humans could be exposed to MTBE through the consumption of polluted water and dermal absorption (Ahmed, 2001). MTBE has been shown in animal tests to cause uterine, testicular, and renal cancer, as well as harm the immune system, liver, kidney, and central nervous system (McGregor, 2008; Ahmed, 2001). The human body could be harmed by MTBE, which is classified as a possible carcinogen. Acute exposure to MTBE has been linked to a number of problems, including headaches, sicchasia, and nasal and optical stimulation.

4.2 Bioaccumulation

In mussel tissues, 43 medicines from various classes were found, with factors of bioaccumulation to the extent of 0.66 (metformin) to 32,022 (sertraline). Vernouillet et al. (2010) found that the antiepileptic medication carbamazepine (CBZ) was bioaccumulated by the algae Raphidocelis subcapitata as well as the crustacea thamnocephalus platyurus, with an extent of bioaccumulation of 2.2 and 12.6, correspondingly. Du et al. (2015) discovered that fluoxetine accumulates in snails with a bioaccumulation value of 3000. One hundred and forty five PPCPs were evaluated in wild and confined mussels from the St. Lawrence River, Ontario, by de Solla et al. (2016). Wang and Gardinali (2013) also found mosquito fish absorption and depuration of pharmaceuticals in reclaimed water (Gambusia holbrooki). Carbamazepine, caffeine, diltiazem, diphenhydramine, and ibuprofen had an extent of bioaccumulation of 1.4, 2.0, 16, 16, and 28, correspondingly. With a bioaccumulation value of 12, oxazepam was found in large amounts in Eurasian perch fish (Brodin et al., 2014). The impact of PPCPs on nontarget creatures, particularly fish, has been studied in a lot of research. The plasma bioconcentration factor of goldfish (Carassius auratus) exposed to waterborne gemfibrozil at an environmentally appropriate concentration for 14 d was 113 (Mimeault et al., 2005). TCS has not been shown to bioaccumulate in aquatic species, but methyl triclosan (M-TCS) (a methyl derivative of TCS) has been found to bioaccumulate in Sesbania herbacea after 28 d (Stevens et al., 2009). Nevertheless, other investigations have shown that bioaccumulation of TCS occurs at a considerably larger level in algae than in M-TCS (Coogan et al., 2007), revealing that bioaccumulation of TCS is much greater in algae than in M-TCS. It is still uncertain if evolutionary variations in physiology throughout trophic groups could impact bioaccumulation or what variables are most relevant for the intake and accumulation. The potential ionization of TCS is one probable reason for the variances in the bioaccumulation of TCS. Based on pKa values of 7.8 at usual pH ranges in SWs, TCS spans from entirely protonated (pH 14.4) to wholly deprotonated (pH 14.2) under typical environmental conditions. These ionization discrepancies result in variances in Dow values, which are related to bioaccumulation differences. However, this is not analyzed for TCS. Investigations with the pharmaceutical fluoxetine (Prozac) show significant changes in bioaccumulation at various pHs (Nakamura et al., 2008). TCS could be predicted to accumulate more at higher pHs due to its pKa value of 7.8, while M-TCS will be assumed to accumulate at greater levels at lower pHs based on these findings.

4.3 Reproductive Hazards

DBPs have recently been hypothesized as infertility risk factors, a lengthy gestational duration, fetal loss, and poor fetal development, as well as fetal malformations, some of which have been investigated in existing records or ongoing research. Several studies looked at DBPs in water from taps and the sperm quality and found that DBP consumption had a detrimental impact on normal sperm morphology and concentration, but there was no effect on motility (Luben et al., 2007). All these studies found no link between the poor quality of sperm in people and the concentration of DBP at or below the regulatory threshold, indicating that further research is needed (Luben et al., 2007; Iszatt et al., 2013). Another study examined the relationship between the menstrual cycle and DBPs and concluded that THM exposure could impact ovary functions, with decreased follicular phase lengths and cycle lengths (Windham et al., 2003). Waller et al. conducted a prospective study and concluded that women who consumed five or more glasses of cold tap water per day containing 75 g/liter of total THMs had a higher risk of spontaneous abortion ("Trihalomethanes in Drinking Water and Spontaneous Abortion : Epidemiology," n.d.). In 1917 (Italy), Righi et al. (2012) conducted a case study on various congenital malformations and discovered abdominal wall defects (OR: 6.88; 95% CI: 1.67–28.33), high renal defects in newborns (OR: 3.30; 95% CI: 1.35–8.09), and cleft palate (OR: 4.1; 95% CI: 0.98–16.8). Babies with cleft palate (OR: 9.60; 95% CI: 1.04–88.9), obstructive urinary abnormalities (OR: 2.88; 95% CI: 1.09–7.63), and spinal bifida (OR: 4.94; 95% CI: 1.10–22) were more likely in women exposed to chlorate levels of 200 g/L. This could be because chlorine dioxide is the most commonly used water disinfectant in Italy.

4.4 Others Effects

According to Joseph and Weiner's (2010) analysis, hypersensitive rhinitis, headaches, coughs, throat stimulation, upper respiratory communicable disease, anxiety, sicchasia, lightheadedness, difficulty in breathing, otitis media, trouble in sleeping, skin infection, irregular heartbeats, malaise, allergy were all linked to MTBE levels in the air. Exposure to NPs in humans occurs primarily during NM synthesis, absorption for medication delivery systems, and possibly inhalation of NPs discharged into the environment physical characteristics of NPs that vary based on their size, origin, and content, with some posing long-term health and environment problems. NPs containing metals could interact with other pollutants and induce cytotoxicity under specific biological and environmental circumstances. Particle-induced reactive oxygen species might well be produced in conjunction with UV radiation, posing a risk of cell damage (Lofrano et al., 2017). Several human studies have looked at the potential negative consequences of UV filters. Currently, the majority of case reports are connected to dermatitis produced by sunscreens. Pharmaceuticals have ecotoxicity impacts on certain nontarget species that share active regions with target species, like cells, tissues, organs, and active molecules. Several studies concentrating on the acute harmful consequences of drug ingestion on aquatic species have recently been published. Propranolol has a significant toxic effect on zooplankton and benthos, with a fatal dosage of 50% (LC50) of around 1 mg/L, while fluoxetine has a greater toxic effect on benthos in comparison to propranolol, with an LC50 of less than 0.5 mg/L (Brooks et al., 2003). The most important representative PFCs are perfluorooctanoic acid (PFOA), perfluorooctanesulfonate (PFOS), and their salts, which can be used in fighting fires, foams, metal spray plating, lubricants, and washing powder products, inks, coating formulations, varnishes, waxes, and textiles. PFCs have great light, heat, and chemical stability, and the microbial metabolism does not readily destroy them (Miralles-Marco and Harrad, 2015). PFCs are therefore considered bioaccumulative, persistent, and possibly harmful to humans and animals (Yadav et al., 2023).

5. Conclusion

This chapter contains a detailed description of ECs and their occurrence sources. There is limited research conducted on the adverse effects of ECs, but this aspect should not be ignored. However, the available reports regarding the adverse effects of ECs have been included in the chapter. ECs are a wide class of pollutants that are continuously affecting the ecosystem with their hazardous effects. The presence of ECs is investigated everywhere in the environment, including water, soil, and air. Although the majority of EC concentrations have been found in water bodies as most human activities are associated with water bodies, which severely affect the lives of aquatic species. Therefore, the direct release of ECs into water streams needs to be avoided.

References

Ahmed, F.E. 2001. Toxicology and human health effects following exposure to oxygenated or reformulated gasoline. Toxicol. Lett. 123: 89–113. https://doi.org/10.1016/S0378-4274(01)00375-7.

Ahmed, M.J.K. and Ahmaruzzaman, M. 2015. A facile synthesis of Fe3O4–charcoal composite for the sorption of a hazardous dye from aquatic environment. J. Environ. Manage. 163: 163–173. https://doi.org/10.1016/J.JENVMAN.2015.08.011.

Allen, L., Cohen, M.J., Abelson, D. and Miller, B. 2012. Fossil Fuels and Water Quality. The World's Water 73–96. https://doi.org/10.5822/978-1-59726-228-6_4.

Balderacchi, M., Benoit, P., Cambier, P., Eklo, O.M., Gargini, A., Gemitzi, A. et al. 2013. Groundwater Pollution and Quality Monitoring Approaches at the European Level. http://dx.doi.org/10.1080/106 43389.2011.604259 43, 323–408. https://doi.org/10.1080/10643389.2011.604259.

Barroso, P.J., Santos, J.L., Martín, J., Aparicio, I. and Alonso, E. 2019. Emerging contaminants in the atmosphere: Analysis, occurrence and future challenges. https://doi.org/10.1080/10643389.2018.15 40761 49: 104–171. https://doi.org/10.1080/10643389.2018.1540761.

Bone, J., Head, M., Barraclough, D., Archer, M., Scheib, C. and Flight, D. et al. 2010. Soil quality assessment under emerging regulatory requirements. Environ. Int. 36: 609–622. https://doi.org/10.1016/J.ENVINT.2010.04.010.

Brausch, J.M. and Rand, G.M. 2011. A review of personal care products in the aquatic environment: Environmental concentrations and toxicity. Chemosphere 82: 1518–1532. https://doi.org/10.1016/J.CHEMOSPHERE.2010.11.018.

Brodin, T., Piovano, S., Fick, J., Klaminder, J., Heynen, M. and Jonsson, M. 2014. Ecological effects of pharmaceuticals in aquatic systems—impacts through behavioural alterations. Philos. Trans. R. Soc. B Biol. Sci. 369. https://doi.org/10.1098/RSTB.2013.0580.

Brooks, B.W., Foran, C.M., Richards, S.M., Weston, J., Turner, P.K., Stanley, J.K. et al. 2003. Aquatic ecotoxicology of fluoxetine. Toxicol. Lett. 142: 169–183. https://doi.org/10.1016/S0378-4274(03)00066-3.

C. Desbrow, †, Routledge, E.J., *,‡, G.C. Brighty, §, J. P. Sumpter, ‡ and Waldock†, M. 1998. Identification of Estrogenic Chemicals in STW Effluent. 1. Chemical fractionation and *in vitro* biological screening. Environ. Sci. Technol. 32: 1549–1558. https://doi.org/10.1021/ES9707973.

Chang, C.C., Ho, S.C., Wang, L.Y. and Yang, C.Y. 2007. Bladder Cancer in Taiwan: Relationship to Trihalomethane Concentrations Present in Drinking-Water Supplies. http://dx.doi.org/10.1080/15287390701459031 70: 1752–1757. https://doi.org/10.1080/15287390701459031.

Cheng, H.H. 1990. Pesticides in the soil environment: processes, impacts, and modeling.

Coogan, M.A., Edziyie, R.E., La Point, T.W. and Venables, B.J. 2007. Algal bioaccumulation of triclocarban, triclosan, and methyl-triclosan in a North Texas wastewater treatment plant receiving stream. Chemosphere 67: 1911–1918. https://doi.org/10.1016/J.CHEMOSPHERE.2006.12.027.

de Solla, S.R., Gilroy, A.M., Klinck, J.S., King, L.E., McInnis, R., Struger, J. et al. 2016. Bioaccumulation of pharmaceuticals and personal care products in the unionid mussel Lasmigona costata in a river receiving wastewater effluent. Chemosphere 146: 486–496. https://doi.org/10.1016/J.CHEMOSPHERE.2015.12.022.

Dimpe, K.M. and Nomngongo, P.N. 2016. Current sample preparation methodologies for analysis of emerging pollutants in different environmental matrices. TrAC Trends Anal. Chem. 82: 199–207. https://doi.org/10.1016/J.TRAC.2016.05.023.

Dong, H., Yuan, X., Wang, W. and Qiang, Z. (2016). Occurrence and removal of antibiotics in ecological and conventional wastewater treatment processes: A field study. Journal of Environmental Management 178: 11–19.

Du, B., Haddad, S.P., Scott, W.C., Chambliss, C.K. and Brooks, B.W. 2015. Pharmaceutical bioaccumulation by periphyton and snails in an effluent-dependent stream during an extreme drought. Chemosphere 119: 927–934. https://doi.org/10.1016/J.CHEMOSPHERE.2014.08.044.

Ema, M., Kobayashi, N., Naya, M., Hanai, S. and Nakanishi, J. 2010. Reproductive and developmental toxicity studies of manufactured nanomaterials. Reprod. Toxicol. 30: 343–352. https://doi.org/10.1016/J.REPROTOX.2010.06.002.

Evgenidou, E.N., Konstantinou, I.K. and Lambropoulou, D.A. 2015. Occurrence and removal of transformation products of PPCPs and illicit drugs in wastewaters: A review. Sci. Total Environ. 505, 905–926. https://doi.org/10.1016/J.SCITOTENV.2014.10.021.

Fabietti, G., Biasioli, M., Barberis, R. and Ajmone-Marsan, F. 2010. Soil contamination by organic and inorganic pollutants at the regional scale: The case of piedmont, Italy. J. Soils Sediments 10: 290–300. https://doi.org/10.1007/S11368-009-0114-9/FIGURES/5.

Fatta-Kassinos, D., Meric, S. and Nikolaou, A. 2011. Pharmaceutical residues in environmental waters and wastewater: Current state of knowledge and future research. Anal. Bioanal. Chem. 399: 251–275. https://doi.org/10.1007/S00216-010-4300-9/TABLES/8.

Fijalkowski, K., Rorat, A., Grobelak, A. and Kacprzak, M.J. 2017. The presence of contaminations in sewage sludge—The current situation. J. Environ. Manage. 203: 1126–1136. https://doi.org/10.1016/J.JENVMAN.2017.05.068.

Fantuzzi, G., Aggazzotti, G., Righi, E., Predieri, G., Giacobazzi, P., Kanitz, S., ... and Triassi, M. (2007). Exposure to organic halogen compounds in drinking water of 9 Italian regions: exposure to chlorites, chlorates, thrihalomethanes, trichloroethylene and tetrachloroethylene. Annali di igiene: medicina preventiva e di comunita, 19(4): 345–354.

Gadore, V. and Ahmaruzzaman, M. 2021. Tailored fly ash materials: A recent progress of their properties and applications for remediation of organic and inorganic contaminants from water. J. Water Process Eng. 41: 101910. https://doi.org/10.1016/J.JWPE.2020.101910.

Gogoi, A., Mazumder, P., Tyagi, V.K., Tushara Chaminda, G.G., An, A.K. and Kumar, M. 2018. Occurrence and fate of emerging contaminants in water environment: A review. Groundw. Sustain. Dev. 6: 169–180. https://doi.org/10.1016/J.GSD.2017.12.009.

Harrad, S. and Hunter, S. 2006. Concentrations of polybrominated diphenyl ethers in air and soil on a rural-urban transect across a major UK conurbation. Environ. Sci. Technol. 40: 4548–4553. https://doi.org/10.1021/ES0606879/SUPPL_FILE/ES0606879SI20060508_100306.PDF.

Iszatt, N., Nieuwenhuijsen, M.J., Bennett, J., Best, N., Povey, A.C., Pacey, A.A. et al. 2013. Chlorination by-products in tap water and semen quality in England and Wales. Occup. Environ. Med. 70: 754–760. https://doi.org/10.1136/OEMED-2012-101339.

Jain, R.B. 2021. Trends in concentrations of selected dioxins and furans across various stages of kidney function for US adults. Environ. Sci. Pollut. Res. 2021 2832 28: 43763–43776. https://doi.org/10.1007/S11356-021-13844-3.

Joseph, P.M. and Weiner, M.G. 2010. Visits to Physicians after the Oxygenation of Gasoline in Philadelphia. https://doi.org/10.1080/00039890209602929 57: 137–154. https://doi.org/10.1080/00039890209602929.

Jurado, A., Gago-Ferrero, P., Vàzquez-Suñé, E., Carrera, J., Pujades, E., Díaz-Cruz, M.S. et al. 2014. Urban groundwater contamination by residues of UV filters. J. Hazard. Mater. 271: 141–149. https://doi.org/10.1016/J.JHAZMAT.2014.01.036.

Jurado, A., Vàzquez-Suñé, E., Carrera, J., López de Alda, M., Pujades, E. and Barceló, D. 2012. Emerging organic contaminants in groundwater in Spain: A review of sources, recent occurrence and fate in a European context. Sci. Total Environ. 440: 82–94. https://doi.org/10.1016/J.SCITOTENV.2012.08.029.

Khan, M.T., Shah, I.A., Ihsanullah, I., Naushad, M., Ali, S., Shah, S.H.A. et al. 2021. Hospital wastewater as a source of environmental contamination: An overview of management practices, environmental risks, and treatment processes. J. Water Process Eng. 41: 101990. https://doi.org/10.1016/J.JWPE.2021.101990.

Kohli, J., Ryan, J.F. and Afghan, B.K. 2017. Phthalate Esters in the Aquatic Environment. Anal. Trace Org. Aquat. Environ. 243–281. https://doi.org/10.1201/9781315149882-7.

Kümmerer, K., Al-Ahmad, A. and Mersch-Sundermann, V. 2000. Biodegradability of some antibiotics, elimination of the genotoxicity and affection of wastewater bacteria in a simple test. Chemosphere 40: 701–710. https://doi.org/10.1016/S0045-6535(99)00439-7.

Lai, K.M., Johnson, K.L., Scrimshaw, M.D. and Lester, J.N. 2000. Binding of Waterborne Steroid Estrogens to Solid Phases in River and Estuarine Systems. Environ. Sci. Technol. 34: 3890–3894. https://doi.org/10.1021/ES9912729.

Li, H., Feng, J., Sheng, G., Lü, S., Fu, J., Peng, P. et al. 2008. The PCDD/F and PBDD/F pollution in the ambient atmosphere of Shanghai, China. Chemosphere 70: 576–583. https://doi.org/10.1016/J.CHEMOSPHERE.2007.07.001.

Lofrano, G., Libralato, G., Sharma, S.K. and Carotenuto, M. 2017. Nano based photocatalytic degradation of pharmaceuticals. Nanotechnologies Environ. Remediat. Appl. Implic. 221–238. https://doi.org/10.1007/978-3-319-53162-5_7/COVER.

Luben, T.J., Olshan, A.F., Herring, A.H., Jeffay, S., Strader, L., Buus, R.M. et al. 2007. The Healthy Men Study: An Evaluation of Exposure to Disinfection By-Products in Tap Water and Sperm Quality. Environ. Health Perspect. 115: 1169–1176. https://doi.org/10.1289/EHP.10120.

Mahmoud, E. and Abd El-Kader, N. 2015. Heavy Metal Immobilization in Contaminated Soils using Phosphogypsum and Rice Straw Compost. L. Degrad. Dev. 26: 819–824. https://doi.org/10.1002/LDR.2288.

Martín-Pozo, L., de Alarcón-Gómez, B., Rodríguez-Gómez, R., García-Córcoles, M.T., Çipa, M. and Zafra-Gómez, A. 2019. Analytical methods for the determination of emerging contaminants in sewage sludge samples. A review. Talanta 192: 508–533. https://doi.org/10.1016/J.TALANTA.2018.09.056.

Mastrup, M., Schäfer, A.I. and Khan, S.J. 2005. Predicting fate of the contraceptive pill in wastewater treatment and discharge. Water Sci. Technol. 52: 279–286. https://doi.org/10.2166/WST.2005.0274.

McGregor, D. 2008. Methyl tertiary-Butyl Ether: Studies for Potential Human Health Hazards. http://dx.doi.org/10.1080/10408440600569938 36, 319–358. https://doi.org/10.1080/10408440600569938.

McMichael, A.J., Woodruff, R.E. and Hales, S. 2006. Climate change and human health: present and future risks. Lancet 367: 859–869. https://doi.org/10.1016/S0140-6736(06)68079-3.

Mimeault, C., Woodhouse, A.J., Miao, X.S., Metcalfe, C.D., Moon, T.W. and Trudeau, V.L. 2005. The human lipid regulator, gemfibrozil bioconcentrates and reduces testosterone in the goldfish, Carassius auratus. Aquat. Toxicol. 73: 44–54. https://doi.org/10.1016/J.AQUATOX.2005.01.009.

Miralles-Marco, A. and Harrad, S. 2015. Perfluorooctane sulfonate: A review of human exposure, biomonitoring and the environmental forensics utility of its chirality and isomer distribution. Environ. Int. 77: 148–159. https://doi.org/10.1016/J.ENVINT.2015.02.002.

Morimoto, Y., Kobayashi, N., Shinohara, N., Myojo, T., Tanaka, I. and Nakanishi, J. 2010. Hazard Assessments of Manufactured Nanomaterials. J. Occup. Health 52: 325–334. https://doi.org/10.1539/JOH.R10003.

Nakamura, Y., Yamamoto, H., Sekizawa, J., Kondo, T., Hirai, N. and Tatarazako, N. 2008. The effects of pH on fluoxetine in Japanese medaka (Oryzias latipes): Acute toxicity in fish larvae and bioaccumulation in juvenile fish. Chemosphere 70: 865–873. https://doi.org/10.1016/J.CHEMOSPHERE.2007.06.089.

Noguera-Oviedo, K. and Aga, D.S. 2016. Lessons learned from more than two decades of research on emerging contaminants in the environment. J. Hazard. Mater. 316: 242–251. https://doi.org/10.1016/J.JHAZMAT.2016.04.058.

Peter Haglund, *,†, Anna Malmvärn, ‡, Sture Bergek, †, Anders Bignert, §, Lena Kautsky, ‖, Takeshi Nakano, ⊥ et al. 2007. Brominated Dibenzo-p-Dioxins: A New Class of Marine Toxins? Environ. Sci. Technol. 41: 3069–3074. https://doi.org/10.1021/ES0624725.

Qi, C., Huang, J., Wang, B., Deng, S., Wang, Y. and Yu, G. 2018. Contaminants of emerging concern in landfill leachate in China: A review. Emerg. Contam. 4: 1–10. https://doi.org/10.1016/J.EMCON.2018.06.001.

Richardson, S., Plewa, M., Wagner, E., Schoeny, R. and Demarini, D. 2007. Occurrence, genotoxicity, and carcinogenicity of regulated and emerging disinfection by-products in drinking water: A review and roadmap for research. Mutat. Res. Mutat. Res. 636: 178–242. https://doi.org/10.1016/J.MRREV.2007.09.001.

Righi, E., Bechtold, P., Tortorici, D., Lauriola, P., Calzolari, E., Astolfi, G. et al. 2012. Trihalomethanes, chlorite, chlorate in drinking water and risk of congenital anomalies: A population-based case-control study in Northern Italy. Environ. Res. 116: 66–73. https://doi.org/10.1016/J.ENVRES.2012.04.014.

Riva, F., Zuccato, E., Pacciani, C., Colombo, A. and Castiglioni, S. 2021. A multi-residue analytical method for extraction and analysis of pharmaceuticals and other selected emerging contaminants in sewage sludge. Anal. Methods 13: 526–535. https://doi.org/10.1039/D0AY02027C.

Rousham, E.K., Unicomb, L. and Islam, M.A. 2018. Human, animal and environmental contributors to antibiotic resistance in low-resource settings: integrating behavioural, epidemiological and One Health approaches. Proc. R. Soc. B Biol. Sci. 285. https://doi.org/10.1098/RSPB.2018.0332.

Rybacka, A. and Andersson, P.L. 2016. Considering ionic state in modeling sorption of pharmaceuticals to sewage sludge. Chemosphere 165: 284–293. https://doi.org/10.1016/J.CHEMOSPHERE.2016.09.014.

Shore, L.S., Gurevitz, M. and Shemesh, M. 1993. Estrogen as an Environmental Pollutant. Bull. Environ. Contam. Toxicol 51: 361–366.

Song, L., Li, L., Yang, S., Lan, J., He, H., McElmurry, S.P. et al. 2016. Sulfamethoxazole, tetracycline and oxytetracycline and related antibiotic resistance genes in a large-scale landfill, China. Sci. Total Environ. 551–552: 9–15. https://doi.org/10.1016/J.SCITOTENV.2016.02.007.

Stevens, K.J., Kim, S.Y., Adhikari, S., Vadapalli, V. and Venables, B.J. 2009. Effects of triclosan on seed germination and seedling development of three wetland plants: Sesbania herbacea, Eclipta prostrata, and Bidens frondosa. Environ. Toxicol. Chem. 28: 2598–2609. https://doi.org/10.1897/08-566.1.

Stuart, M., Lapworth, D., Crane, E. and Hart, A. 2012. Review of risk from potential emerging contaminants in UK groundwater. Sci. Total Environ. 416: 1–21. https://doi.org/10.1016/J.SCITOTENV.2011.11.072.

Suzuki, G., Someya, M., Takahashi, S., Tanabe, S., Sakai, S. and Takigami, H. 2010. Dioxin-like Activity in Japanese Indoor Dusts Evaluated by Means of *in Vitro* Bioassay and Instrumental Analysis:

Brominated Dibenzofurans Are an Important Contributor. Environ. Sci. Technol. 44: 8330–8336. https://doi.org/10.1021/ES102021C.

Ternes, T.A., Kreckel, P. and Mueller, J. 1999. Behaviour and occurrence of estrogens in municipal sewage treatment plants—II. Aerobic batch experiments with activated sludge. Sci. Total Environ. 225: 91–99. https://doi.org/10.1016/S0048-9697(98)00335-0.

Trihalomethanes in Drinking Water and Spontaneous Abortion : Epidemiology [WWW Document], n.d.

Vernouillet, G., Eullaffroy, P., Lajeunesse, A., Blaise, C., Gagné, F. and Juneau, P. 2010. Toxic effects and bioaccumulation of carbamazepine evaluated by biomarkers measured in organisms of different trophic levels. Chemosphere 80: 1062–1068. https://doi.org/10.1016/J.CHEMOSPHERE.2010.05.010.

Villanueva, C.M., Cantor, K.P., Grimalt, J.O., Malats, N., Silverman, D., Tardon, A. et al. 2007. Bladder Cancer and Exposure to Water Disinfection By-Products through Ingestion, Bathing, Showering, and Swimming in Pools. Am. J. Epidemiol. 165: 148–156. https://doi.org/10.1093/AJE/KWJ364.

Villanueva, C.M., Cordier, Sylvaine, Font-Ribera, L., Salas, Lucas A, Levallois, Patrick, Salas, L A, Cordier et al. 2015. Overview of Disinfection By-products and Associated Health Effects. Curr. Environ. Heal. Reports 2015 21 2, 107–115. https://doi.org/10.1007/S40572-014-0032-X.

Waller, K., Swan, S.H., DeLorenze, G. and Hopkins, B. 1998. Trihalomethanes in drinking water and spontaneous abortion. Epidemiology 9(2): 134–140.

Wang, J. and Gardinali, P.R. 2013. Uptake and depuration of pharmaceuticals in reclaimed water by mosquito fish (Gambusia holbrooki): A worst-case, multiple-exposure scenario. Environ. Toxicol. Chem. 32: 1752–1758. https://doi.org/10.1002/ETC.2238.

Windham, G.C., Waller, K., Anderson, M., Fenster, L., Mendola, P. and Swan, S. 2003. Chlorination by-products in drinking water and menstrual cycle function. Environ. Health Perspect. 111: 935–941. https://doi.org/10.1289/EHP.5922.

Wu, M.H., Que, C.J., Xu, G., Sun, Y.F., Ma, J., Xu, H. et al. 2016. Occurrence, fate and interrelation of selected antibiotics in sewage treatment plants and their receiving surface water. Ecotoxicol. Environ. Saf. 132: 132–139. https://doi.org/10.1016/J.ECOENV.2016.06.006.

Yadav, G. and Ahmaruzzaman, M. 2022a. New generation advanced nanomaterials for photocatalytic abatement of phenolic compounds. Chemosphere 304: 135297. https://doi.org/10.1016/J.CHEMOSPHERE.2022.135297.

Yadav, G. and Ahmaruzzaman, M. 2022b. Sustainable Development of Nanomaterials for Removal of Dyes from Water and Wastewater 167–188. https://doi.org/10.1007/978-981-19-0987-0_8.

Yadav, G., Yadav, N., Sultana, M. and Ahmaruzzaman, M. 2023. A comprehensive review on low-cost waste-derived catalysts for environmental remediation. Mater. Res. Bull. 164: 112261. https://doi.org/10.1016/J.MATERRESBULL.2023.112261.

Yadav, G.K. and Ahmaruzzaman, M. 2021. Recent advances in the development of nanocomposites for effective removal of pesticides from aqueous stream. J. Nanoparticle Res. 23: 1–31. https://doi.org/10.1007/S11051-021-05290-6/TABLES/5.

Yadav, N. and Ahmaruzzaman, M. 2023. Recent advancements in CaFe2O4-based composite: Properties, synthesis, and multiple applications. https://doi.org/10.1177/0958305X231155491. https://doi.org/10.1177/0958305X231155491.

Yang, Y., Zheng, M., Yang, L., Jin, R., Li, C., Liu, X. and Liu, G. 2021. Profiles, spatial distributions and inventory of brominated dioxin and furan emissions from secondary nonferrous smelting industries in China. J. Hazard. Mater. 419: 126415. https://doi.org/10.1016/J.JHAZMAT.2021.126415.

CHAPTER 3

Occurrence of Emerging Contaminants in the Environment
Causes and Effects

Maria Alice Prado Cechinel,[1] *Domingos Lusitâneo Pier Macuvele,*[1,2] *Natan Padoin,*[1] *Humberto Gracher Riella*[1] and *Cíntia Soares*[1,]*

1. Introduction

Contaminants of Emerging Concerns (CECs) represent a larger group of toxic chemicals found in environmental matrices and are not currently regulated. These contaminants include pesticides, microplastics, nano plastics, personal-care additives, drugs, etc. More recently, microplastics have attracted the attention of the scientific community and governments.

Microplastic contamination has become an environmental concern, as they have been found in various sources, including water, soil, and air. The term microplastic describes plastic particles with dimensions smaller than 5 mm and includes nano plastics smaller than 1 μm (World Health Organization, 2019). These particles can be generated from the fragmentation of large plastic objects, such as bottles and packaging, or directly originate from the composition of certain products, such as pharmaceutical and personal hygiene products (Wang et al., 2021a).

[1] Laboratory of Materials and Scientific Computing (LabMAC), Department of Chemical and Food Engineering, Federal University of Santa Catarina, Florianópolis, Santa Catarina, Brazil.
[2] Center for Studies in Science and Technology (NECET), Department of Science, Engineering, Technology and Mathematics, University of Rovuma, Lichinga, Niassa, Mozambique.
* Corresponding author: cintia.soares@ufsc.br

The degradation of microplastics in natural environments is a relatively slow process, which causes these particles to accumulate in the environment and can gradually be absorbed, directly or indirectly, by living organisms. In addition, microplastics can be transported through water and wind currents, spreading throughout the environment and affecting the soil, air, and water quality (Wang et al., 2021b). The effects of microplastic exposure vary according to the particle type, size, and concentration, as well as the characteristics of the exposed organisms. Studies indicate that the presence of microplastics can interfere with the health and behavior of organisms, as well as cause modifications to the physicochemical characteristics of the environment (Ahmed et al., 2022). Microplastics interact with other contaminants with emerging concerns (CECs), reducing and decreasing their toxicity (?Zhu et al., 2019; Sheng et al., 2021; Araújo et al., 2022; Rubin and Zucker, 2022; Lu et al., 2023a).

In recent years, there has been an evident growth in the number of studies related to microplastics. A search for the keywords "microplastic" and "occurrence" carried out on a journal platform (Scopus Elsevier) found 1,053 works that addressed the theme in their title or abstract, and of these, 681 were published between 2021 and 2023. Initially, the studies were restricted to analyzing aquatic environments and the impact of microplastic presence on marine life. Still, the scope of investigations has recently expanded to terrestrial habitats and the atmosphere. It is essential to highlight that the results published so far emphasize the urgency of actions to mitigate microplastic pollution, safeguard the biological integrity of terrestrial and aquatic ecosystems, and preserve human health.

This chapter aims to provide a broader understanding of the occurrence of microplastics in different media and the main effects caused by the presence of these particles in living organisms, to serve as a reason for future research related to monitoring and reducing the presence of microplastics in natural environments.

2. Occurrence of Microplastics in Different Media

As a result of the widespread use and increased inappropriate disposal of plastic materials, the occurrence of microplastics is seen more frequently in a wide number of environments, including oceans (Abelouah et al., 2022), rivers (Zhang et al., 2021), soil (Cao et al., 2021), foods (Bai et al., 2022), drinking water for human consumption (Tong et al., 2020) and even in the air we breathe (Chen et al., 2020). Recent works also report the occurrence of microplastics in animal and human tissues (Yan et al., 2021; Zhu et al., 2023). Occurrences of microplastics have been confirmed even in inhospitable places, such as the Mariana Trench (Peng et al., 2018) and Mount Everest (Napper et al., 2020).

Regardless of the environment where they are observed, microplastics can come from two main sources, known as primary and secondary. Primary microplastics are those directly manufactured in micrometric size for specific applications in products such as toothpastes, cosmetics, cleaning products, clothing, textiles, and plastics, among others (Wang et al., 2021a). These materials are frequently released into the aquatic environment through industrial and human activities, domestic sewage systems and wastewater treatment plants. In addition, primary microplastics can be dispersed onto surface water and shorelines by runoff and wind, eventually reaching the oceans (Li et al., 2018a). On the other hand, secondary microplastics are those obtained from the degradation of larger plastic items, such as trawl nets, industrial resin pellets, and household supplies, into small plastic debris. This breakdown can occur because of ultraviolet radiation, mechanical abrasion, and wave impact, and also by biological sources of wear (Kasmuri et al., 2022; Zhang et al., 2022). Secondary microplastics are more commonly found in the ecosystem since they come from different polluting sources. The main sources for the occurrence of microplastics in the environment are shown in Fig. 3.1.

Therefore, it is important to understand the occurrence and possible impacts of microplastics in different environments to develop effective measures to mitigate

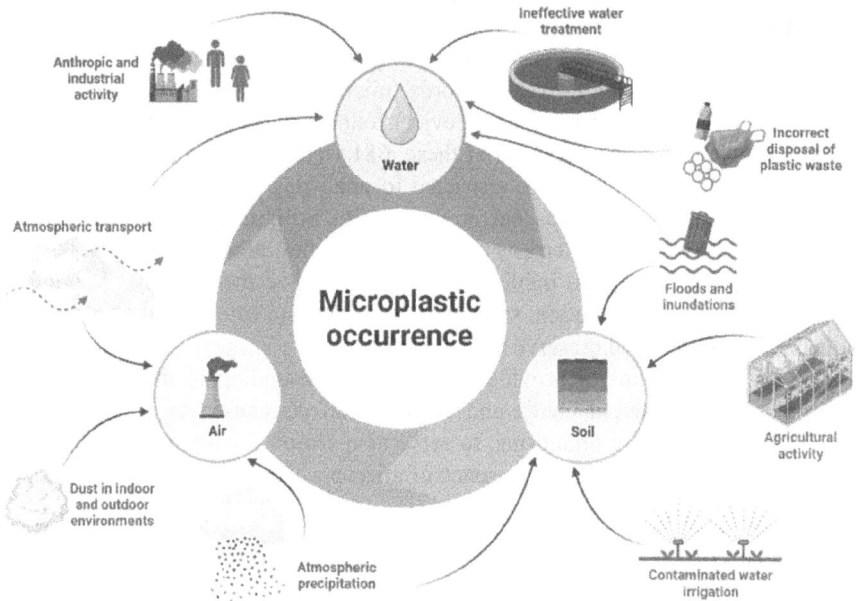

Figure 3.1 The main forms of occurrence of microplastics in air, soil and water.

their spread and minimize their negative effects. This occurrence will be discussed in three main environments: water, soils, and air.

2.1 Occurrence of Microplastic in Water

In recent years, microplastics have been confirmed in a wide number of water environments. Its dispersion is even more amplified since the flow of water from rivers, lakes and oceans transports small-sized particles that are highly resistant to degradation over long distances (Alimi et al., 2018). Several studies have pointed out that its occurrence in aquatic environments is mainly derived from anthropogenic activities, such as everyday life and industrial production.

The ocean is the main recipient of plastic waste, with most of this waste (80%) coming from land-based plastic debris (Li et al., 2016). This debris can indirectly reach the ocean through improper disposal on beaches, atmospheric transport, and rivers, or through human activities such as fishing, shipping, and aquaculture. Among these sources, plastic marine litter from landfills or poorly managed collection points stands out; which is transported to the ocean due to adverse weather conditions or the direct dumping of large pieces of plastic into the sea (Wang et al., 2021a).

Microplastics have been found worldwide in surface and deep marine waters, from polar to tropical regions. This is because the ocean is the destination for much of the plastic that is discarded on land and in rivers. In a study conducted by Cózar et al. (2023), a broad sampling of the Arctic Ocean was carried out in search of floating plastic debris, finding high concentrations of plastic materials in the northern and eastern areas of Greenland and the Barents Seas. The plastic fragmentation and typology identified in the study suggested that the debris originates from distant and aged sources. Furthermore, more than 70% of the plastic material collected was small-sized plastic fragments. Russel and Webster (2021) identified the presence of microplastics in surface waters of all Scottish Marine Regions and Offshore Maritime Regions, with concentrations of up to 91,128 particles/km² of the marine surface. As was identified in the aforementioned study, shredded plastics accounted for almost 50% of the material analyzed, indicating that microplastics in Scotland's seas predominantly come from the decomposition of larger items.

The occurrence of microplastics is also reported in equatorial marine regions. Garcés-Ordonez et al. (2022) identified the presence of microplastics in fragile ecosystems on the Caribbean coast of Colombia, both in surface waters and in sediments. 0.3 particles/L of surface water were identified, mainly in the form of fibers and fragments. These microplastics were more intense in areas with high fishing and aquaculture activities, as well as near river mouths, which were identified as significant polluting sources. Another study carried out by Suteja et al. (2021) identified the presence of microplastics in the Benoa Bay, Bali, Indonesia. The study indicated that 26.4% of the microplastic particles collected were smaller than 500 and that 73.2% were in the form of fragments (73.19%). As it is possible to identify, microplastic pollution in oceans is not restricted to a specific region of the planet and, once plastic fragments and debris are released into the ocean, sea currents can transport them to other distant areas, contaminating other places.

Rivers and lakes are reported to be important sources of transporting microplastics from the continent to the ocean. The hydrological characteristics of these water bodies, such as water flow velocity, depth, water flow variation, and level of contamination, are relevant parameters for the occurrence of microplastics in these aquatic environments and freshwater oceans (Campanale et al., 2020; Alfonso et al., 2021). However, rivers and lakes can also be the destination for these pollutants. The main sources of microplastics in freshwater are those of industrial origins, such as plastic resin dust, spillage of pellets from air blasting machines, microbeads present in personal care products, as well as raw materials used to produce plastic products. The decomposition of large plastic debris and the presence of anthropological activity also contribute to the increase in the number of microplastics present in aquatic systems (Horton et al., 2017).

The study by Jian et al. (2020) revealed high levels of microplastics (up to 1,064 ± 90 particles/m³) in China's largest freshwater lake, Lake Poyang. The dominant form of microparticles was as fragments and fibers, with characteristics associated with weathering and fragmentation, in addition to significant fractions of particles with a size between 0.03 and 0.1 mm. Baldwin et al. (2020) conducted a study to evaluate the amount and morphology of microplastics in two large reservoirs along the Colorado River within the Lake Mead National Recreation Area. The identified microplastic concentrations ranged from 0.44 to 9.7 particles/m^3 on the surface of the water, being higher in areas of greater anthropic impact. Regarding the shape, fibers were the most abundant type of microplastic in all samples collected. The occurrence of microplastics flowing into Port Phillip Bay from urban rivers outside Melbourne, Australia was investigated by Samandra et al. (2023). Rivers contained, on average, 9 ± 15 microplastics/L and the flow of microplastics into the Port Philip Bay was estimated to be 7.5 × 10^6 microplastics/day. Finally, another study finds the occurrence of microplastics in the surface waters of the Densu River in West Africa (Blankson et al., 2022). The results indicated widespread pollution of the river with microplastics in all evaluated samples. The average number of microplastic particles varied between 83 and 150 particles/L. Again, microplastic pollution affects freshwater environments globally. Furthermore, it was observed that the differences in concentrations (which range from several to millions of tons) are mainly influenced by locations, natural conditions, and anthropogenic activity (Pivokonsky et al., 2018).

Underground aquifers, which are the main source of drinking water for thousands of people around the world, are also subject to the occurrence of microplastics, as these pollutants can seep into the soil and be transported to groundwater. However, there are still only a few studies that evaluate this occurrence. Samandra et al. (2022b) identified different types of microplastics in all groundwater samples collected from seven groundwater monitoring boreholes in Bacchus Marsh (Victoria, Australia). The study revealed that the average size of the microplastics found ranged between 18 and 491 μm. Furthermore, the mean number of microplastics detected at all sites was 38 ± 8 microplastics/L. Considering that the samples were collected from plugged groundwater holes, the authors stated that the main route of entry for microplastics was permeation through the soil. Another study conducted by Panno et al. (2019) found the presence of microplastics in springs and wells of two karst aquifers in Illinois, USA. All identified microplastics were in the form of fibers, with an average concentration of 6.4 particles/L. The authors also related the presence of microplastics in groundwater to the hydrogeological connections of the surface to the underlying aquifers. Although the contamination of groundwater with microplastics may seem less evident and present greater difficulty in detection, which results in less attention from the scientific community compared to marine or surface water bodies, it is crucial that the monitoring of water quality in these places is not neglected.

Recent studies have identified microplastics in the water ready for human consumption. Samandra et al. (2022a) documented the presence of microplastics in 15 brands of bottled water sold in Australia, with an average of 13 particles/L and an average size of 77 ± 22 μm. Based on these numbers, the study estimated that Australians are exposed to 400 microplastics annually through drinking bottled water. A study carried out by Schymanski et al. (2018) found microplastics in water samples collected from 22 reusable and disposable plastic packages and nine types of glass bottles, all purchased in supermarkets in Germany. Microplastic fragments of size 1 and 500 μm were detected in all analyzed samples. The study also revealed the occurrence of larger numbers of plastic particles in some of the glass bottled water samples, with an average of 50 ± 52 particles/L, which was unexpected considering the type of material used to make these bottles. The authors suggested that this contamination may be a result of wear on the soft plastic cap and bottle seal and/or the bottle cleaning process. The average content of microplastics was 118 ± 88 particles/L in returnables and 14 ± 14 particles/L in disposable plastic bottles, which reinforces that the continuous use and cleaning process of returnable material can enhance the accumulation of microplastics in materials. The effect of daily use and abrasion of a plastic material after a series of bottle openings/closings on the occurrence of microplastics in bottled water was evaluated by Winkler et al. (2019), indicating that the chances of microplastic ingestion by humans increase with frequent use of the same single-use plastic bottle.

The occurrence of microplastics is also observed in tap water distributed for consumption. Tong et al. (2020) identified microplastics in 36 samples of tap water collected in different cities in China, with an average concentration of 440 ± 275 particles/L and a predominance of particles smaller than 50 μm and in the form of fragments. The identification of microplastics in 41 tap water samples collected in the Metropolitan Area of Barcelona, Spain, was also described by Vega-Herrera et al. (2022). Since tap water undergoes treatment before distribution, it is possible to infer that the microplastics present in these samples come from the transport process from the water treatment plant to the collection point and/or are remnants of the treatment, that is, there may be an inadequacy in conventional water and sewage treatment processes for removing these plastic particles, due to their extremely small size. The interconnection between discharges from wastewater treatment plants and the occurrence of plastics and microplastics in the ocean and freshwater has been supported in other studies (Sun et al., 2019; Wong et al., 2020). Although the levels of microplastics in the studies reported here can still be considered low, the presence of these pollutants is a growing concern for public health and poses a challenge for waste management and water treatment.

2.2 Occurrence of Microplastic in Soils

Given that most plastic waste is generated and discharged in terrestrial ecosystems and that the soil is still considered an important long-term reservoir for this waste, it is not surprising that the occurrence of microplastic is proved in this environment in different parts of the world. However, research that seeks to identify the presence of microplastics in soils is still limited compared to investigations carried out in aquatic environments, and only recently has it begun to be published.

It is estimated that the annual release of microplastics into the soil is 4 to 23 times greater than into the oceans (Nizzetto et al., 2016) and that massive volume far exceeds all aquatic environments (Qiu et al., 2022). Furthermore, the existence of plastic debris in the soil enables its easy transfer to rivers, lakes, and oceans through surface runoff, and to groundwater through the infiltration of particles through the soil. Factors such as direct ultraviolet irradiation, relatively high temperatures and direct contact with oxygen make topsoil a potential degradation environment for microplastics (Chae and An, 2018). These particles can also migrate to deeper soils due to agricultural cultivation, soil cracking, disturbance of soil organisms (e.g., earthworms) and leaching, this being the most common process by which microplastics are usually moved, and may reach groundwater reservoirs (Rillig et al., 2017). Koutnik et al. (2021) reported that most microplastics settled in urban soils are fragmented, while fibers are more easily transported by water due to their shape. The presence of vegetation, especially in the rhizosphere, influences the content of microplastics in the soil since the roots of plantations can intercept the particles, adhering them to their surface and causing more microplastics to remain in the soil around them (Huang et al., 2021). A clear relationship between the concentration of microplastics and the chemical composition of the soil (pH, dissolved organic matter content and total iron content) has not been found, so far, by the scientific community (Yang et al., 2021).

The main sources of microplastic in soils are agricultural coverings (such as plastic mulching), irregularly discarded garbage, sewage sludge, irrigation, flooding episodes and water runoff in streets and through atmospheric entry (Guo et al., 2020; Yang et al., 2021). Agricultural mulching covers, which have been widely used as an alternative for soil and plant protection against erosion (Steinmetz et al., 2016), can be made with plastic materials that slowly degrade into smaller particles, releasing microplastics into the soil. In a study conducted by Huang et al. (2020), the occurrence of microplastics was confirmed in 384 soil samples used for agricultural activities, collected in 19 provinces of China. Macroplastic concentrations in soil samples ranged from 0.1 to 324.5 kg/ha. The study indicated that the presence of plastic coverings can be an important source of macroplastics and that the abundance of microplastic particles has increased over time in places where this covering has been used continuously for years. Feng et al. (2021) investigated the presence of microplastics in 35 samples of farmland and grasslands in an area of the Qinghai-Tibet Plateau in East Asia. The authors identified a notable difference in the abundance of microplastics between soil samples from agricultural lands with and without mulch, in addition to proving that the presence of microplastics is more substantial with the increase in the time of use of these mulches in agricultural installations. The highest concentration of microplastics identified in this study (260 particles/kg in shallow soil and 193.3 particles/kg in deep soil) was obtained for a soil sample with more than 15 yr of use of plastic coverings.

Improper waste disposal is also a major source of microplastic contamination in soil. It is estimated that between 1950 and 2015, around 6.3 billion tons of plastic waste were generated globally, of which 4.97 billion tons are accumulated in landfills and natural environments (Geyer et al., 2023). Plastic waste, when disposed of incorrectly, can fragment and turn into microplastics, which end up being carried by the wind or water to the soil, where they contaminate the environment (Yang et al., 2021). However, to date, no study has quantified the amount of microplastic that reaches the soil through garbage or illegal dumping.

Sewage sludge, which is the solid residue resulting from the treatment of water and sanitary and industrial sewage, is often applied as fertilizers in agricultural soils. However, recent research has indicated that sludge may contain microplastics, especially from personal care products and cosmetics, which are not retained in the

WWTP due to their small size. Horton et al. (2021) identified the occurrence of microplastic in 20 sludge samples collected from wastewater treatment plants in the United Kingdom. The study found high concentrations of microplastics in the analyzed samples, with values between 301 and 10,380 microplastics per gram of dry weight. Based on the average of the results obtained, it is estimated that the addition of sludge to the soil can generate an annual contribution of approximately $2,7 \times 10^{15}$ microplastics. A study conducted by Van den Berg et al. (2020) identified the presence of microplastics in 97% of the samples collected in four wastewater treatment plants and 16 agricultural fields that apply sewage sludge to correct the soil. The data obtained by the researchers indicated a significant concentration of microplastics in sewage sludge (on average, 32,070 ± 19,080 microplastics/kg). In addition, fields with sewage sludge application had an average of 5,190 microplastics/kg, while fields without application had an average of 2,030 microplastics/kg. The study also identified that the microplastic content increased with the number of applications of sewage sludge to the soil. The data presented in both studies indicated that the application of sewage sludge could be contributing to the increase in the presence of microplastics in soils and that the frequency of this practice should be monitored and regulated to minimize its environmental impacts.

Irrigation can also carry microplastics from the aquatic environment to the soil, since the main sources of water for irrigation include rivers, lakes, reservoirs, and groundwater and, as already reported in this chapter, water bodies have suffered successive and persistent contamination by microplastics. Furthermore, it is common to use treated wastewater for irrigation in regions with scarce water resources and, as previously detailed, the occurrence of microplastics is reported in water after the treatment process in WWTPs (Li et al., 2018b). A study conducted by Ahmad et al. (2022) identified microplastics in water and soil samples collected in the peri-urban area of Faisalabad, Pakistan, where municipal and industrial waste is used to irrigate agricultural land. The average abundance of microplastics found in the soil was 2,790.75 items/kg. Another study conducted by Pérez-Reverón et al. (2022) evaluated the occurrence of microplastics in wastewater used in soil irrigation in Fuerteventura (Canary Islands, Spain) and in samples of these irrigated soils. The results showed the prevalence of the presence of cellulosic and polyester microfibers (between 84.4 and 100%), with concentrations of up to 40.0 ± 19.0 items/L. The presence of microplastics in samples of irrigated soils was up to 159 ± 338 items/kg, on the other hand, non-irrigated/non-cultivated soils did not show a detectable concentration of microplastics, suggesting that agricultural activities are the only source of microplastics in soils of this area.

The occurrence of floods and water runoff in streets can also cause soil contamination by microplastics since these particles can be dragged by the water and deposited on the surface during the sedimentation process. Illegal dumping of waste near roads and tire abrasion also enhances the occurrence of microplastics in soils (He et al., 2018) through the runoff of rainwater or in cases of flooding. Despite this, studies evaluating the impact of floods and water runoff on increasing the concentration of microplastics in the soil are rare. A study conducted by Yukioka et al. (2020) evaluated the occurrence of microplastics in road surface dust in three cities: Kusatsu (Japan), Da Nang (Vietnam) and Kathmandu (Nepal). The results demonstrated the presence of up to 19.7 items/m², mostly composed of fragments of containers and/or plastic packaging. The study also highlighted the importance of evaluating the flow of microplastics present on roads into the aquatic environment.

Considering the information presented, it is important to adopt measures aimed at reducing soil contamination by microplastics, including raising awareness about proper waste disposal, the use of more sustainable agricultural techniques, the application of waste treatment practices, more efficient sewage systems, in addition to controlling and monitoring of water flow in urban areas. Furthermore, it was possible to prove the scarcity of studies about the distribution, transport and degradation of microplastics in terrestrial environments, reinforcing the need to carry out more research to deepen the knowledge about the potential environmental impacts of these materials.

2.3 Occurrence of Microplastic in the Atmosphere

Currently, research on the occurrence of microplastics in the air and atmosphere is still scarce, but there is growing evidence that these materials may be present in different environments in the atmosphere. Due to the limited number of studies, many issues related to the sources of emission of microplastics in the air and the mechanisms of transport and deposition of these materials in different environments are still in the initial stages but already indicate possible impacts on human health and the ecosystems. However, there are still few reports on the concentration of microplastics in the air. This shortage of studies is primarily due to limitations in the capabilities for detecting atmospheric microplastics (Prata, 2018).

However, there is a growing number of studies that seek to analyze air samples in different regions of the world and that have detected the presence of microplastics. The atmosphere is an important pathway by which many suspended materials are transported regionally or globally. With their small size and low density, microplastics are easily transported by wind over long distances, including ocean surface air and remote areas (Allen et al., 2019; Liu et al., 2019a; Wang et al., 2021a). Particle characteristics, such as their density, length, and diameter, along with environmental factors including wind, air currents and precipitation, influence both the dispersion and fate of microplastics in the airborne environment (Zhang and Wu, 2023). Its abundance in the air also seems to be mostly associated with human activities, population density and level of industrialization (Zhao et al., 2023).

Among the main sources of microplastics in the air, worn synthetic textiles, erosion of synthetic rubber tires and urban dust, construction waste, industrial emissions, landfills, waste incineration, exhaust from dryer machines and synthetic particles. used in horticultural soils (O'Brien et al., 2023) are highlighted. In addition, the occurrence of microplastics is identified in rural and urban areas, as well as in outdoor and indoor environments (Cui et al., 2022; Rosso et al., 2023).

In urban areas, the main source is human activity, including vehicle traffic and the burning of fossil fuels. In addition, plastic products discarded incorrectly on the streets and the lack of adequate solid waste management systems can contribute to microplastic pollution. Abbasi et al. (2019) collected 15 samples of street dust and 16 samples of dust in suspension from urban and industrial areas of Asaluyeh

Municipality, Iran. The study identified an average of 60 microplastics/g of street dust sample, with varying colors and sizes (between 100 and 1,000 μm). Airborne dust revealed the ubiquity of fibrous microplastics of sizes ranging from around 2 μm to 100 μm and an abundance of around 1 item/m³. The concentration of microplastics in atmospheric deposition in the metropolitan region of Hamburg, Germany, was investigated by Klein and Fischer (2019). Samples were collected from different points in the countryside south of Hamburg and selected locations within the city and all showed detectable concentrations of microplastics. Average concentrations of up to 512.0 microplastic/m² in 1 d were found during the sampling period. Interestingly, higher concentrations of microplastics were obtained in samples collected in rural areas. The authors pointed out the influence of proximity to highways and emissions associated with vehicular traffic (dust and abrasion from tires, paint and road surfaces) as a factor for increasing the concentration of particles in this region. The presence of vegetation also enhances the fixation of microplastics in the leaf area, leaving them susceptible to atmospheric dispersion through processes of drag by wind and transport by rain. Another study conducted by Liu et al. (2019a) estimated that approximately 120.7 kg of microplastics are carried annually by air from Shanghai, China and that 21 particles of microplastics are inhaled daily by people in Shanghai outdoors. The occurrence of microplastics in the collected samples ranged from 0 to 4.18 items/m³.

In addition to rural and urban areas with significant population densities, studies also presented evidence that microplastics can also be transported to remote mountainous areas or marine air through atmospheric transport. A study conducted by González-Pleiter et al. (2021) detected microplastics in the troposphere, approximately 2,800 m above ground level in an agricultural town in Spain, at average concentrations of 1 microplastic/m³ in rural areas and an average of 1.71 microplastics/m³ above urban areas. The distribution of suspended atmospheric microplastics in the western Pacific Ocean was investigated by Liu et al. (2019b) through continuous sampling during a ship cruise. The microplastics amount ranged from 0 to 1.37 items/m³, with higher concentrations being recorded in the coastal area (0.13 ± 0.24 items/m³) compared to the pelagic area, where the amount detected was smaller (0.01 ± 0.01 items/m³). The study verified that the collection carried out during the day presented twice the amount of suspended atmospheric microplastics collected during the night. Allen et al. (2019) observed the presence of microplastics in atmospheric depositions collected in a remote basin located in the mountainous region of the French Pyrenees. The microplastic counts from field samples illustrate an average daily particle deposition of 365 items/m². The authors also presented an air mass trajectory analysis, which identified the transport of microplastics through the atmosphere over up to 95 km.

Indoor environments such as homes, offices and schools can also contain microplastics in the air. This can occur when using products that contain microplastics, such as cleaning products and cosmetics, as well as household dust that contains plastic particles. Cui et al. (2022) identified the presence of microplastics in all samples collected by atmospheric deposition in various rooms of 20 homes in Yangzhou, China. A total of 23,889 microplastics were identified from 100 sampling sites in 20 households, with average daily concentrations of up to 96,367 items/m². The study conducted by Zhu et al. (2022) also investigated the abundance and characteristics of microplastics in 242 dust samples collected from different indoor environments (residential apartments, offices, hotels, classrooms, and university dormitories) in Hangzhou City, China. From the data, the authors estimated an average daily intake through indoor dust inhalation by adults of 0.23 microplastics/kg.

The professional activity carried out indoors also enhances the presence of microplastics in suspension. The study conducted by Chen et al. (2022) compared the presence of microplastics in indoor and outdoor air samples from beauty salons in Taiwan. The researchers estimated that the average annual exposure to microplastics in these environments is 67,567 particles per year. In addition, the physical characteristics and polymeric compositions of these particles showed differences between beauty salons and other indoor spaces studied earlier. It was found that the use of air conditioning increased the emission of particles, and nail salons with plastic ceilings and floors, as well as those with a larger number of occupants, had higher concentrations. Finally, another significant source of microplastic emissions into the environment is the washing of clothes and synthetic fabrics due to the mechanical drying step. O'Brien et al. (2020) evaluated the effect of mechanical drying of synthetic textiles, carried out by a ventilated domestic dryer, on the release of microplastic fibers into the surrounding air or captured by the built-in filtration system. The data presented by the study indicated that mechanical drying contributes approximately 2 fibers/m^3 to the surrounding atmospheric environment and estimate, based on the operation data and dry fabric mass, that the annual contribution can reach 3×10^3 fibers transported by the air in households and/or atmospheric environment per household.

In summary, the presence of microplastics in the air is a global reality that occurs regardless of where they are collected, whether due to the direct influence of human activities or the transport of these particles to places that are not very accessible through air masses such as in oceanic waters. This finding should encourage the scientific community to expand research on the subject, to investigate the concentration, fate, and effects of these particles on human health and the environment. In addition, it is essential to promote sustainable practices that minimize the presence of these suspended materials and, thus, mitigate the negative impacts that their presence can cause. In summary, this issue requires a holistic, multi-disciplinary, and collaborative approach to minimize the risks associated with microplastics in the atmosphere.

3. Effects and Interaction of Microplastics in Organisms

Microplastic toxicity is a complex and paramount hot topic for academicians and governments. This issue is analyzed in terms of pure MPs' interaction with organisms. The toxicity of microplastics is also analyzed in terms of the interaction of MPs with other contaminants and how these interactions intensify the toxicology (Fig. 3.2).

3.1 Pristine MPs Interaction with Organisms

The toxicology of microplastics depends on several factors, such as type of polymeric resin, size, shape, etc. For these reasons, is very complicated to generalize the

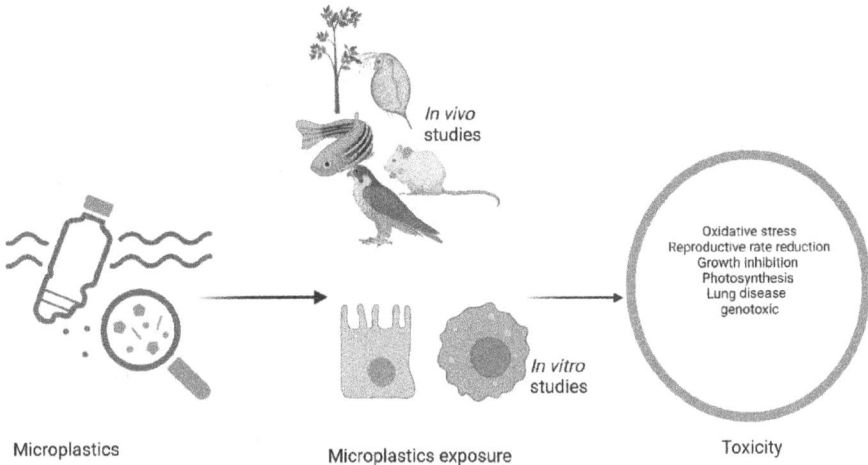

Figure 3.2 Schematic illustration of microplastics' interaction with organisms.

toxicology aspects for all microplastics. In this section, the analysis will be directed to a specific type of microplastics.

3.1.1 PS Polystyrene base Microplastics (MP-PS)

Polystyrene-based microplastics are one of the most found particles in the environment. This trend is due to the various application of PS-based materials. Several works showed that microplastics with PS resin present specific toxicity in various cells and selected organisms. The toxicity of PS microplastics depends on several factors. These factors include UV-light exposition, pH, presence, or absence of other contaminants, etc. Due to the aquatic environment's relevance and trophic interactions, this subject will first highlight the toxicological aspects of aquatic organisms.

Euglena gracilis, is a microzooplankton mainly used in toxicological assays due to their sensitivity to contaminants. For this reason, it is extremely important to assess the toxicology of PS-based microplastics using *Euglena gracilis* as a model. MP-PS was evaluated in terms of toxicity against *Euglena gracilis*. The results showed that low concentrations of MP-PS did not affect the essential parameters of *E. gracilis*. However, concentrations at 5 and 25 mg/L, affected the growth, motility, and photosynthesis.

Interestingly, the size of MP-PS affected the toxicity against *E. gracilis*. MP-PS with a size of 75 nm was more toxic than MP-PS with 1000 nm (Liao et al., 2020). In the same perspective, MP-PS (5 um and 0.1 um) toxicology behavior was evaluated against *E. gracilis*. However, in this work, the authors focused on photosynthesis pigment contents, antioxidant enzyme activities, and transcriptomic response variations after MP-PS exposure. MP-PS at 1 mg/L induced *E. gracilis* vacuoles after 24 hr of exposure. All photosynthetic pigments were significantly reduced after MP-PS exposure. However, chlorophyll b (Chl b) and carotenoids (CAR) were the most affected compared to Chl b. The oxidative stress was also induced by MP-PS exposure; this was confirmed by the peroxidase activity.

Interestingly, the results of molecular level assays to understand the pathways involved in MP-PS toxicity showed that the 3-Ketoacyl-CoA Synthase (KCS) genes and copper uptake protein 1 (CTR1) are probably key mechanisms to induce *E. gracilis* adverse effects after MP-PS exposure (Xiao et al., 2020). Other studies evaluated the MP-PS adverse effects against *E. gracilis*. The results showed that nanoplastics were more toxic than microplastics, only in high concentrations. While MP-PS did not significantly affect. *E. gracilis* growth (Liao et al., 2020).

MP-PS, due to their adverse effects and easy penetration into several organs, led the scientific communities to suggest that these particles can interfere in reproductive performance of the organisms. *In vitro* and *in vivo* studies, indicated that MP-PS can cause adverse effects in organisms significantly reducing their reproductive performance. MP-PS ($0.5\,\mu m$, $4\,\mu m$, and $10\,\mu m$) toxicity was evaluated in mice aiming to understand how the MP-exposure affects the male reproductive system. The results showed that all evaluated reproductive parameters were significantly affected by MP-PS. MP-PS caused abscission and disordered arrangement in spermatogenic cells. The MP-PS exposure reduced a testosterone level compared to the control. In addition, a sperm reduction was also observed after MP-PS exposure. MP-PS size did not affect the testosterone levels and sperm reduction (Jin et al., 2021). However, molecular studies are still needed to understand the pathways that govern toxicity of MP-PS in mice. MP-PS were also evaluated for their toxicity in female mice to understand their role in the reproductive system. MP-PS induced a reduction of antral follicles of female mice. The quality and developmental competence of mouse oocytes was also affected by MP-PS (Liu et al., 2022). Research was performed to understand the pathways that govern the MP-PS toxicity in the reproductive systems of male mice. The results regarding the toxicity corroborated strongly with earlier work (Jin et al., 2021). After MP-PS exposure into male mice, Reactive Oxygen Ppecies (ROS), glutathione-reduced form (GSH) and malondialdehyde (MDA) increased significantly compared to the control. In addition, MP-PS increased the degree p38 phosphorylation. Conversely, adding SB203580, the p38 mitogen-activated protein kinases (p38 MAPK) specific inhibitor, attenuated the p38 phosphorylation. These results indicate that the MP-PS reproductive toxicity pathways involve oxidative stress and p38 MAPK activation (Xie et al., 2020).

MP-PS were detected in the atmosphere in several samples collected in various countries. Therefore, the study of pulmonary-related toxicity is of paramount importance. In *an in vitro* assay, MP-PS were assessed for their toxicity against human lung epithelial cell line BEAS-2B. Cell viability of cell line BEAS-2B reduced significantly in higher concentrations ($1000\ \mu g/cm^2$) of MPs, after 24 hr of exposure. Interestingly, after 48 hr of MPs exposure, the cell viability reduction is significant in MPs concentration $\geq 10\,\mu g/cm^2$. The toxicity of MP-PS pathways was related to reactive oxygen species productions as a main pathway (Dong et al., 2020). In an *in vivo* study, the MP-PS (0, 0.5, 1, and 2 mg/200 µL 100 nm for 14 d) exposure to rats caused lung injury and inflammation. A genetic sequence was performed to understand the mechanism and pathways involved in MP-PS toxicity in rats. The results revealed that pro-inflammatory cytokines IL-6, TNF-α and IL-1β were upregulated after MP-PS exposure. In addition, the work also showed that circular RNA (circRNAs) and long non-coding RNAs (lncRNAs) play a paramount role on in development of lung inflammation (Fan et al., 2022). MP-PS induced nasal and lung microbial dysbiosis in mice. Interestingly, the work also evaluated the effect of nanoplastics (NP-PS) on nasal and lung microbial dysbiosis. The results showed that microplastics (MP-PS) presented a stronger influence than NP-PS on lung microbial dysbiosis (Zha et al., 2023).

More recently, the MP-PS toxicity (0, 1, 10, and 100 mg/L) was assessed in birds. The MP-PS exposure was obtained for 6 wk to understand the toxicity in the bird's lung. The MP-PS exposure caused the alteration of oxidative stress makers such as malonaldehyde (MDA)—content increased, and catalase (CAT) and glutathione (GSH) activity. These alterations confirm that oxidative stress is a crucial pathway of the bird's lung injury. Additionally, the MP-PS exposure caused apoptosis, confirmed by increasing cytochrome (Cytc), Bcl-2 Associated X-protein (Bax), B-cell lymphoma 2 (Bcl-2), dynamin-related protein 1 (DRP1), Caspase-3, Caspase-8, and Caspase-9 in mitochondria. The Myofibroblasts 2 (MF2) lever was also decreased, indicating that MPs exposure led to mitochondrial dysfunction (Lu et al., 2023b). In the same perspective, MP-PS was studied for its effects on pulmonary diseases, a type of lung disease. MP-PS was administered in mice via intratracheal for 3 wk. MP-PS was administered via the trachea-induced pulmonary fibrosis, this was confirmed by fibrotic features and increased expression of collagen and α-SMA. Curiously, the oral administration of MP-PS did not cause pulmonary fibrosis in mice. MP-PS also, induces alveolar epithelial injury. This was confirmed by the increased expression of KL-6, IL-1β, and TNFα (Li et al., 2022).

In another study, MP-PS (size ≤ 450 nm and 500 µg/mL) induced a cellular response of THP-1 macrophages. Macrophages are major components of the immune system, therefore, understanding their reaction in the presence of microplastics is essential. MP-PS decreased the cell viability of THP-1 macrophages. Interestingly, the MP-PS dose, exposure time, size, and shape affected cell viability reduction. Microplastics with an irregular shape and size less or equal to 450 nm reduced the cell viability significantly at maximum at 500 µg/mL. However, spherical MP-PS with size of 0.5 nm, did not affect the viability of $CaCO_3$ cells at concentration up to 50 µg/mL.

Furthermore, MP-PS decreased the cell's proliferation that was time exposure dependent. Nonetheless, the role of shape and size in cell proliferation is not clear. In addition, MP-PS affected the cell morphology, from rounded to flattened. This alteration morphology of macrophages was correlated to immune activation (Koner et al., 2023).

3.1.2 PET (PolyEthylene Terephthalate) based Microplastics (MP-PET)

PET (PolyEthylene Terephthalate) is one of the most applied polymers. This polymer is cheap and has a low production cost. Furthermore, PET presents a higher wear resistance; however, in marine and other, environments can release microplastics. For these reasons, understanding the toxicological aspects related to PET-based microplastics is essential.

MP-PET (5-300µm) was evaluated for its effects against bacteria (*V. fischeri*; UNI EN ISO 11348-3:2009), algae (*P. tricornutum*; UNI EN ISO 10253:2016E), and echinoderms (*P. lividus*; EPA 600/R-95-136/Section 15) species in different pH

(8 and 7.5). Bacteria and algae were not sensitive to MP-PET. However, *P. lividus* larval stage was sensitive to MP-PET leachates and suspensions (Piccardo et al., 2020). Another work exposed MP-PET (p-PET, approximately 150 µm in diameter) into zebrafish embryos. MP-PET does not induce the death or deformity of zebrafish embryos. However, after MP-PET exposure, an increase in heart rate was observed. In addition, the blood flow significantly increased only in 48 hr (Cheng et al., 2021).

Cell lines derived from *Oncorhynchus mykiss*, were used as a model to assess the toxicity of MP-PET (25-µm and 90-µm particles). MP-PET did not alter the cell viability, and no significant ROS was produced during the essay (Boháčková et al., 2023). In *Gammarus pulex*, MP-PET (chronic exposure over 48 d and PET, 10–150 µm; 0.8–4,000 particles/mL) did not modify survival, feeding activity, energy reserves and molting (Weber et al., 2018). Biomolecular responses of mussels (*Mytilus galloprovincialis*) were recorded after MP-PET (small 5–60 µm, S-PET; medium 61–499 µm, M-PET; large 500–3,000 µm, L-PET) exposure at 0.1g/mL. The results showed that mussels were stressed under MP-PET exposure. However, only Lipid peroxidation (LPO) and Glutathione Peroxidase (GPx) biomarkers were significantly increased after MP-PET exposure. Non-significant alteration of superoxide dismutase (SOD) and adrenocorticotropic hormone (AtCh) activity was recorded (Provenza et al., 2020).

3.1.3 Polyethylene-based Microplastic (MP-PE)

PE is another polymer widely used in modern societies. This polymer is applied in several fields. Therefore, understanding the toxicological aspect related to PE debris in different environments is extremely important.

Several studies related to MP-PE toxicology have been reported. MP-PE exposure was studied in the presence of zebrafish, *Danio rerio* under static and semi-static aquatic systems. The MP-PE affected the hatching rate of embryos under static and semi-static methods. Under a semi-static system, the hatching rating was reduced to 60%, but the mechanism that governs the process is still unknown. Conversely, under a static method, the embryo hatching rate increased after 48 hr. This trend was not observed under a semi-static system. In addition, MP-PE (exposure to zebra fish caused morphological and teratogenic alterations (Malafaia et al., 2020). Similarly, zebrafish was exposed to MP-PE (0–22 µm, 45–53 µm, 90–106 µm, 212–250 µm, and 500–600 µm and moderate (110 particles/L), and high concentrations (1,100 particles/L) to evaluate their toxicity. The results showed that MP-PE (high-density polyethylene) exposure affected the behavior of zebra fish. This alteration includes sickness (seizures), indicating that MP-PE exposure causes neurotoxicity. Furthermore, gene expression performed by real-time PCR in zebra fish organs (the intestines and liver), revealed overexpression of cytochrome P450 family 1 subfamily A member (1cyp 1a) in moderate concentrations of MP-PE (in the intestines). In the liver, moderate and higher concentrations of MP-PE induced vitellogenin 1 (vtg 1) significantly, but not cyp 1a (Mak et al., 2019). In another study zebra fish (*Danio rerio* and *Perca fluviatilis)* were exposed to MP-PE (sized 10–45 µm and 106–125 µm) for 21 d. The results showed microparticles with 10-45 µm were more toxic than those sized 106–125 µm (Fig. 3.3). This result indicated that microplastic toxicity is

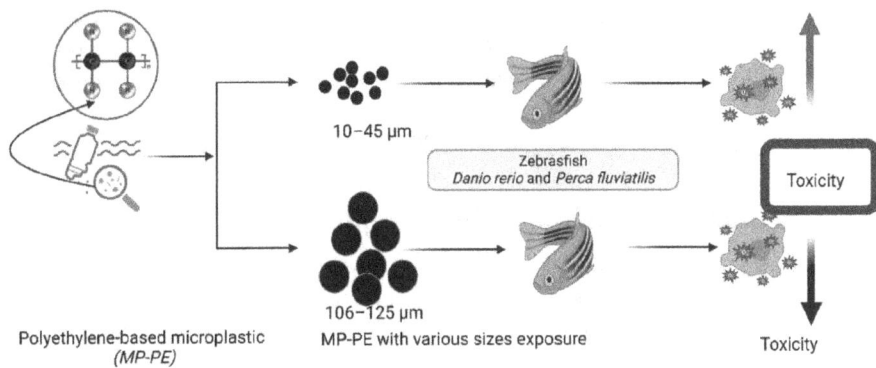

Figure 3.3 MP-PE exposure to zebrafish—the role of MP-PE in toxicity.

size dependent. In addition, the oxidative stress was intensified by small particles rather than larger particles (Bobori et al., 2022).

Recently, an in-depth study was performed to understand the antioxidant response and Na^+-K^+-ATPase activity in zebrafish exposed to polyethylene microplastics (5 and 50 µg/L). MP-PE induced significant alterations in antioxidants enzyme and Na^+/K^+-ATPase activity in zebrafish. Interestingly, in all studied organs time and MP-PE treatment interacted, led to CAT and GST activity alteration. For example, a greater than 20% reduction of CAT activity was observed in the liver after 10 d. While, in the brain, a CAT activity was reduced in 30% on the 10th d. In addition, Na^+/K^+-ATPase activity increased linearly with MP-PE concentrations (Rangasamy et al., 2022).

Carps are fish that belong to the family *Cyprinidae*. This fish is mainly consumed in Asia and Europe, and understanding the toxicological aspects of MP-PE exposure is essential. Several works reported that MP-PE caused injury in carps, damaging gills, and disrupting their immune function; however, a mechanism that governs the process is still unclear (Cao et al., 2023). A recent work demonstrated that exposure to MP-PE (1,000 ng/L; 8 µm) induces severe oxidative stress in carp gills. Furthermore, the results revealed that MP-PE exposure intensified the expression of the nuclear factor-κB (NF-κB) pathway (NF-κB p65, IKKα, IKKβ) and apoptosis biomarkers (p53, caspase-3, caspase-9, and Bax). The results filled the gap regarding understanding the pathways involved in the toxicity of MP-PE in carps.

Humans are susceptible to being exposed to MP-PE microplastics. Therefore, understanding the toxicological aspects of this type of microplastic is imperative. MP-PE (1 µg/mL to 1,000 µg/mL; and 30.5 ± 10.5 and 6.2 ± 2.0 µm) was evaluated in terms of their toxicity in epithelial cells. All concentrations of MP-PE tested did not cause cytotoxicity. However, the highest concentration reduced the cell viability slightly. In addition, various MP-PE concentration were used to understand an oxidative stress profile in defined cell lines. All treatments upregulated the nitrite level and depended on MP-PE dose and size. ROS was dose-dependent; however, the MP-PE size did not affect the ROS profile. MP-PE affected the level of the pro-inflammatory cytokine, altering HaCaT, THP-1, and U937 cell lines (Gautam et al., 2022). The genotoxic and cytotoxic effects of polyethylene microplastics

(25–500 μg/mL; 10–45 μm), on the human peripheral blood lymphocytes was reported for the first time. In all treatments, MP-PE exposure increased the frequency of Micronucleus Cytome (MN). However, a significant increase in MN was observed in 250 and 500 μg/mL MP-PE concentrations, indicating that MN frequency is MP-PE dose-dependent (Çobanoğlu et al., 2021).

3.2 MPs-other CECs Interaction with Organisms

Microplastics represent significant concerns in the modern society. The problem of microplastics is related to their direct toxicity and their interaction with other CECs altering its effects on the environment (Fig. 3.4). For these reasons, is essential to understand the interactions of microplastics with other CECs, and how it affects their toxicology.

Chlorpyrifos [O,O-diethyl O-(3,5,6-trichloro-2-pyridinyl) phosphorothioate (CPF) is an insecticide widely used in agriculture to control insect pests. Therefore, studying its toxicology in the presence of microplastics is essential. MP-PE increased the toxicity of chlorpyrifos to the marine copepod *Acartia tonsa*. CPF loaded in MP-PE decreased by 94% the egg production compared to CPF in the absence of microplastics. The recruitment rate was largely affected by the CPF and MP-PE combination. This behavior was associated to the desorption and adsorption of CPF in the MP surface (Bellas and Gil, 2020).

Health care and personal care products are the main CECs that were detected in environmental matrices. Understanding the interaction of this with MP particles is essential. Ketoconazole (KZ) is an antifungal mainly used to treat most infections. However, KZ was primarily detected in the environmental samples in the last few years. A recent study evaluated a toxic interaction between KZ and microplastics (0.1 +0.1, 1 +1, 10 +10, and 100 +100 μg/g) on *Limnodrilus hoffmeistteri*. MP-PS intensified the KZ bioaccumulation five times compared to KZ alone. Other parameters such as enzyme activity, weight, and offspring were affected negatively by KZ, MP-PS, and KZ+MP-PS; however, the work did not report a synergism or antagonism (Lu et al., 2023a). In the same perspective, MP-PE with a mix of 15 CECs were evaluated for their potential genotoxicity, mutagenicity, and redox imbalance in zebra fish. The results showed that microplastics alone or in combination induced an accumulation in zebrafish. In addition, there was no antagonism or synergism between MP-PE and 15 emerging contaminants (Araújo et al., 2022).

Triclosan (TCS) is an antimicrobial widely used in personal care products such as toothpastes, mouthwashes, hand sanitizers, and surgical soaps. Therefore, the

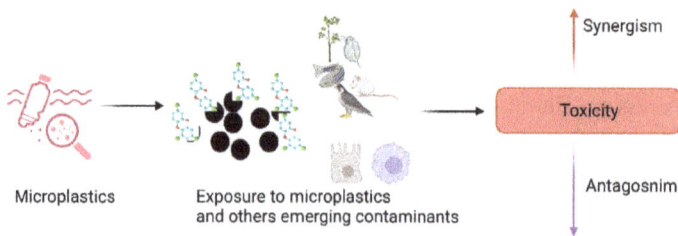

Figure 3.4 Toxicity of microplastics in combination with CECs.

study of the interaction of TCS and microplastics is essential. The joint toxicity of TCZ and MP-PS was evaluated using CaCo-2 cell as a model. First, the data showed that TCZ is adsorbed on MP-PS. The adsorption capacity of MP-PS increased as a function of surface functionalization. The functionalization study and their role on TCS adsorption are essential as the MP-PS are exposed to several biotic and abiotic conditions that lead to their surface modification in the environment. Pristine MP-PS did not affect the CaCo-2 cell viability, while functionalized MP-PS and TCS decreased the cell viability. To understand the joint toxicity of MP-PS+TCS, various proportions with TCS+MP-P (pristine), TCS+MPS-functionalized by-NH_2 and -COOH were performed. The results showed that the TCS and functionalized MP-PS combination illustrated a joint toxicity higher than other formulations (Fig. 3.5). This result was ascribed to the interaction of CaCo-2 with MP-PS and the increased adsorption capacity (Rubin and Zucker, 2022). Similarly, a joint toxicity of TCS and MPs (polyethylene (PE, 74 μm), polystyrene (PS, 74 μm), polyvinyl chloride (PVC, 74 μm), and PVC800 (1 μm) on microalgae *Skeletonema costatum* were evaluated. All MPs inhibited the growth of *Skeletonema costatum* and followed the order PS < PE < PVC < PVC800. This trend, corroborating with other works, showed that the toxicology of MPs depends on the polymer type and size. Surprisingly, the TCS increased the growth of *Skeletonema costatum* until 96 hr. After 96 hr, the inhibition growth of *Skeletonema costatum* was observed and was associated to the hormesis effect. The joint toxicity experiment showed an antagonism effect. The presence of MPs resulted in a reduction of toxicity and corroborated with the adsorption capacity of MPs. The adsorption of TCS by MPs reduced the availability of TCS in the solution, reducing its direct contact with *Skeletonema costatum* (Zhu et al., 2019).

The influence of different polymer types of microplastics on the adsorption, accumulation, and toxicity of triclosan (TCS) in zebrafish was evaluated. The adsorption capacity of MP-PP was higher than MP-PVC and MP-PE. All MPs types

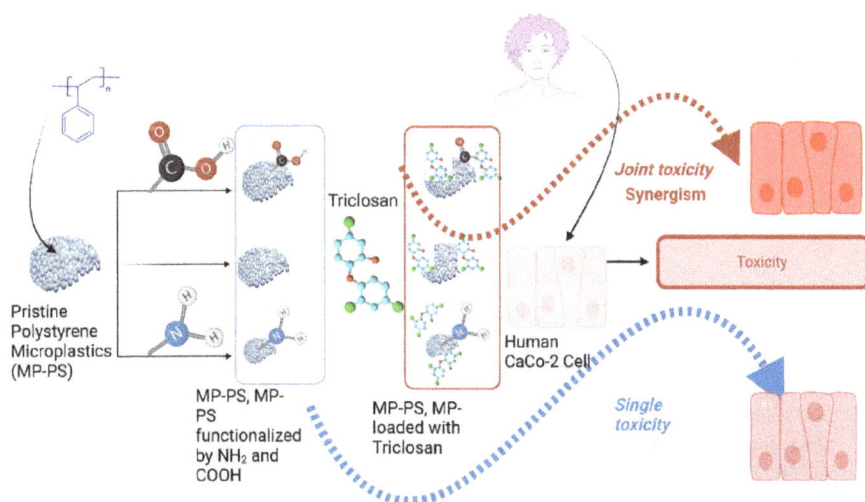

Figure 3.5 Co exposure of MP-PS functionalized by NH_2 and COOH, and Triclosan on the Human CaCo-2 cell.

increased the accumulation of TCS in zebra fish. However, different TCS accumulation patterns were observed depending on the areas of the body. For example, the effect of MPs on TCS accumulation was more intensive in the gut compared to other body parts. As expected, TCS +MP-PP combination accumulated more TCS compared to other types of MPs evaluated, and this was consistent to the adsorption capacity of PP. Furthermore, TCS+MP-PP aggravated hepatic oxidative stress and enhanced the cerebral toxicity in zebrafish. MP-PVC and MP-PE in combination with TCS do not increase toxicity compared to TCS, MP-PE, and MP-PVC alone (Sheng et al., 2021).

4. Conclusion

Contaminants of Emergent Concerns (CECs) are the main environmental challenges in society. Among several types of CECs, microplastics represent an environmental concern due to higher sources availability and interactions with other contaminants. The present chapter focused on the occurrence of microplastics and toxicological aspects. Microplastics have been found in several media, such as soil, air, and water. More recently, microplastics were found in the organisms' tissues. The most impactful detection occurred in the human placenta, raising concerns regarding the role of microplastics in the development of the embryo. The toxicology of MPs in various organisms was also found in literature. Most organisms were evaluated in the presence of microplastics, and the results showed specific toxicity. However, this toxicity cannot be generalized because, in some organisms, the MPs did not cause toxicity to the organism. In addition, the MPs toxicity depends on several factors such as pH, size, shape, etc. Microplastics interact with CECs increasing/decreasing their toxicity. Although because of these advances, more works aiming to the standardization of toxicology protocols are still needed. Furthermore, the pathways involved in toxic processes are not clearly understood due to the variability of organisms and microplastics (type, size, and shape).

Acknowledgments

The authors acknowledge the National Council for Scientific and Technological Development (CNPq), the Coordination for the Improvement of Higher Education Personnel (CAPES), the Federal University of Santa Catarina (UFSC), and the Support Foundation for Research and Innovation of Santa Catarina (FAPESC).

References

Abbasi, S., Keshavarzi, B., Moore, F., Turner, A., Kelly, F.J., Dominguez, A.O. et al. 2019. Distribution and potential health impacts of microplastics and microrubbers in air and street dusts from Asaluyeh County, Iran: Environmental Pollution 244: 153–164.

Abelouah, M.R., Ben-Haddad, M., Hajji, S., De-la-Torre, G.E., Aziz, T., Oualid, J.A. et al. 2022. Floating microplastics pollution in the Central Atlantic Ocean of Morocco: Insights into the occurrence, characterization, and fate: Marine Pollution Bulletin 182: 113969.

Ahmad, T., Amjad, M., Iqbal, Q., Batool, A., Noor, A., Jafir, M. et al. 2022. Occurrence of Microplastics and Heavy Metals in Aquatic and Agroecosystem: A Case Study: Bulletin of Environmental Contamination and Toxicology 109: 266–271.

Ahmed, R., Hamid, A.K., Krebsbach, S.A., He, J. and Wang, D. 2022. Critical review of microplastics removal from the environment: Chemosphere 293: 133557.

Alfonso, M.B., Arias, A.H., Ronda, A.C. and Piccolo, M.C. 2021. Continental microplastics: Presence, features, and environmental transport pathways: Science of The Total Environment 799: 149447.

Alimi, O.S., Farner Budarz, J., Hernandez, L.M. and Tufenkji, N. 2018. Microplastics and Nanoplastics in Aquatic Environments: Aggregation, Deposition, and Enhanced Contaminant Transport: Environmental Science & Technology 52: 1704–1724.

Allen, S., Allen, D., Phoenix, V.R., Le Roux, G., Jiménez, P.D., Simonneau, A. et al. 2019. Author Correction: Atmospheric transport and deposition of microplastics in a remote mountain catchment: Nature Geoscience 12: 679.

Araújo, A.P. da C., Luz, T.M. da, Rocha, T.L., Ahmed, M.A.I., Silva, D. de M., Rahman, M.M. et al. 2022. Toxicity evaluation of the combination of emerging pollutants with polyethylene microplastics in zebrafish: Perspective study of genotoxicity, mutagenicity, and redox unbalance: Journal of Hazardous Materials 432: 128691.

Bai, C.-L., Liu, L.-Y., Hu, Y.-B., Zeng, E.Y. and Guo, Y. 2022. Microplastics: A review of analytical methods, occurrence and characteristics in food, and potential toxicities to biota: Science of The Total Environment 806: 150263.

Baldwin, A.K., Spanjer, A.R., Rosen, M.R. and Thom, T. 2020. Microplastics in Lake Mead National Recreation Area, USA: Occurrence and biological uptake: PLOS ONE 15: e0228896.

Bellas, J. and Gil, I. 2020, Polyethylene microplastics increase the toxicity of chlorpyrifos to the marine copepod Acartia tonsa: Environmental Pollution 260: 114059.

van den Berg, P., Huerta-Lwanga, E., Corradini, F. and Geissen, V. 2020. Sewage sludge application as a vehicle for microplastics in eastern Spanish agricultural soils: Environmental Pollution 261: 114198.

Blankson, E.R., Tetteh, P.N., Oppong, P. and Gbogbo, F. 2022. Microplastics prevalence in water, sediment and two economically important species of fish in an urban riverine system in Ghana: PLOS ONE 17: e0263196.

Bobori, D.C., Dimitriadi, A., Feidantsis, K., Samiotaki, A., Fafouti, D., Sampsonidis, I. et al. 2022, Differentiation in the expression of toxic effects of polyethylene-microplastics on two freshwater fish species: Size matters: Science of The Total Environment 830: 154603.

Boháčková, J., Havlíčková, L., Semerád, J., Titov, I., Trhlíková, O., Beneš, H. et al. 2023. *In vitro* toxicity assessment of polyethylene terephthalate and polyvinyl chloride microplastics using three cell lines from rainbow trout (Oncorhynchus mykiss): Chemosphere 312: 136996.

Campanale, C., Stock, F., Massarelli, C., Kochleus, C., Bagnuolo, G., Reifferscheid, G. et al. 2020. Microplastics and their possible sources: The example of Ofanto river in southeast Italy: Environmental Pollution 258: 113284.

Cao, J., Xu, R., Wang, F., Geng, Y., Xu, T., Zhu, M. et al. 2023. Polyethylene microplastics trigger cell apoptosis and inflammation via inducing oxidative stress and activation of the NLRP3 inflammasome in carp gills: Fish & Shellfish Immunology 132: 108470.

Cao, L., Wu, D., Liu, P., Hu, W., Xu, L., Sun, Y. et al. 2021. Occurrence, distribution and affecting factors of microplastics in agricultural soils along the lower reaches of Yangtze River, China: Science of The Total Environment 794: 148694.

Chae, Y. and An, Y.-J. 2018. Current research trends on plastic pollution and ecological impacts on the soil ecosystem: A review: Environmental Pollution 240: 387–395.

Chen, E.-Y., Lin, K.-T., Jung, C.-C., Chang, C.-L. and Chen, C.-Y. 2022. Characteristics and influencing factors of airborne microplastics in nail salons: Science of The Total Environment 806: 151472.

Chen, G., Feng, Q. and Wang, J. 2020. Mini-review of microplastics in the atmosphere and their risks to humans: Science of The Total Environment 703: 135504.

Cheng, H., Feng, Y., Duan, Z., Duan, X., Zhao, S., Wang, Y. et al. 2021. Toxicities of microplastic fibers and granules on the development of zebrafish embryos and their combined effects with cadmium: Chemosphere 269: 128677.

Çobanoğlu, H., Belivermiş, M., Sıkdokur, E., Kılıç, Ö. and Çayır, A. 2021. Genotoxic and cytotoxic effects of polyethylene microplastics on human peripheral blood lymphocytes: Chemosphere 272: 129805.

Cózar, A., Martí, E., Duarte, C.M., García-de-Lomas, J., van Sebille, E., Ballatore, T.J. et al. 2023. The Arctic Ocean as a dead end for floating plastics in the North Atlantic branch of the Thermohaline Circulation: Science Advances 3: e1600582.

Cui, J., Chen, C., Gan, Q., Wang, T., Li, W., Zeng, W. et al. 2022. Indoor microplastics and bacteria in the atmospheric fallout in urban homes: Science of The Total Environment 852: 158233.

Dong, C.-D., Chen, C.-W., Chen, Y.-C., Chen, H.-H., Lee, J.-S. and Lin, C.-H. 2020. Polystyrene microplastic particles: *In vitro* pulmonary toxicity assessment: Journal of Hazardous Materials 385: 121575.

Fan, Z., Xiao, T., Luo, H., Chen, D., Lu, K., Shi, W. et al. 2022. A study on the roles of long non-coding RNA and circular RNA in the pulmonary injuries induced by polystyrene microplastics: Environment International 163: 107223.

Feng, S., Lu, H. and Liu, Y. 2021. The occurrence of microplastics in farmland and grassland soils in the Qinghai-Tibet plateau: Different land use and mulching time in facility agriculture: Environmental Pollution 279: 116939.

Garcés-Ordóñez, O., Saldarriaga-Vélez, J.F., Espinosa-Díaz, L.F., Patiño, A.D., Cusba, J., Canals, M. et al. 2022. Microplastic pollution in water, sediments and commercial fish species from Ciénaga Grande de Santa Marta lagoon complex, Colombian Caribbean: Science of The Total Environment 829: 154643.

Gautam, Ravi, Jo, JiHun, Acharya, Manju, Maharjan, Anju, Lee, DaEun, Bahadur K.C. et al. 2022. Evaluation of potential toxicity of polyethylene microplastics on human derived cell lines: Science of The Total Environment 838: 156089.

Geyer, R., Jambeck, J.R. and Law, K.L. 2023. Production, use, and fate of all plastics ever made: Science Advances 3: 1700782.

González-Pleiter, M., Edo, C., Aguilera, Á., Viúdez-Moreiras, D., Pulido-Reyes, G., González-Toril, E. et al. 2021. Occurrence and transport of microplastics sampled within and above the planetary boundary layer: Science of The Total Environment 761: 143213.

Guo, J.-J., Huang, X.-P., Xiang, L., Wang, Y.-Z., Li, Y.-W., Li, H. et al. 2020. Source, migration and toxicology of microplastics in soil: Environment International 137: 105263.

He, D., Luo, Y., Lu, S., Liu, M., Song, Y. and Lei, L. 2018. Microplastics in soils: Analytical methods, pollution characteristics and ecological risks: TrAC Trends in Analytical Chemistry 109: 163–172.

Horton, A.A., Walton, A., Spurgeon, D.J., Lahive, E., and Svendsen, C. 2017. Microplastics in freshwater and terrestrial environments: Evaluating the current understanding to identify the knowledge gaps and future research priorities: Science of The Total Environment 586: 127–141.

Horton, A.A., Cross, R.K., Read, D.S., Jürgens, M.D., Ball, H.L., Svendsen, C. et al. 2021. Semi-automated analysis of microplastics in complex wastewater samples: Environmental Pollution 268: 115841.

Huang, B., Sun, L., Liu, M., Huang, H., He, H., Han, F. et al. 2021. Abundance and distribution characteristics of microplastic in plateau cultivated land of Yunnan Province, China: Environmental Science and Pollution Research 28: 1675–1688.

Huang, Y., Liu, Q., Jia, W., Yan, C. and Wang, J. 2020. Agricultural plastic mulching as a source of microplastics in the terrestrial environment: Environmental Pollution 260: 114096.

Jian, M., Zhang, Y., Yang, W., Zhou, L., Liu, S. and Xu, E.G. 2020. Occurrence and distribution of microplastics in China's largest freshwater lake system: Chemosphere 261: 128186.

Jin, H., Ma, T., Sha, X., Liu, Z., Zhou, Y., Meng, X. et al. 2021. Polystyrene microplastics induced male reproductive toxicity in mice: Journal of Hazardous Materials 401: 123430.

Kasmuri, N., Tarmizi, N.A.A. and Mojiri, A. 2022. Occurrence, impact, toxicity, and degradation methods of microplastics in environment—a review: Environmental Science and Pollution Research 29: 30820–30836.

Klein, M. and Fischer, E.K. 2019. Microplastic abundance in atmospheric deposition within the Metropolitan area of Hamburg, Germany: Science of The Total Environment 685: 96–103.

Koner, S., Florance, I., Mukherjee, A. and Chandrasekaran, N. 2023. Cellular response of THP-1 macrophages to polystyrene microplastics exposure: Toxicology 483: 153385.

Koutnik, V.S., Leonard, J., Alkidim, S., DePrima, F.J., Ravi, S., Hoek, E.M.V. et al. 2021. Distribution of microplastics in soil and freshwater environments: Global analysis and framework for transport modeling: Environmental Pollution 274: 116552.

Li, W.C., TSE, H.F. and FOK, L. 2016. Plastic waste in the marine environment: A review of sources, occurrence and effects: Science of The Total Environment 566–567: 333–349.

Li, J., Liu, H. and Paul Chen, J. 2018a. Microplastics in freshwater systems: A review on occurrence, environmental effects, and methods for microplastics detection: Water Research 137: 362–374.

Li, X., Chen, L., Mei, Q., Dong, B., Dai, X., Ding, G. et al. 2018b. Microplastics in sewage sludge from the wastewater treatment plants in China: Water Research 142: 75–85.

Li, X., Zhang, T., Lv, W., Wang, H., Chen, H., Xu, Q. et al. 2022. Intratracheal administration of polystyrene microplastics induces pulmonary fibrosis by activating oxidative stress and Wnt/β-catenin signaling pathway in mice: Ecotoxicology and Environmental Safety 232: 113238.

Liao, Y., Jiang, X., Xiao, Y. and Li, M. 2020. Exposure of microalgae Euglena gracilis to polystyrene microbeads and cadmium: Perspective from the physiological and transcriptional responses: Aquatic Toxicology 228: 105650.

Liu, K., Wang, X., Fang, T., Xu, P., Zhu, L. and Li, D. 2019a. Source and potential risk assessment of suspended atmospheric microplastics in Shanghai: Science of The Total Environment 675: 462–471.

Liu, K., Wu, T., Wang, X., Song, Z., Zong, C., Wei, N. et al. 2019b. Consistent Transport of Terrestrial Microplastics to the Ocean through Atmosphere: Environmental Science & Technology 53: 10612–10619.

Liu, Z., Zhuan, Q., Zhang, L., Meng, L., Fu, X. and Hou, Y. 2022. Polystyrene microplastics induced female reproductive toxicity in mice: Journal of Hazardous Materials 424: 127629.

Lu, G., Xue, Q., Ling, X. and Zheng, X. 2023a. Toxic interactions between microplastics and the antifungal agent ketoconazole in sediments on Limnodrilus hoffmeistteri: Process Safety and Environmental Protection 172: 250–261.

Lu, H., Yin, K., Su, H., Wang, D., Zhang, Y., Hou, L. et al. 2023. Polystyrene microplastics induce autophagy and apoptosis in birds lungs via PTEN/PI3K/AKT/mTOR: Environmental Toxicology 38: 78–89.

Mak, C.W., Ching-Fong Yeung, K. and Chan, K.M. 2019. Acute toxic effects of polyethylene microplastic on adult zebrafish: Ecotoxicology and Environmental Safety 182: 109442.

Malafaia, G., de Souza, A.M., Pereira, A.C., Gonçalves, S., da Costa Araújo, A.P., Ribeiro, R.X. et al. 2020. Developmental toxicity in zebrafish exposed to polyethylene microplastics under static and semi-static aquatic systems: Science of The Total Environment 700: 134867.

Napper, I.E., Davies, B.F.R., Clifford, H., Elvin, S., Koldewey, H.J., Mayewski, P.A. et al. 2020. Reaching New Heights in Plastic Pollution—Preliminary Findings of Microplastics on Mount Everest: One Earth 3: 621–630.

Nizzetto, L., Langaas, S. and Futter, M. 2016. Pollution: Do microplastics spill on to farm soils? Nature 537: 488.

O'Brien, S., Okoffo, E.D., O'Brien, J.W., Ribeiro, F., Wang, X., Wright, S.L. et al. 2020. Airborne emissions of microplastic fibres from domestic laundry dryers: Science of The Total Environment 747: 141175.

O'Brien, S., Rauert, C., Ribeiro, F., Okoffo, E.D., Burrows, S.D., O'Brien, J.W. et al. 2023. There's something in the air: A review of sources, prevalence and behaviour of microplastics in the atmosphere: Science of The Total Environment 874: 162193.

Panno, S.V., Kelly, W.R., Scott, J., Zheng, W., McNeish, R.E., Holm, N. et al. 2019. Microplastic Contamination in Karst Groundwater Systems: Groundwater 57: 189–196.

Peng, X., Chen, M., Chen, S., Dasgupta, S., Xu, H., Ta, K. et al. 2018. Microplastics contaminate the deepest part of the world's ocean: Geochemical Perspectives Letters 9: 1–5.

Pérez-Reverón, R., González-Sálamo, J., Hernández-Sánchez, C., González-Pleiter, M., Hernández-Borges, J. and Díaz-Peña, F.J. 2022. Recycled wastewater as a potential source of microplastics in irrigated soils from an arid-insular territory (Fuerteventura, Spain): Science of The Total Environment 817: 152830.

Piccardo, M., Provenza, F., Grazioli, E., Cavallo, A., Terlizzi, A. and Renzi, M. 2020. PET microplastics toxicity on marine key species is influenced by pH, particle size and food variations: Science of The Total Environment 715: 136947.

Pivokonsky, M., Cermakova, L., Novotna, K., Peer, P., Cajthaml, T. and Janda, V. 2018. Occurrence of microplastics in raw and treated drinking water: Science of The Total Environment 643: 1644–1651.

Prata, J.C. 2018. Airborne microplastics: Consequences to human health? Environmental Pollution 234: 115–126.

Provenza, F., Piccardo, M., Terlizzi, A. and Renzi, M. 2020. Exposure to pet-made microplastics: Particle size and pH effects on biomolecular responses in mussels: Marine Pollution Bulletin 156: 111228.

Qiu, Y., Zhou, S., Zhang, C., Zhou, Y. and Qin, W. 2022. Soil microplastic characteristics and the effects on soil properties and biota: A systematic review and meta-analysis: Environmental Pollution 313: 120183.

Rangasamy, B., Malafaia, G. and Maheswaran, R. 2022. Evaluation of antioxidant response and Na+-K+-ATPase activity in zebrafish exposed to polyethylene microplastics: Shedding light on a physiological adaptation: Journal of Hazardous Materials 426: 127789.

Rillig, M.C., Ziersch, L. and Hempel, S. 2017. Microplastic transport in soil by earthworms: Scientific Reports 7: 1362.

Rosso, B., Corami, F., Barbante, C. and Gambaro, A. 2023, Quantification and identification of airborne small microplastics (< 100 µm) and other microlitter components in atmospheric aerosol via a novel elutriation and oleo-extraction method: Environmental Pollution 318: 120889.

Rubin, A.E. and Zucker, I. 2022. Interactions of microplastics and organic compounds in aquatic environments: A case study of augmented joint toxicity: Chemosphere 289: 133212.

Russell, M. and Webster, L. 2021. Microplastics in sea surface waters around Scotland: Marine Pollution Bulletin 166: 112210.

Samandra, S., Mescall, O.J., Plaisted, K., Symons, B., Xie, S., Ellis, A.V. et al. 2022a. Assessing exposure of the Australian population to microplastics through bottled water consumption: Science of The Total Environment 837: 155329.

Samandra, S., Johnston, J.M., Jaeger, J.E., Symons, B., Xie, S., Currell, M. et al. 2022b. Microplastic contamination of an unconfined groundwater aquifer in Victoria, Australia: Science of The Total Environment 802: 149727.

Samandra, S., Singh, J., Plaisted, K., Mescall, O.J., Symons, B., Xie, S. et al. 2023. Quantifying environmental emissions of microplastics from urban rivers in Melbourne, Australia: Marine Pollution Bulletin 189: 114709.

Schymanski, D., Goldbeck, C., Humpf, H.-U. and Fürst, P. 2018, Analysis of microplastics in water by micro-Raman spectroscopy: Release of plastic particles from different packaging into mineral water: Water Research 129: 154–162.

Sheng, C., Zhang, S. and Zhang, Y. 2021. The influence of different polymer types of microplastics on adsorption, accumulation, and toxicity of triclosan in zebrafish: Journal of Hazardous Materials 402: 123733.

Steinmetz, Z., Wollmann, C., Schaefer, M., Buchmann, C., David, J., Tröger, J. et al. 2016. Plastic mulching in agriculture. Trading short-term agronomic benefits for long-term soil degradation? Science of The Total Environment 550: 690–705.

Sun, J., Dai, X., Wang, Q., van Loosdrecht, M.C.M. and Ni, B.-J. 2019. Microplastics in wastewater treatment plants: Detection, occurrence and removal: Water Research 152: 21–37.

Suteja, Y., Atmadipoera, A.S., Riani, E., Nurjaya, I.W., Nugroho, D. and Cordova, M.R. 2021. Spatial and temporal distribution of microplastic in surface water of tropical estuary: Case study in Benoa Bay, Bali, Indonesia: Marine Pollution Bulletin 163: 111979.

Tong, H., Jiang, Q., Hu, X. and Zhong, X. 2020. Occurrence and identification of microplastics in tap water from China: Chemosphere 252: 126493.

van den Berg, P., Huerta-Lwanga, E., Corradini, F. and Geissen, V. 2020. Sewage sludge application as a vehicle for microplastics in eastern Spanish agricultural soils: Environmental Pollution 261: 114198.

Vega-Herrera, A., Llorca, M., Borrell-Diaz, X., Redondo-Hasselerharm, P.E., Abad, E., Villanueva, C.M. et al. 2022. Polymers of micro(nano) plastic in household tap water of the Barcelona Metropolitan Area: Water Research 220: 118645.

Wang, C., Zhao, J. and Xing, B. 2021a. Environmental source, fate, and toxicity of microplastics: Journal of Hazardous Materials 407: 124357.

Wang, Y., Huang, J., Zhu, F. and Zhou, S. 2021b. Airborne Microplastics: A Review on the Occurrence, Migration and Risks to Humans: Bulletin of Environmental Contamination and Toxicology 107: 657–664.

Weber, A., Scherer, C., Brennholt, N., Reifferscheid, G. and Wagner, M. 2018. PET microplastics do not negatively affect the survival, development, metabolism and feeding activity of the freshwater invertebrate Gammarus pulex: Environmental Pollution 234: 181–189.

Winkler, A., Santo, N., Ortenzi, M.A., Bolzoni, E., Bacchetta, R. and Tremolada, P. 2019. Does mechanical stress cause microplastic release from plastic water bottles? Water Research 166: 115082.

Wong, J.K.H., Lee, K.K., Tang, K.H.D. and Yap, P.-S. 2020. Microplastics in the freshwater and terrestrial environments: Prevalence, fates, impacts and sustainable solutions: Science of The Total Environment 719: 137512.

World Health Organization. 2019. Microplastics in Drinking-Water, Licence: C.

Xiao, Y., Jiang, X., Liao, Y., Zhao, W., Zhao, P. and Li, M. 2020. Adverse physiological and molecular level effects of polystyrene microplastics on freshwater microalgae: Chemosphere 255: 126914.

Xie, X., Deng, T., Duan, J., Xie, J., Yuan, J. and Chen, M. 2020. Exposure to polystyrene microplastics causes reproductive toxicity through oxidative stress and activation of the p38 MAPK signaling pathway: Ecotoxicology and Environmental Safety 190: 110133.

Yan, M., Li, W., Chen, X., He, Y., Zhang, X. and Gong, H. 2021. A preliminary study of the association between colonization of microorganism on microplastics and intestinal microbiota in shrimp under natural conditions: Journal of Hazardous Materials 408: 124882.

Yang, L., Zhang, Y., Kang, S., Wang, Z. and Wu, C. 2021. Microplastics in soil: A review on methods, occurrence, sources, and potential risk: Science of The Total Environment 780: 146546.

Yukioka, S., Tanaka, S., Nabetani, Y., Suzuki, Y., Ushijima, T., Fujii, S. et al. 2020. Occurrence and characteristics of microplastics in surface road dust in Kusatsu (Japan), Da Nang (Vietnam), and Kathmandu (Nepal): Environmental Pollution 256: 113447.

Zha, H., Xia, J., Li, S., Lv, J., Zhuge, A., Tang, R. et al. 2023. Airborne polystyrene microplastics and nanoplastics induce nasal and lung microbial dysbiosis in mice: Chemosphere 310: 136764.

Zhang, K., and Wu, C., 2023, Chapter One - Formation of airborne microplastics. *In*: Wang, J.B.T.-C.A.C. (ed.). Airborne Microplastics: Analysis, Fate And Human Health Effects 100. Elsevier: 1–16.

Zhang, Z., Gao, S.-H., Luo, G., Kang, Y., Zhang, L., Pan, Y. et al. 2022. The contamination of microplastics in China's aquatic environment: Occurrence, detection and implications for ecological risk: Environmental Pollution 296: 118737.

Zhang, Z., Deng, C., Dong, L., Liu, L., Li, H., Wu, J. et al. 2021, Microplastic pollution in the Yangtze River Basin: Heterogeneity of abundances and characteristics in different environments: Environmental Pollution 287: 117580.

Zhao, X., Zhou, Y., Liang, C., Song, J., Yu, S., Liao, G. et al. 2023, Airborne microplastics: Occurrence, sources, fate, risks and mitigation: Science of The Total Environment 858: 159943.

Zhu, J., Zhang, X., Liao, K., Wu, P. and Jin, H. 2022. Microplastics in dust from different indoor environments: Science of The Total Environment 833: 155256.

Zhu, Long, Zhu, Jingying, Zuo, Rui, Xu, Qiujin, Qian, Yanhua and An, Lihui. 2023. Identification of microplastics in human placenta using laser direct infrared spectroscopy: Science of The Total Environment 856: 159060.

Zhu, Z., Wang, S., Zhao, F., Wang, S., Liu, F. and Liu, G. 2019. Joint toxicity of microplastics with triclosan to marine microalgae Skeletonema costatum: Environmental Pollution 246: 509–517.

CHAPTER 4

Occurrence and Fate of Emerging Contaminants with Microplastics

Current Scenario, Sources and Effects

Pankaj Kumar, Savita Chaudhary* and Aman Bhalla*

1. Introduction

Plastic has been made ubiquitous in the environment by humans (Sajjad et al., 2022). Man has made plastics available in the soil, rivers, lakes, oceans, and air. Based on the current production of plastics, it is estimated that by the year 2050, its global production will reach 33 billion tonnes (Chen et al., 2020). The main reason for the large production of plastic is that it is durable, light weight, easy to transport, easy to synthesize, and cheap in cost. Today plastic is used in almost every sector including, agricultural, automobiles, clothing, building materials, chemicals, telecommunication, medical equipment, and most importantly the packaging industry (Singh and Sharma, 2016). The main sources of plastic production are crude oil and petroleum products (Harding et al., 2007). Although the discovery of plastic has revolutionized every industry, it has given rise to environmental pollution. It increases the CO_2 concentration in the environment and is the major source of air, soil, and water pollution. Its presence in the environment makes it available to the biota resulting in the death of animals due to ingestion and entanglement (López-

Department of Chemistry & Centre of Advanced Studies in Chemistry, Panjab University, Chandigarh-160014

* Corresponding authors: amanbhalla@pu.ac.in; schaudhary@pu.ac.in

Martínez et al., 2021). However, with the discovery of "microplastics" an even more harmful and dangerous situation than plastics has been predicted by scientists (Campanale et al., 2020).

The term "microplastics" was introduced by Richard Thompson, who identified them on British beaches in 2004 (Napper and Thompson, 2020). Microplastics (MPs) are defined by plastic particles of 100 nm –5 mm in size. They can be of two kinds primary MPs or secondary MPs. Primary MPs are plastic beads that are synthesized for use in Personal Care and Cosmetic Products (PCCPs) such as shaving creams, scrubs, face washes, toothpastes, etc. (Laskar and Kumar, 2019). While secondary MPs are those that are formed due to the breaking down of larger plastics under environmental conditions such as weathering, heat, sunlight, etc., to the size limit of MPs.

Microbeads generally are of the size 0.1 μm –1 mm and made of different polymers such as polyethylene (PE), polypropylene (PP), polyethylene terephthalate (PET), polyamides (PA), polymethyl methacrylate (PMMA), polyester, and polyurethanes (McDevitt et al., 2017). Primary MPs are discarded from homes into the Waste Water Treatment Plants (WWTP) and directly to the environment by incorrect waste disposal. However, the WWTP are not that much more efficient for the removal of microbeads completely and hence remain untreated leading to their effluents into the water bodies including rivers, lakes, and sea (Ngo et al., 2019). Based on a survey done by Cosmetics Europe it was reported that 4360 tons of microbeads were used in the European Union countries, Switzerland, and Norway in 2012 (Juliano and Magrini, 2017). It was estimated that a single use of an exfoliant results in the release of 4594–94,500 microbeads in the environment (Jiang et al., 2022). About 4000 microbeads are released from toothpaste in a single wash (Prata, 2018). It was reported that 260 tons of microbeads were released as a result from using liquid soup per year in the United States (Nawalage and Bellanthudawa, 2022). Figure 4.1 shows the global release of microbeads in the environment. Other than PCCPs the sources of primary MPs are plastic pellets, paints, plastic running tracks in school, artificial turfs, rubber roads in cities and vehicle tire wears (An et al., 2020).

Sources of secondary MPs include landfills, municipal debris, fishing waste and agricultural films. Plastic waste in the environment undergoes various changes under environmental conditions. Along with the change in their physical properties such as the conversion to MPs, their chemical properties also change (Efimova et al., 2018). These alterations are the result from weathering conditions, such as, heat, sunlight, and the presence of various microorganisms. Municipal debris are mainly composed of plastic bags, plastic packaging and plastic bottles. Other electronic plastic waste and automobile plastic waste also exist; however, these wastes are organized because of their post-use value. Agricultural film composed of PVC and PE are also very important sources of MPs in the environment (Jia et al., 2022). Fishing gears are the main source of MPs in the marine environment (Li et al., 2021). However, rain, wind, and flooding are also responsible for increased MPs in the ocean by swiping plastics from landfills and beaches.

MPs are widely known as emerging contaminants today and pose a very alarming situation to the environment. The bioavailability to biota due to ingestion by various organisms has made their entry into the food chain (Sharma and Chatterjee, 2017).

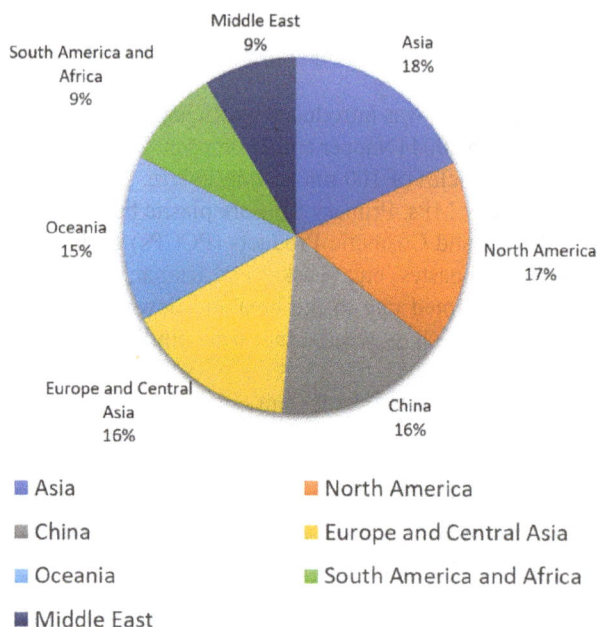

Figure 4.1 Global contribution in the release of microbeads in the environment.

The sea which is the eventual fate of MPs made their availability to aquatic life into serious toxic effects to aquatic life, plants, and organisms. The removal of MPs from the sea water is very costly (Schuhen et al., 2019). Therefore , banning their use and the proper disposal are some of the solutions to control the release of MPs in environment. In this regard many countries have enacted a legal ban on the production and sales of PCCPs having MPs (Tiller and Nyman, 2018). The United States was the first country to ban microbeads in cosmetics under "Microbead-Free Water Act" on July 1, 2017. Other countries including Canada, United Kingdom, New Zealand, Denmark, and South Korea have issued various policies to prohibit the use of microbeads.

An estimated 5.25 trillion plastic particles weighing approximately 268,000 tonnes have been released into oceans (Khan et al., 2019). MPs are also present in the terrestrial environment, soil, rivers, and lakes. The surface of MPs contains various polar and non-polar groups, alkyl chains, and aromatic groups (Sturm et al., 2021). Furthermore, the surface properties of the MPs are altered with environmental conditions, which is scientifically known as ageing which results in aged MPs (Fig. 4.3). These conditions result in Reactive Organic Species (ROS) and increases surface area, pores and other carbonyl-containing functionalities. These reactive groups and surface properties of pristine and aged MPs results in various interactions with other emerging contaminants, including pharmaceuticals, fertilizers, pesticides, insecticides, metals, polychlorinated biphenyls (PCBs) and polybrominated ethers (PBDEs) (Martinho et al., 2022). These interactions are hydrophobic interactions, hydrogen bonding, π–π interactions, electrostatic interactions, van der Waal forces and other weak forces. Not only do these interactions result in surface adsorption of these emerging contaminants but also make MPs vehicles for these contaminants to biota. As the size of MPs ranges with the planktons, they are consumed by various aquatic organisms (di Mauro et al., 2017). The co-effect of MPs and emerging contaminants results in even more toxicity. Their desorption is also effected by pH, salinity, temperature and other environmental conditions. In this chapter, the recent investigations done by various researchers on the adsorption and desorption of emerging contaminants on various MPs and their toxic effects alone and in combination on various biota, aquatic and terrestrial life have been presented.

2. Pharmaceuticals as Emerging Contaminants

Due to the extensive production and requirement of pharmaceuticals, they are continuously discarded in the environment making them emerging pollutants (Valdez-Carrillo et al., 2020). As a result, various antibiotics, antidepressants, beta-blockers, and other Pharmacologically Active Compounds (PhACs) are being detected in aquatic and terrestrial environment. Their presence in the environment makes their availability to non-target organisms causing alterations in their physiology and behaviours. In addition, they give rise to antibiotic resistance. Furthermore, these pharmaceuticals have different properties, such as polarity, ionizability, and chirality. In the past decade, their interactions with pristine and aged MPs via adsorption has posed a serious threat.

The presence of MPs in the environment can affect the fate of pharmaceuticals by altering their transfer and distribution (Puckowski et al., 2021). MPs can act as a sink for pharmaceuticals, as the plastic particles can adsorb and accumulate these compounds from the surrounding environment. This can lead to the concentration of pharmaceuticals on the surface of MPs, potentially increasing their toxicity. Furthermore, the presence of MPs can also affect the degradation of pharmaceuticals in the environment (Santos et al., 2021). MPs can provide a protective barrier that can prevent the degradation of pharmaceuticals by microorganisms or UV radiation. This can lead to the persistence of pharmaceuticals in the environment and potentially increase their toxicity over time. Additionally, MPs can also affect the behaviour and transfer of pharmaceuticals in aquatic environments such as their entry into the food chain (Fig. 4.2). MPs can alter the hydrodynamics and sedimentation patterns of aquatic environments, which can affect the deposition and accumulation of pharmaceuticals in sediments. Overall, the interaction between pharmaceuticals and MPs is an important area of research, as it can have significant implications for the health of ecosystems and the human population. In this regard, different research groups have conducted studies on the adsorption/desorption behaviour of different pharmaceuticals on MPs and their individual and co-effects on plants and organisms. Some recent important studies (2018–2022) has been included here.

2.1 Pharmaceutical Adsorption/Desorption from Pristine MPs

Pristine MPs refer to newly-produced or recently-introduced MPs in the environment that have not undergone any significant degradation or weathering. The adsorption

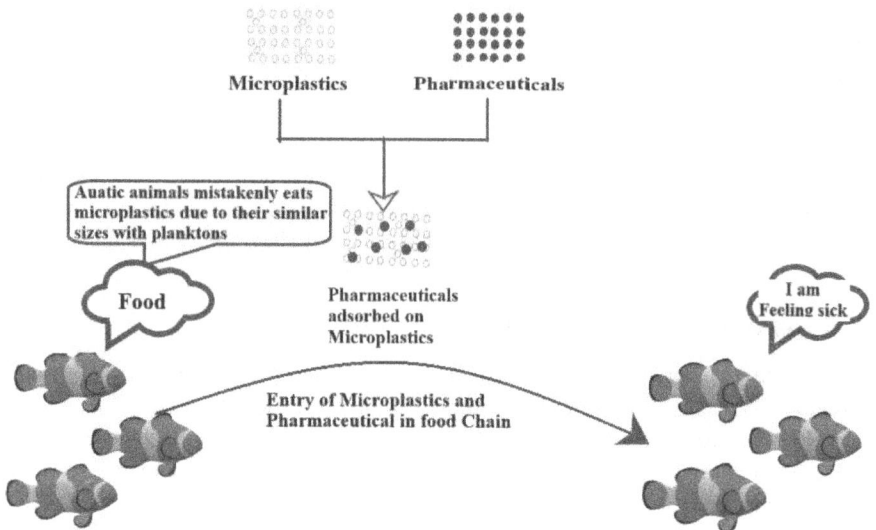

Figure 4.2 Entry of pharmaceuticals with MPs in the food chain.

and desorption of pharmaceuticals on pristine MPs has been a topic of interest for several research groups which is addressed here.

Xu and co-workers (2018) carried out investigations on sorption of tetracycline (TC) on PE, PP, and polystyrene (PS) based MPs. They further tested the influence of various environmental factors, i.e., pH, Dissolved Organic Matter (DOM) and salinity on the sorption behaviour of the TC. Langmuir model which was used to fit the sorption isotherms. The results indicated that the hydrophobic and electrostatic interactions were responsible for the sorption. Based on the data they concluded that the sorption capacity of the TC is more towards PS out of the three. The reason for the higher sorption is polar interactions and π–π interactions of the benzene ring present on both TC and PS. Furthermore, at lower and higher pH the sorption of the TC on MPs decreases due to electrostatic repulsion and is found to be higher at pH 6. In the presence of fulvic acid (DOM), the sorption capacity decreases on the MPs with increasing affinity of the TC towards it. Furthermore, salinity does not affect the adsorption process. They finally concluded, that in a non-neutral environment and the presence of the DOM, MPs have a lesser ability to transfer the TC.

Razanajatovo et al. (2018) conducted experiments to study both sorption and desorption process of pharmaceuticals agents on PE MPs in water. For this they chose, sulphamethoxazole (SMX) (antibiotic), propranolol (PRP) (beta-blocker), and sertraline (SER) (antidepressant) as sorbates. They used Linear and Freundlich models to interpret the sorption isotherm of the three pharmaceuticals. The result disclosed sorption of all pharmaceuticals on PE MPs via hydrophobicity and electrostatic interactions with the sorption percentage of 28.61, 21.61, and 15.31% for SER, PRP and SMX, respectively. Furthermore, the desorption results reveal irreversible sorption of SMX and, 4 and 8% for SER and PRP from PE MPs within 48 hr. This also increased the threat of bioaccumulation of SER and PRP in aquatic environments.

Guo (2019a) reported the findings of sorption studies of SMX and sulphamethazine (SMT) (Guo, 2019b) on six MPs including polyamide (PA), PE, PET, polystyrene (PS), polyvinyl chloride (PVC) and PP. The linear and Freundlich isotherms were used to interpret the SMX sorption equilibrium on six MPs. The sorption of SMX on MPs was found to have increased during the 1st hr, while it slows down till 16th hr when the sorption reached equilibrium. The results revealed greater affinity of SMX towards PA MPs via hydrogen bonding. Furthermore, they disclosed the decrease in sorption capacity of SMX with increasing salinity and pH of the solution. The sorption of SMT was rapid in the first 2 hr while the equilibrium was achieved at 16 hr. Pseudo first-order model was used to interpret the sorption kinetics of SMT sorption on PA, PVC, PE, and PP, while sorption of SMT on PA and PET was interpreted by Pseudo Second-Order Model (PSOM). Furthermore, they revealed Van der Waals and electrostatic interactions involved in the sorption process which decreased on increasing pH and salinity. The sorption capacity of MPs towards SMX and SMT follows the order as PA > PET > PE > PS > PVC > PP.

Li and co-workers (2019) investigated the sorption isotherm of Triclosan (TCS) (an antimicrobial) and PS MPs along with the effect of pH, temperature, ionic strength, and heavy metals (HMs), i.e., Cu (II) and Zn (II). They used PSOM to describe the sorption kinetics of TCS affinity towards PS. They revealed higher sorption of TCS on smaller PS MPs on accessing the sorption kinetics at different sizes of PS MPs. Furthermore, they disclosed rapid sorption in first 24 hr followed by a slower decrease in the sorption process and sorption equilibrium was achieved after 72 hr. On studying the effect of pH on sorption, they revealed TCS^0 sorption via that the hydrophobic interaction was much higher at pH 3–6 while the sorption decreases after pH > 6. Furthermore, the investigation of temperature and ionic strength effect on the sorption of TCS on PS did not reveal any significant effects. In addition to this, Ma's group (2019) studied the sorption of TCS on different sizes of PVC MPs. To proceed with the investigations, they chose two sizes of PVC, i.e., PVC-S (size < 1 µm) and PVC-L (74 µm). They used PSOM and the D-R model to interpret the sorption kinetics and sorption isotherms. Based on their findings they concluded higher sorption of TCS on PVC-S (12.7 mg/g) as compared to PVC-L (8.98 mg/g). Furthermore, they disclosed decreased sorption of TCS on PVC MPs on increasing pH values while salinity has opposite effect.

Elizalde-Velázquez et al. (2020) investigated the sorption behaviour of Non Steroidal Anti-Inflammatory Drugs (NSAIDs), i.e., ibuprofen, naproxen, and diclofenac (DCF) on four different types of MPs including PS, ultra-high molecular weight polyethylene (UHMWPE), the average molecular weight medium density polyethylene (AMWPE), and PP.The particle size diameter of PP were (1 mm) > PS (600–800 µm) > AMWPE (300–400 µm) > UHMWPE (2–10 µm). They revealed that the highest sorption of NSAIDs on UHMWPE was due to the lower size and more surface area of MPs, while the sorption was lowest in the case of PP. Based on their findings, they revealed lower sorption of NSAIDs over various MPs at pH: 6.9 (freshwater), pH: 8.1 (synthetic seawater conditions) and pH: 10.0 (alkaline conditions). Furthermore, they disclosed the highest sorption of NSAIDs on MPs only at pH: 2 (acidic conditions) due to the hydrophobic interactions. Finally, they concluded that the highest sorption coefficient of DCF was for MPs, while the NSAIDs sorption was highest among PE MPs.

Studies on adsorption and desorption of amoxicillin (AMX), atrazine (ATZ), diuron (DIR), paracetamol (PAC), phenol (PHN) and vancomycin (VAC) with PE, PET, PP, PS and PVC was carried out by Godoy and co-workers (2020). They disclosed higher affinity of AMX and PHN towards MPs. Among all the investigated MPs, PHN sorption was higher for PET and PS, while AMX sorption was higher for PE and PP. Sorption of PAC was also found significant for PE, PET, and PVC. Based on these experiments they revealed increased desorption of contaminants with increased temperature and pH.

McDougall and coworkers (2022) investigated the bioavailability of pharmaceuticals adsorbed on PE MPs in municipal wastewater. Three different types of pharmaceuticals with cationic, anionic, and neutral speciation were chosen for the investigations. Based on their experiments they revealed more adsorption of the cationic form of pharmaceuticals due to their hydrophobicity, while the anionic pharmaceuticals adsorption was blocked due to repulsion caused by negative charges surface of MPs. Neutral pharmaceuticals also got adsorbed on PE MPs. A further range of pH dependant octanol-water distribution coefficients (log D_{ow}) were also used to study the adsorption of pharmaceuticals on MPs. It was disclosed that no adsorption was observed for all pharmaceuticals with log $D_{ow} < 1$. Also, a negligible effect on sorption was found on varying the pH and composition of wastewater (diluting with NaCl and stormwater). Next they studied the desorption of pharmaceuticals in rivers and simulated intestinal and gastric fluids. They reported higher desorption of cationic pharmaceuticals in gastric fluids which was triggered in low pH due to the reduced surface charge of MPs. However, low desorption of pharmaceuticals was observed in river water.

> Sorption of DCF on PS MPs and the effects of various factors such as size, pH and ionic strength were explored by Li group's (2022). The studies revealed 24 hr of equilibrium sorption time for the DCF on 0.5 μm PS MPs through the process of chemisorption. They also revealed the increase in the sorption capacity with increasing particle size of MPs. Hydrophobic and electrostatic interactions were the main mechanism for the sorption. They revealed the dependency of sorption capacity on pH and ionic strength. The sorption capacity decreased with increasing pH, which increased with increasing ionic strength.

Wagstaff and Petrie (2022) investigated the desorption of fluoxetine (an antidepressant drug) from PET MPs in simulated gastric and intestinal fluid, river waters and the sea. Based on various experiments they revealed rapid desorption in a few hours after exposure. They disclosed higher desorption in sea water (18–23%) as compared to river water (4–11%). The desorption follows the decreasing order as: gastric fluid at 20°C and 37°C; sea water at 20°C; intestinal fluid at 20°C and 37°C; then river water at 20°C.

Chen's group (2022) reported the effects of chirality in the adsorption of pharmaceuticals on the MPs. For this they chose ofloxacin (OFL) and levofloxacin (LEV) and investigated their enantioselectivity in the adsorption on PE MPs. Based on their findings they disclosed 3–5% higher adsorption of OFL as compared to LEV. On increasing the pH from 7 to 10, they showed 17.9% less adsorption of OFL with respect to LEV having 21.3% less adsorption. However, no effect of PE particle size was observed on adsorption of OFL and LEV. While increasing the Natural Organic Matter (NOM) concentration and UV irradiation time of 8.7–9.9% was a higher adsorption of OFL on PE MPs than was observed in LEV.

2.2 Pharmaceutical Adsorption/Desorption from Aged MPs

Aged MPs refer to MPs that have undergone degradation over time due to environmental factors such as sunlight, weathering, and microbial activity. As a result of this degradation, aged MPs have different surface and chemical properties compared to pristine MPs. Specifically, aged MPs have a higher surface area, more pore volume, and more oxygen-containing functional groups on their surface (Fig. 4.3). These changes in surface and chemical properties of aged MPs can affect their interaction with pharmaceuticals and recent investigations are discussed here.

Liu's group (2019) studied and reported the sorption interaction of hydrophilic organic pollutant, ciprofloxacin (CIP) with aged PS and PVC MPs in both freshwater and seawater. To achieve this, they artificially irradiated the PS and PVC pristine MPs under UV light. The surface studies reported small wrinkles and cracks on the surface of old MPs. The Cl and oxygen containing functional group were also determined which were not there in the pristine MPs. The generation of free radicals and hydroperoxides were also observed which further accelerated due to ageing. Based on the sorption isotherm for both pristine and aged MPs, it was found that aged PS and PVC has 123.3 and 20.4% higher adsorption capacity for CIP than pristine PS and PVC. This can be explained due to the increase in oxygen containing functional groups reducing the hydrophobicity of the MPs. They further reported pristine PVC has higher sorption capacity as compared to pristine PS which was vice versa in the case of aged PS and PVC. The hydrogen bonding, partition and electrostatic interactions were the main mechanism for adsorption which decreased on increasing salinity.

Guo and Wang (2019c) explored the sorption of SMX, SMT, and cephalosporin C (CEP-C) on aged PS and PE MPs collected from coast of East China Sea and Yellow Sea, China. To understand the sorption kinetics of antibiotics on aged MPs they used mixed-order kinetics models. The surface studies of the aged MPs revealed oxygen containing functional groups such as hydroxy and carbonyl, and free radicals. The investigations revealed, the requirement of 8 hr to reach the sorption equilibrium for SMT on MPs while 16 hr equilibrium time was observed for SMX and CEP-C

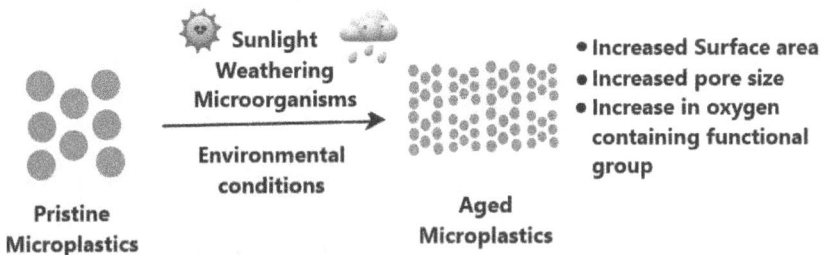

Figure 4.3 Transformation of pristine MPs to aged MPs under environmental conditions.

sorption. The sorption amount was highest in SMX which follows the order SMX > SMT > CEP-C. However, only CEP-C sorption was obtained in stimulated seawater system within the 24 hr equilibrium time. Finally, they revealed, Van der Waals, hydrophobic and electrostatic interactions as the main mechanisms for sorption.

Liu's group (2020) studied the effects of aged PS on the adsorption of atorvastatin (ATV) and amlodipine (AML). They enhanced the ageing of PS by the photo-Fenton process. Based on the results of various experiments they concluded hydrophobic and π–π interactions as major processes for adsorption in pristine PS, while the electrostatic interaction and hydrogen bonding played the main role in the case of aged PS. Furthermore, they disclosed decreased adsorption of ATV and increased adsorption of AML on PS MPs due to the interference of intermediates released during ageing process.

Wang and coworkers (2021) investigated the effect of ROS produced during ageing of MPs on the phototransformation of pharmaceuticals. They chose PS MPs and ATV to continue with their studies under simulated sunlight. Based on their studies they revealed an increased rate of degradation of ATV with ageing of PS MPs. This is due to the oxygen containing-functional groups present on PS MPs which on absorbing sunlight results in the generation of ROS. Further they revealed $^1O_2$ (singlet oxygen state) as the main supporter for increasing the phototransformation of ATV. Moreover, with ageing the number of degraded products of ATV also increases.

Fan's group (2021) explored and compared the adsorption-desorption of chlortetracycline (CTC) and AMX on aged Tire Wear Particles (TWP) and PE MPs. To fasten the ageing process a potassium persulphate solution was added to TWP and PE contained in a brown conical flask. The flask then was heated with constant shaking. After ageing, cracks and small holes were produced at the surface of both MPs. While specific surface areas (SBET) of both TWP and PE increased but more SBET was observed for TWP. Further they revealed 1.13–23.40 times more adsorption of antibiotics on aged TWP and 1.08–14.24 times on aged PE. The desorption data disclosed higher desorption rate of antibiotics from TWP and PE in case of simulated gastric fluid as compared to ultrapure water while the desorption of antibiotics from TWP was greater than PE.

Munoz and coworkers (2021) explored the role of ageing, size, nature, and Natural Organic Matter (NOM) fouling of various MPs including PS, PET, PP and high-density polyethylene (HDPE) on adsorption of DCF and metronidazole (MNZ). The results showed a higher adsorption capacity of DCF than MNZ due to aromatic character and its strong hydrophobic interactions. More adsorption was also observed in the case of PS and PET containing aromatic MPs. Moreover, adsorption increased on decreasing the size of MPs. Ageing of the MPs cause less sorption of hydrophobic DCF and increased the adsorption of hydrophilic MNZ, while the NOM fouling decreases the sorption capacity of MPs as blocking of sorption sites. Finally, they tested the effects of pH and salinity which results in more desorption of DCF and MNZ.

Wang's group (2022) investigated the effects of photoaged PS MPs on the photodegradation of cimetidine. The ageing of PS MPs was enhanced under simulated sunlight irradiation. It was noticed that the 5-d aged PS-MPs degraded the cimetidine with a faster rate than pristine PS MPs. The reason was again the formation of free radicals or ROS in PS MPs on ageing which interfere with the photodegradation of cimetidine. They also reported both $^1O^3$ and $^3PS^*$ states of ROS affecting the degradation. Moreover, they also studied and reported a significant increase in degradation of other pharmaceutical such as codeine and morphine due to aged PS MPs.

Later Zhang's group (2022) explored effect of aged PS on the photodegradation of SMX in simulated sunlight conditions. Based on various experiments they disclosed blocking of photodegradation of SMX in the presence of PS MPs. However, with ageing of PS MPs the photodegradation also decreases in sunlit water due to the light-screening effect. To get more knowledge about the mechanism of inhibition for photolysis of SMX they did a radical quenching experiment which revealed the decreased triplet excited state ($^3SMX^*$) of SMX. Further they disclosed a smaller number of degradation products with less yield of SMX upon photodegradation in presence of aged PS MPs.

To get more knowledge about the competition for sorption of various pharmaceuticals (SMX, PRP and SER), role of pH and ageing of MPs (PE) Wang's group (2022) carried out various experiments. They performed the ageing of PE MPs in presence of UV irradiation. They revealed hydrophobic interaction as the main mechanism for adsorption, while electrostatic interaction played the main role for adsorption on varying the pH from 2.00–12.00. Further they disclosed higher adsorption of pharmaceutical on aged PE than the pristine one. It was also reported that SER showed the highest sorption percentage (23.0%) followed by PRP (17.6%) on 6 d aged MPs, while SMX show lowest sorption (5.4%) even for 10 d aged MPs. At pH 7 the sorption follows the order as SER cations > SMX anions > PRP cations.

Miranda's group (2022) disclosed the results of sorption experiments of various pharmaceuticals (trimethoprim, florfenicol, tramadol, citalopram, DCF, venlafaxine (VFX)) and pesticides (clofibric acid, alachlor, DIR, pentachlorophenol) on virgin and aged MPs of low-density polyethylene (LDPE), poly (ethylene terephthalate) (PET) and unplasticized poly (vinyl chloride) (uPVC). They used the ozone treatment and rooftop weathering process to convert pristine MPs to aged ones. The results revealed increased sorption of pharmaceuticals and pesticides on aged MPs as compared to pristine MPs. No or very little affinity of pharmaceutical and pesticides was obtained in the case of pristine MPs. Further they disclosed a type of polymer and ageing treatment as the main factor to govern the adsorption. Based on their studies they reported an increase in LDPE sorption on ageing followed by uPVC and the least for PET.

2.3 Individual and co-effects of MPs and Pharmaceuticals on Biota

The presence of pharmaceuticals in the environment, particularly in the presence of MPs, can have complex and toxic effects on plants and animals (Fig. 4.4). There has been a growing body of research on this topic in recent years, as scientists seek to understand the potential risks and develop strategies to mitigate them. This section covers the recent literature in this research area (Table 4.1).

Prata and coworkers (2018) investigated the potential influence of MPs on the toxicity of the pharmaceutical's procainamide and doxycycline to the marine

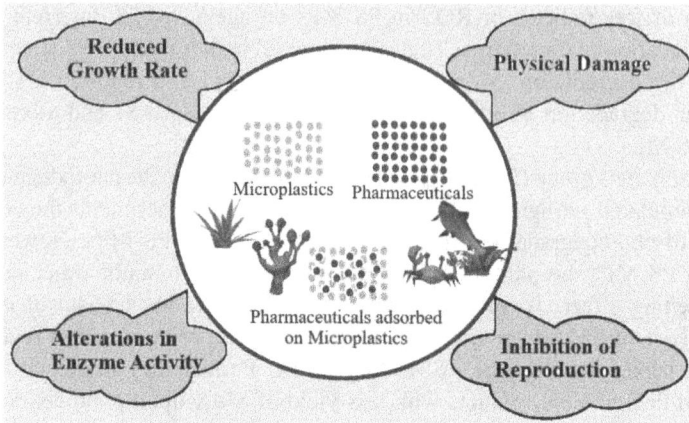

Figure 4.4 Individual and co-effect of MPs and pharmaceuticals on biota.

microalga *Tetraselmis chuii*. The researchers conducted bioassays to examine the toxicity of these substances alone and in mixtures, using the criteria such as growth rate and chlorophyll concentration. They determined EC10, EC20, and EC50 values for each of the substances. The results showed that MPs alone did not significantly affect the growth rate at concentrations up to 41.5 mg/l, but the chlorophyll content was significantly reduced at lower concentrations. Both pharmaceuticals were found to be toxic to *T. chuii* in the low ppm range, and co-effect of both MPs and pharmaceuticals were more toxic than the pharmaceuticals individually. The researchers also found that doxycycline concentrations in the test media decreased significantly, indicating degradation of the antibiotic.

Zhu's group (2019) studied the joint toxicity of TCS and four types of MPs including PE, PS, PVC, and PVC800 on microalgae species *Skeletonema costatum*. They examined both the growth inhibition and oxidative stress. The studies found that TCS and single MPs had a significant inhibitory effect on microalgae growth, with PVC800 showing the most significant reduction in growth followed by PVC, PS, and PE. However, the toxicity of PVC and PVC800 jointly with TCS decreased more than that of PS and PE. This suggests that when TCS is adsorbed onto the MPs, it reduces the concentration of both TCS and MPs in the surrounding solution, which in turn reduces their joint toxicity to microalgae. The study also found that the reduction of superoxide dismutase (SOD) was higher than malondialdehyde (MDA). Based on SEM images they revealed that physical damage caused by the MPs could be the primary factor responsible for this joint toxicity.

The ecotoxicological effects of two beta-blockers including Diltiazem and Bisoprolol, and PVC MPs on marine meiofauna, particularly nematode communities individually and in-combination were investigated by Allouche (2022). They exposed the invertebrates to different concentrations of the two beta-blockers and MPs, separately and mixed. The results showed that all treatments led to a significant reduction in overall meiofauna abundance, particularly polychaetes and amphipods. Further treatment with Bisoprolol and with a 1.8 mgL^{-1} mixture of Diltiazem and MPs, revealed variable influence on the maturity status of the species at different dosages. The demographic results are supported by the *in-silico* approach with

Table 4.1 Studies towards individual and combined effects of MPs and Pharmaceuticals to various species of Biota.

Sr No.	Research group	Pharmaceutical	Microplastics (MPs)	Species (Common name)	Effects (single and in combination)
1.	Prata and coworkers, 2018	Procainamide and doxycycline	Not mentioned	*Tetraselmis chuii* (Microalgae)	MPs reduces chlorophyll content.
2.	Zhu and coworkers	Triclorosan (TCS)	PE, PS, PVC, and PVC800	*Skeletonema costatum* (Microalgae)	Reduction in microalgae growth. Reduction of SOD and MDA. Physical damage.
3.	Allouche and coworkers, 2022	Diltiazem and Bisoprolol	PVC	*Coenorhabditis elegans* (Nematode)	Reduction in overall meiofauna abundance and influence on the maturity status of the species.
4.	Qu and coworkers, 2019	Venlafaxine (VFX) with its metabolite O-desmethylvenlafaxine	PVC	*Misgurnus anguillicaudatus* (loach)	Oxidative stress in the loach liver. Increased levels of SOD and MDA. Distribution of VFX and O-desmethylvenlafaxine in the loach liver and tissues. Contaminants get accumulated in the liver. MPs amplified bioaccumulation factor (BAF) of venlafaxine and O-desmethylvenlafaxine in the loach tissue by 10 times.
5.	Qu and coworkers, 2018	Venlafaxine (VFX)	PVC	*Misgurnus. anguillicaudatus* (M. anguillicaudatus)	High levels of MPs accumulated more VFX and O-desmethylvenlafaxine in the loach liver.
6.	Zhang and coworkers	Roxithromycin (ROX)	PS	*Oreochromis niloticus* (Red tilapia)	MPs enhanced the bioaccumulation of ROX in fish tissues and decreased the neurotoxicity caused by ROX. Activities of cytochrome P450 (CYP) enzymes get affected. Activity of the antioxidant enzyme SOD increased significantly.

Table 4.1 contd. ...

... Table 4.1 contd.

Sr No.	Research group	Pharmaceutical	Microplastics (MPs)	Species (Common name)	Effects (single and in combination)
7.	Shi and coworkers, 2020	Sertraline (SER)	PS	*Tegillarca granosa* (Bivalve mollusc)	MPs can increase the immunotoxicity of SER.
8.	Huang and coworkers, 2021	Sulphamethoxazole (SMX) and β-blocker Propranolol (PRP)	PS	*Oreochromis niloticusas* (Red tilapia)	Bioaccumulations of MPs and pharmaceuticals. Changes in enzyme activities and gene expressions. Inhibitions of cytochrome P450 enzymes activities.
9.	Takai and coworkers, 2022	Diazepam	PS	*Oryzias latipes* (Medaka)	Change in social behaviours.
10.	Nugnes and coworkers, 2022	Acyclovir	PS	*Ceriodaphnia dubia* (Crustacean)	Antagonistic genotoxic effect on *C. dubia* neonates. An additive chronic toxic effect. Inhibition of reproduction and damage to the DNA of the crustacean.

acceptable binding affinities and interactions in *Coenorhabditis elegans* having key residues in the germ line development protein 3 and sex-determining protein.

The study conducted by Qu et al. (2018) provided insights into the behaviour of the chiral antidepressant VFX in aquatic ecosystems in the presence of PVC MPs. They explored the distribution, bioaccumulation, and metabolism of VFX and effects of MPs on aquatic ecosystems. The researchers set up laboratory-scale aquatic ecosystems, including water-sediment, water-Lemna.minor (L.minor), water-*Misgurnus.anguillicaudatus* (M.anguillicaudatus), and water-sediment-L.minor-M.anguillicaudatus, and exposed them to VFX and two levels of MPs. The study found that PVC MPs could adsorb 58–96% of VFX and its metabolite O-desmethylvenlafaxine. R-venlafaxine degraded preferentially in four ecosystems. Additionally, high levels of MPs accumulated more VFX and O-desmethylvenlafaxine in the loach's liver, indicating that MPs could magnify the bioaccumulation of contaminants in aquatic ecosystems.

Later the same group investigated the toxicity, distribution, and metabolism of VFX with its metabolite O-desmethylvenlafaxine in *Misgurnus anguillicaudatus* (loach) in the presence of PVC MPs (Qu et al., 2019). The findings of the study suggested that exposure to these compounds causes oxidative stress in the loach's liver, as evidenced by the increased levels of SOD and MDA. Co-exposure with MPs exacerbates these adverse effects. The study also detected the distribution of VFX and O-desmethylvenlafaxine in the liver and tissues of loaches. The concentrations of these compounds were lower in water in the presence of MPs, while more contaminants get accumulated in the liver with MPs, which can act as a carrier for the transport of contaminants into the organisms. Based on their studies they further disclosed that with the MPs present, the bioaccumulation factor (BAF) of VFX and O-desmethylvenlafaxine in loach tissue amplified by 10 times.

Zhang's group (2019) examined the impact of PS-MPs on the bioaccumulation of roxithromycin (ROX) in *Oreochromis niloticus* (Red tilapia) and their interactive biochemical effects. The results showed that the presence of MPs enhanced the bioaccumulation of ROX in fish tissues and decreased the neurotoxicity caused by ROX after 14 d of exposure. The activities of cytochrome P450 (CYP) enzymes in fish livers exposed to co-exposure treatments exhibited great variability compared to ROX alone after 14 d of exposure, indicating that the presence of MPs may affect the metabolism of ROX in tilapia. Additionally, when fish were exposed to both MPs and ROX, the activity of the antioxidant enzyme SOD increased significantly compared to when exposed to ROX alone. These results indicate that the oxidative damage caused by ROX was mitigated in the fish livers after 14 d of exposure to both MPs and ROX.

Desorption mechanisms of pharmaceuticals from PS MPs under simulated gastric and intestinal conditions of marine organisms was investigated by Liu's group (2020). The study found that the type of digestive system defines the transportation mechanism of pharmaceuticals in the organism. Specifically, the increased desorption in the stomach primarily relies on pepsin solubilization to pharmaceuticals and the competition between adsorbent and sorption sites on MPs via π–π and hydrophobic interactions. They further reported the dependency of high desorption in the gut on the intestinal components solubilizations, such as Bovine Serum Albumin (BSA) and bile salts (NaT), and the competitive adsorption of NaT, as the enhancement of solubility increases the pharmaceuticals partition in an aqueous phase. The study also investigated the effect of ageing on the transportation mechanism of pharmaceuticals and found that the ageing process suppresses the desorption of pharmaceuticals. Further they revealed reduced hydrophobic and π-π interactions, but increased electrostatic interaction between MPs and pharmaceuticals on ageing, which became less affected by gastrointestinal components. The risk assessment of the MP-associated pharmaceuticals indicated that they posed low risks to organisms. Still, warm-blooded organisms suffered higher risks than cold-blooded organisms relatively.

Shi's group (2020) investigated the synergistic impact of SER and PS MPs on the immune responses of *Tegillarca granosa*, a bivalve mollusc. The results showed that both MPs and SER had a significant immunosuppressive effect on the mollusc. Interestingly, a size-dependent interaction between SER and nanoscale MPs was observed, which enhanced the immunotoxicity of SER. Further they explored the underlying toxication mechanisms by analyzing the intracellular content of reactive oxygen species, ATP content, apoptosis status, pyruvate kinase activity, plasma cortisol level. In addition, the *in vivo* concentrations of cytochrome P450 1A1 and neurotransmitters was determined. Based on their investigations they disclosed nanoscale MPs can increase the immunotoxicity of SER indicating nanoscale MPs alone or in combination to various toxicants have more toxic effects on marine organisms than those by the larger MPs.

Huang's group (2021) reported the effects of aged and virgin MPs (MPs) and the antibiotics SMX and β-blocker PRP on red tilapia (*Oreochromis niloticusas*). The researchers simulated MP ageing in the environment using ultraviolet irradiation and evaluated the accumulations of MPs and pharmaceuticals, the changes in enzyme activities and gene expressions in tilapia. With the ageing process, MPs specific surface area and average pore volume was increased including more carbonyl formation on the surface. It was observed that aged MPs increased the PRP accumulation in the brain by 82.3% while concentration of SMX in the gills reduced by 46.1%. Further they revealed alleviated stress on tilapia caused by PRP due to aged MPs, lipid peroxidation damages and lower neurotoxicity. They also disclosed higher inhibitions of cytochrome P450 enzymes activities under co exposure of aged MPs and SMX.

Wang's group (2021) explored the impact of MPs (PE and PVC) and pharmaceuticals on Antibiotic Resistance Genes (ARGs) and microbial communities in sewage. Three commonly used antibiotics (TC, ampicillin (AMP), and TCS) were selected for the study, and two types of MPs (PVC and PE) were used. The study found that the adsorption capacity of the three antibiotics on the MPs decreased in the order of AMP > TC > TCS. PE was found to be more conducive to microbial attachment than polyvinyl chloride. The presence of MPs led to an increase in the total number of ARGs and Mobile Genetic Elements (MGEs) in the sewage, with the multidrugs ARGs and MGEs get enriched on the plastisphere. Based on the results they revealed co-occurrence of TC and MPs can pose a higher risk of spreading ARGs and MGEs. Furthermore, the study found that potential pathogenic bacteria including Mycobacterium, Legionella, Arcobacter and Neisseria were more found on the plastisphere than sewage making them hosts for ARGs and MGEs.

Liu's group (2022) investigated the co-effect of aged PS MPs to the bioaccumulation of ATV and AML in the environment. Based on their studies they revealed that pharmaceuticals were more easily released from aged MPs in gastrointestinal fluids than in simulated seawater. Hydrophobic pharmaceuticals were also found to be more bio accessible than hydrophilic ones. The model analysis results indicated ingestion of water and food as important uptake routes for pharmaceuticals in the aquatic environment (fish and seabirds). Interestingly, the study found that aged PS MPs actually resulted in reduced bioaccumulation of pharmaceuticals in marine organisms. Further they disclosed warm-blooded organisms to be more susceptible to exposure to pharmaceuticals via ingested MPs than cold-blooded organisms.

Takai's group (2022) evaluated the combined effect of PS MPs and diazepam on the social behaviour of medaka (*Oryzias latipes*), a freshwater fish species. The researchers used a shoaling behaviour test to assess the effect of the different treatments on the fish's social behaviour. The study had five treatment groups: solvent control, polystyrene MPs exposure, low-concentration diazepam exposure, high-concentration diazepam exposure, and polystyrene MPs and low-concentration diazepam co-exposure. The results of the study showed that exposure to high concentrations of diazepam (0.3 mg/L) and co-exposure to MPs and low concentrations of diazepam (0.04 mg/L and 0.03 mg/L, respectively) significantly increased the shoal-leaving behaviour of medaka compared to the solvent control group. The shoal-leaving behaviour was assessed by counting the number of times the fish left the shoal during a 10-min observation period. The study also found that even after 5 d of recovery, the medaka in the co-exposure group still left the shoal more often than those in the control group, suggesting that the combined effect of MPs and diazepam on the social behaviour was prolonged. Overall, the study suggested that the presence of MPs can enhance the adverse effects of pollutants such as diazepam on the social behaviour of aquatic organisms.

Nugnes's group (2022) investigated the chronic and sub-chronic effects of polystyrene beads (1.0 μm) as an individual xenobiotic and in combination with the antiviral drug acyclovir and the insecticide imidacloprid (IMI) on the freshwater crustacean *Ceriodaphnia dubia*. They used Bliss independence as a reference model for the research. The study found that exposure to mixtures of polystyrene MPs, acyclovir, and IMI for 24 hr mostly resulted in an antagonistic genotoxic effect on C. dubia neonates. However, an additive chronic toxic effect was observed for 7 d exposure to the mixtures which occurred at concentrations near to or overlapping with the environmental concentrations of the pollutants. They disclosed inhibition of reproduction and damage to the DNA of the crustacean C. dubia caused by mixtures of the selected xenobiotics.

The adsorption properties of triazole based fungicides (hexaconazole (HEX), myclobutanil (MYC), triadimenol (TRI)) on PS MPs including particle size and environmental factors (pH, salinity) was investigated by Fang's group (2019). The adsorption kinetics and isotherm. fitted with the Pseudo-second-order and Freundlich model. The adsorption capacity of these fungicides follows the order as HEX > MYC > TRI. They further revealed an increase in the adsorption capacity on increasing salinity, decreasing particle size of PS MPs, and changing pH. They disclosed hydrophobic and electrostatic interactions as the main adsorption mechanisms.

Jiang and coworkers (2020) explored the sorption of two fungicides (triadimefon and difenoconazole (DIFE)) on the biodegradable polybutylene succinate (PBS) and, conventional PE and PVC MPs in aqueous solution. The data revealed the highest sorption of triadimefon (104.2 ± 4.8 mg/g) and DIFE (192.8 ± 2.3 mg/g) on PBS MPs. However, no significant impact of pH, salinity and DOM was observed on sorption capacity of PBS. But the PE and PVC sorption capacity got decreased with increased salinity and DOM. The results suggested that the sorption capacity of biodegradable MPs and conventional MPs are different and environment factors also have a variable influence on both type of MPs.

3. Pesticides as Emerging Contaminants

Pesticides are chemicals used in agriculture to control pests, diseases, and weeds (Nicolopoulou-Stamati et al., 2016). At present they are known as emerging contaminants due to their excessive use and persistence in the environment (Fig. 4.5) (Smital et al., 2004). They can enter the environment bodies such as soil, rivers, lakes, sea, etc., by different routes such as runoff from agricultural fields, leaching into groundwater, and spray drift. Pesticides are known to have adverse effects on non-target organisms, including humans, as they can accumulate in the food chain and causes many health problems. Additionally, they are also known to be toxic to aquatic life (Mensah et al., 2014). The persistence of pesticides in the environment also poses a long-term threat to ecosystems, as they can persist in soil and water for many years and continue to impact the environment and human health. Therefore, it is important to regulate the use of pesticides and minimize their environmental impact. This can be achieved by using alternative methods of

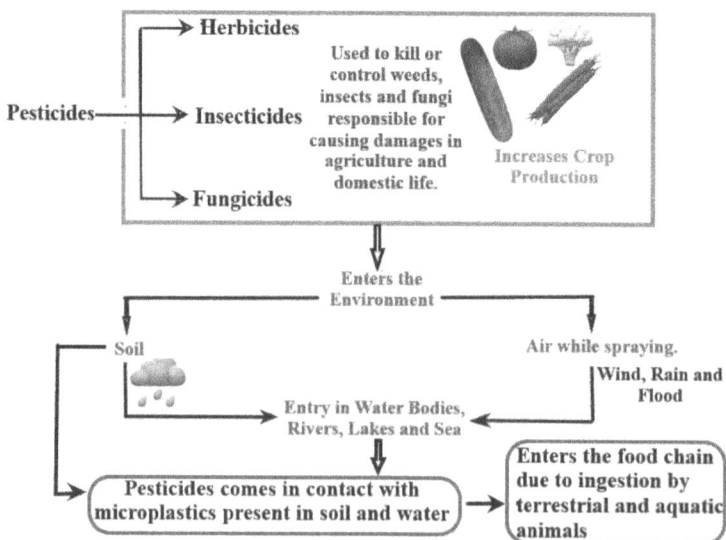

Figure 4.5 Fate of pesticides from their use to entry in food chain.

pest control such as integrated pest management, as well as the implementation of measures to reduce the use of pesticides and their impact on the environment.

When pesticides are released into the environment, they can bind to the surface of MPs, which can affect their distribution and toxicity (Verla et al., 2019). The presence of MPs can increase the persistence of some pesticides in the environment, as plastic particles can provide a protective barrier that prevents the pesticides from degrading or being degraded by microorganisms. On the other hand, MPs can also enhance the uptake of pesticides by organisms, as the plastic particles can act as a transfer mechanism for the pesticides (Peña et al., 2023). This can lead to increased exposure and potential toxicity in organisms that ingest MPs. The current investigations on the uptake and desorption of pesticides with MPS and their harmful effects on biota are discussed here.

3.1 Pesticides Adsorption/Desorption from Pristine and Aged MPs

MPs can adsorb and desorb pesticides, which can affect the way these chemicals are distributed in the environment. Researchers are investigating the adsorption and desorption of pesticides from MPs and the recent work has been reported here.

Wang's group (2020) studied the adsorption process, capacity, and mechanisms of five pesticides including carbendazim (CAR), dipterex (DIP), diflubenzuron (DIF), malathion (MAL), DIFE on to PE MPs. They revealed adsorption of all five pesticide on the MPs via hydrophobic interactions with DIF and DIFE having more adsorption. Further they disclosed a 2 hr equilibrium time for the adsorption process. Adsorption at surface sites of MPs, mass transfer and intraparticle diffusion were the main mechanisms involved in adsorption process. Based on the data from their experiments they concluded adsorption of pesticides on PE MPs follow the order as DIF > DIFE > MAL > CAR > DIP.

> Lan's group (2021) explored the adsorption of commonly used pesticides including CAR, DIF, MAL, DIFE on aged PE MPs (APE), Wheat Straw segment (WSS), Wheat Straw Powder (WSP), and Straw Crystalline Cellulose (SCC). They revealed that hydrophobic diffusion was the main adsorption force, while pesticides adsorption on WSP and WSS also occurred through π-π conjugation and electrostatic interaction. They disclosed adsorption of pesticides follows the order as WSP > WSS > APE > SCC. Later they compared the adsorption among these four pesticides on pristine PE MPs and aged PE MPs (APE MPs). Based on the surface studies they reported layered structures, flakes, and cracks in aged PE. The presence of larger surface area and polar O-containing functional groups due to oxidation occurred during ageing and weathering making APE MPs a better adsorbent as compared to PE MPs.

The interaction of pesticides including acetamiprid, chlorantraniliprole and flubendiamide on polyester fibres and PP MPs were studied by Šunta et al., 2020. They chose various concentrations of (1, 5 and 10 mg L^{-1}) of pesticides in 1 and 5% (w/w) MPs content in soil. They revealed that the adsorption of pesticides depends on the octanol/water partition coefficient. Most hydrophobic pesticides have more adsorption on MPs. They disclosed the adsorption of pesticides on the MPs decrease their availability in soil. MPs can also act as good vectors for organic pollutants which can transfer them to underground water.

Li's group (2021) studies the adsorption of three pesticides including IMI, Buprofezin, DIFE on PE MPs and investigated the effects of pH, NaCl salinity, MPs dosages and the reaction time on the adsorption characteristics in the aqueous solution. The adsorption data was interpreted by the Freundlich isotherm model and the pseudo-first-order kinetics revealing physical functions controlling the adsorption. Based on the results they concluded at high pH and low salinity adsorption capacity increases. They disclosed that adsorption occurred through intermolecular Van der Waals force and the microporous filling mechanism. Elevated temperature also favours the adsorption. The adsorption capacity of DIFE on PE MPs was highest followed by buprofezin and the lowest for IMI.

Wang's group (2020) investigated the effect of PE MPs on pesticides concentration and degradation in the aquatic environment. To carry out their studies they chose eight pesticides including tebuconazole, epoxiconazole, MYC, simazine, azoxystrobin, terbuthylazine, metolachlor and atrazine. They disclosed that 2–50 g/l concentration of MPs in water can reduce the pesticides' residue in water. However, the ageing process of MPs does not show effects on the pesticide's interaction with MPs. Moreover, due to adsorption of pesticides on the MPs increases the half-life of pesticides in water.

Zhang and coworkers (2021) studied the interaction between soil PE MPs and organochlorine pesticides (OCPs) including hexachlorocyclohexanes (HCHs) and dichlorodiphenyltrichloroethanes (DDTs). They used sorption kinetics and isotherm models for their studies in soil suspension. They also explored the solution/soil ratio and size of MPs. Based on their studies they revealed a good adsorption when 75:1 to 100:1 solution/soil ratio was taken. Further on increasing the size of MPs the sorption decreased. They also disclosed Van der Waals forces and hydrophobicity as the main mechanism for sorption.

Concha-Graña et al. (2022) investigated the adsorption behaviour of two pesticides (α-endosulfan and chlorpyrifos (CPF)) and six musk fragrances (celestolide, galaxolide, tonalide, musk moskene, musk ketone, and musk xylene) on polyamide (PA6) and on polyhydroxybutyrate (PHB) MPs in the marine environment. They also explored the effect of temperature and ageing of MPs due to weathering. They revealed that more adsorption was obtained in the case of PHB than PA. Almost in all cases the highest percentage of adsorption was attained in 24 hr. Moreover, they reported the inhibition of α-endosulfan degradation in the aquatic environment due to adsorption. However, no difference was obtained in adsorption percentage of the pollutants on aged and pristine MPs.

Liu's group (2023) investigated the alteration in adsorption properties of neonicotinoid insecticide IMI on the photo-aged biodegradable polylactic acid (PLA) MPs and polar PA MPs. To increase the ageing process of the MPs they exposed both to ultraviolet irradiation. With ageing the surface and chemical properties of MPs changed as the C-N bond in PA MPs and functional groups with the oxygen of PLA MPs got disrupted resulting in more carbonyl groups and other small molecules. This result was caused by decreasing the adsorption of IMI on the aged PA MPs by 19.2% and increased adsorption of IMI on aged PLA MPs by 41.2%. They revealed hydrogen bonding, Van der Waals interactions, electrostatic interactions, and polar-polar interactions mechanism responsible for adsorption. They also found high pH and lower ionic strength favours the adsorption of IMI on the MPs surface. In water humic acid-IMI complexes formation occurs which reduces the IMI concentration in water and as a result adsorption decreases.

3.2 Individual and Co-effects of MPs and Pesticides on Biota

It is known that the use of pesticides can have both positive and negative impacts on the soil and water biota. In recent years, there has been growing concern about the potential impact of MPs on soil and water ecosystems. Studies have shown that the presence of MPs in soil and water can interact with pesticides, leading to increased toxicity to biota which was discussed here (Table 4.2).

Table 4.2 Investigations done by various research groups on the toxic effects of MPs and pesticides to various species of organisms.

Sr No.	Research group	Pesticides	Microplastics (MPs)	Species (Common name)	Effects (single and in combination)
1.	Rios-Fuster and coworkers, 2021	Lindane, HCB, aHCH, Aldrin, p,p′ DDE, p,p′ DDD	LDPE	*Sparus aurata* (Juvenile gilthead seabream)	Higher concentration of pesticides in the liver as compared to the muscles
2.	Zhu and coworkers	Chlorpyrifos (CPF)	LDPE, PBAT	*Lumbricus terrestris* (Earthworm)	Decrease in earthworm weight and increase in mortality by exposing the earthworms to 28% Bio-MPs MPs decreases the bioaccumulation of pesticide in earthworm bodies.
3.	Villegas and coworkers, 2022	Malathion (MAL)	PE	*Minuca ecuadoriensis* (Fiddler crab)	Increased mortality. Increased bioaccumulation of MPs in tissues. MPs bioaccumulation increased MLT toxicity.
4.	Hanachi and coworkers, 2021	Chlorpyrifos (CPF)	PE	*Onchorhynchus mykiss* (Rainbow trout)	Alteration in fatty acid and amino acid composition. Significant variations in all the contents of fish muscle on exposure to combination of CPF and PE MPs.

Rios-Fuster's group (2021) investigated the levels of OCPs and PCBs in the liver and muscle of juvenile gilthead seabream (*Sparus aurata*) in the presence of virgin and weathered MPs. For this they exposed S aurata to MPs along with OCPs and PCBs for 4 mon in their diets. They revealed higher concentration of both pollutants in their livers as compared to muscles.

Ju's group (2023) explored the co-effects of MPs (low density polyethylene MPs (LDPE-MPs) and biodegradable MPs (Bio-MPs) consisting of poly (butylene adipate co-terephthalate) (PBAT)) with the pesticide CPF on the mortality, growth, and reproduction in earthworms (*Lumbricus terrestris*). In addition, they investigated the biogenic transfer of MPs and pesticides through earthworm burrows. Based on their studies they revealed a decrease in the earthworm's weight by 17.6% and an increase in mortality by 62.5% by exposing the earthworms to 28% Bio-MPs. Further on treatment with a combination of Bio-MPs (28%) and CPF have less toxicity, while the treatment with both pesticides including 28% LDME-MPs and 7% Bio-MPs in combination with CPF showed the highest toxicity. On treatment with 28% Bio-MPs and CPF resulted in a significant decrease in bioaccumulation of CPF in the earthworms' bodies in comparison to exposing them with CPF alone. They also found that the addition of CPF results in the transfer of 8% LDME-MPs into the earthworms' burrows but no such transfer of Bio-MPs was observed. It was disclosed that MPs also inhibited the CPF degradation. However, no effect was observed in the earthworms' reproduction.

Villegas and coworkers (2022) assessed the effects of PE MPs in combination with MAL on the survival of *Minuca ecuadoriensis* (fiddler crab). They expose the crab for 120 hr to investigate the MP tissue bioaccumulation in four treatments which includes: T1) control, T2) MLT 50 mg/L, T3) MP 200 mg/L, and T4) MLT (50 mg/L) + MP (200 mg/L). The results of the study showed that the 80% mortality in T4, 28% in T2 while no mortality was observed in T3. They revealed higher MPs bioaccumulation in case of T4 (572 items g/tissue) following T3 (70 items g/tissue). They also disclosed increased MAL toxicity in the presence of MP, and the MLT + MP mixture increased MP tissue bioaccumulation.

Hanachi's group (2021) explored the changes in nutritional parameters in the muscle of *Onchorhynchus mykiss* (rainbow trout) with the effect of PE MPs alone or in combination with CPF insecticide. On exposing the fish to PE MPs with 30 or 300 µg/L concentration showed minimal effects on fatty acid and amino acid composition while no effects were observed for muscle protein content. They disclosed significant variations in all the contents of the fishes muscle on exposure to combination of CPF concentrations (2 or 6 µg/L) and PE MPs concentrations (30 or 300 µg/L). This suggests that PE MPs can cause significant toxicity in combination with CPF on the fish muscle.

4. Metals as Emerging Contaminants

Metals are considered as emerging contaminants due to their increasing presence in the environment and potential adverse effects on human and ecological health (Francisco et al., 2019). Metals such as lead, arsenic, mercury, and cadmium are naturally occurring and can also be released into the environment through industrial activities, mining, and agricultural practices (Fig. 4.6). Exposure to metals can cause a range of health effects, including developmental and neurological disorders, cardiovascular disease, and cancer (Balali-Mood et al., 2021). In addition, metals can accumulate in the environment and the food chain, leading to long-term impacts on the ecosystem. The presence of metals in drinking water sources has become a major public health concern, particularly in areas with contaminated groundwater or industrial activities. Metals can also contaminate the soil, leading to reduced soil fertility and potential crop contamination (Alengebawy et al., 2021). To mitigate the risks associated with metal contamination, it is important to identify and monitor sources of contamination, develop effective remediation strategies, and implement regulations and policies to reduce exposure to metals.

MPs can act as carriers or adsorbents for metals, and can increase their bioavailability and mobility in the environment (Liu et al., 2021). The small size and high surface area of MPs can lead to a greater accumulation of metals on their surface, which can potentially increase their toxicity. In addition, the presence of MPs can also affect the behavior and transfer of metals in aquatic environments. MPs can alter the hydrodynamics and sedimentation patterns of aquatic environments, which can affect the deposition and accumulation of metals in sediments (He et al., 2023). Furthermore, MPs can also act as a vector for the transfer of metals over long distances, as they can be transported by ocean currents and wind. This can lead to the dispersal of metals over large areas and potentially affect the health of the ecosystem and organisms in remote regions. Here the fate and current scenarios of effects of metals in combination with MPs on the biota along with the adsorption and desorption studies in environment are described.

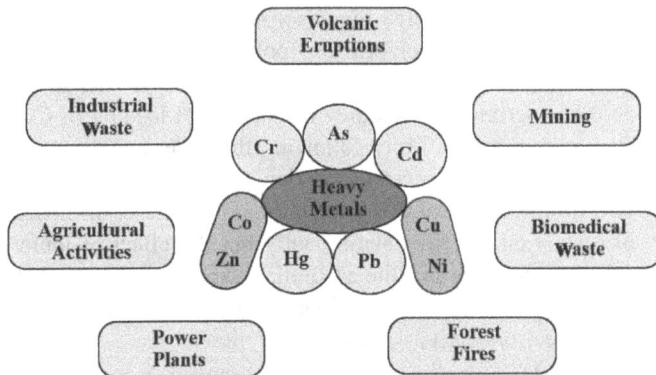

Figure 4.6 Sources of heavy metals in the environment.

4.1 Metals Adsorption/Desorption from Pristine and Aged MPs

Metals and MPs can interact with each other in various ways, and researchers are studying these interactions to better understand their effects on the environment and human health. The adsorption and desorption of metals from MPs depends on numerous factors including the type of MPs, interactions, natural and environment factors (Fig. 4.7). When metals and MPs come into contact with each other, several processes can occur. For example, metals can adsorb onto the surface of MPs,

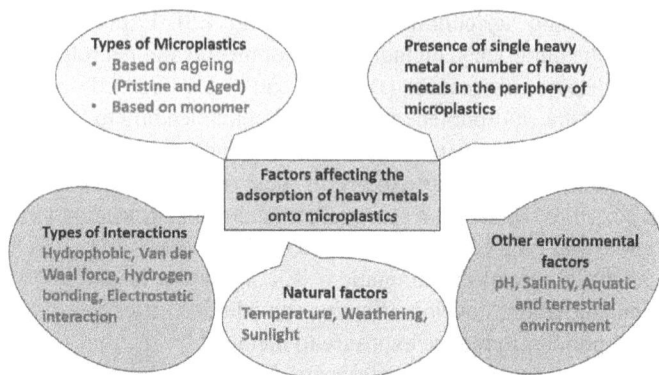

Figure 4.7 Factors affecting adsorption of HMs onto MPs.

which can increase their toxicity and make them more bioavailable to organisms that ingest them. Additionally, MPs can act as a transfer mechanism for metals, carrying them to new locations and potentially spreading them throughout the environment. Researchers are studying the interactions between metals and MPs to better understand their potential effects on the environment and human health.

Wang's group (2020) studied the sorption of metal ions including Cu^{2+} and Zn^{2+} on virgin and aged PET MPs. Photo-ageing of PET MPs were simulated by UV radiation exposure. They revealed more adsorption of HMs on the aged MPs as compared to the pristine ones due to the increased surface area and oxygen containing functional groups after the ageing process. Moreover, the adsorption capacity keeps on increasing when the MPs are kept under UV radiation for a prolonged period.

Zou's group (2020) investigated the adsorption of three HMs including Cu^{2+}, Cd^{2+} and Pb^{2+} on four pristine chlorinated polyethylene (CPE), PVC, and two polyethylene plastic particles (LPE (low crystallinity) and HPE (high crystallinity)) MPs. Based on the experimental data they revealed high loading of Cu^{2+}, Cd^{2+} and Pb^{2+} on to MPs with sorption affinity following the order as CPE > PVC > HPE > LPE. However, Pb^{2+} have stronger adsorption to MPs as compared to Cu^{2+} and Cd^{2+} because of a stronger electrostatic interaction. They further disclosed pH affects the sorption of metals on the MPs significantly but the effect of ionic strength is relatively less. With increasing pH, the adsorption capacity of metals gets increased as on increasing the pH the surface functional group of MPs become deprotonated resulting in higher electronegativity and more sorption sites.

Fu's group (2021) studied the adsorption mechanism of lead (Pb^{2+}) on pristine and natural-aged MPs in an aqueous medium. They revealed higher adsorption on to aged MPs as compared to pristine MPs. Adsorption kinetics and data was interpreted with Langmuir model and obtained 13.60 mg/g as maximum adsorption of Pb^{2+} on the natural-aged MPs. Based on the surface studies of the aged MPs they revealed more oxygen containing functional groups after ageing (the carboxyl and hydroxyl groups) are responsible for undergoing electrostatic interaction with Pb^{2+} resulting in more adsorption.

Yu's group (2021) investigated the effects of PE MPs on the chemical uptake of HMs including Zn, Cu, Ni, Cd, Cr, As, and Pb through incubation experiments and soil fractionation for 5 mon. The results showed reduced heavy metal concentrations in bioavailability and increased their content in organic-bound fractions on the addition of PE MP with 100 μM at 28% concentration. They concluded that MPs results in transformation of HMs from their bioavailability to organic bound species. They further revealed that the physiochemical factors of soil also affected varied HMs in different pathways. MPs also have a varied extent of influence on adsorption and complexation properties for various heavy metals. Due to these processes different HMs resulted in discrepant responses to the MPs addition. They also disclosed that within the incubation period, larger-sized aggregate fractions show greater response as compared to the smaller-sized aggregate fractions.

4.2 Individual and co-effects of MPs and Metals on Biota

The combination of metals and MPs can have devastating effects on the biota, including both plants and animals. When MPs and metals are present in aquatic and terrestrial environments, they can interact with each other and create synergistic effects that can increase the toxicity of both substances and recent studies have been described here (Table 4.3).

Zong's group (2021) investigated the adsorption capacity of HMs on PS MPs along with their influence on bioavailability and toxicity by hydroponic wheat (*Triticum aestivum* L.) seedlings experiment. Based on the results they revealed no significant effect on seedling growth of wheat, photosynthesis, and ROS content on exposure to PS MPs of the size of 0.5 μm and 100 mg/L concentration. Furthermore, they disclosed adsorption of Cu and Cd on PS MPs via chemisorption which results in reducing their accumulation mitigating their toxic effects. On exposure of both PS MPs and HMs increased the chlorophyll content-enhancing photosynthesis and decreased the ROS accumulation.

Pinto-Poblete and coworkers (2022) evaluated the coeffect of HDPE MPs and Cd on soil microbiological activity and yield, growth parameters and roots of strawberries (*Fragaria x ananassa* Duch) planted in clay pots. They exposed the soil and strawberry plant to MPs, Cd, MPs + Cd and observed significant changes in the plant's growth, biomass, fruits number, characteristics of roots, acid phosphate, dehydrogenase activity. An increased Cd accumulation was observed in roots and soil with changes in microbial biomass due to increased bioavailability of the Cd.

The only and combined effects of PS MPs and metals (Cu, Zn and Mn) on the growth rate and chlorophyll content of microalgae (*Chlorella vulgaris*) at different concentrations was investigated by Tunali (et al., 2020). For their studies they chose PS of 0.5 μm size and at varied concentrations (1, 5, 50, 100 and 1000 mg/l) of the metals. Based on the data of their experiments they concluded that there was no impact on the growth rate and chlorophyll content at lower concentrations of MPS (1, 5 mg/l). However, significant reduction in both was obtained at higher concentrations of MPs (50, 100 and 1000 mg/l) with the growth rate decreasing by 15.71–28.86% and chlorophyll concentration decreasing by 9.2–21.3%. They further reported greater reduction in growth (47.83–49.57%) and chlorophyll (44.75–50.25%) content when microalgae was exposed to a combination of MPs with the metal. In addition, more impact was obtained when algae were exposed to a

Table 4.3 Studies towards effects of MPs and metals alone and in combination to various species of Biota.

Sr. No.	Research group	Metals	Microplastics (MPs)	Species (Common name)	Effects (single and in combination)
1.	Zong and coworkers, 2021	Cu^{2+}, Cd^{2+}	PS	*Triticum aestivum* L. (Wheat)	No toxic effect
2.	Pinto-Poblete and coworkers, 2022	Cd	PE	*Fragaria x ananassa* Duch (Strawberry)	Increase in Cd accumulation was observed in roots and soil with changes in microbial biomass due to increased bioavailability of Cd. Significant changes in plant growth, biomass, fruits number, characteristics of roots, acid phosphate, dehydrogenase activity.
3.	Tunali and coworkers, 2020	Cu, Zn and Mn	PS	(*Chlorella vulgaris*) (Microalgae)	Significant reduction in growth rate (15.71–28.86%) and chlorophyll content (9.2–21.3%) at higher concentration of MPs. Reduction in growth (47.83–49.57%) and chlorophyll (44.75–50.25%) content when microalgae was exposed to a combination of MPs with metal.
4.	Yuan and coworkers, 2020	Pb^{2+}, Cu^{2+}, Cd^{2+}, and Ni^{2+}	PS	*Daphnia magna* (Water fleas)	Decreased immobilization of D. magna at lower MPs concentration. Combined toxicity increased rapidly at high MPs concentrations.
5.	Zhou and coworkers, 2020	Cd	PP	*Eisenia foetida* (Earthworm)	Reduced growth rate and increased mortality of *E. foetida* after co-exposure. Oxidative damage was also observed in the presence of MPs, whereas the presence of Cd enhanced the adverse effects of MPs. Cd accumulation of 9.7–161.3% in the earthworm bodies.

6.	Yan and coworkers, 2020	Cd, Pb, and Zn	PS	*Oryzias melastigma* (Marine medaka)	Alterations in intestinal microbioata (enhanced alterations in co-exposure). Alterations in specific bacterial species such as betaproteobacteria, burkholderiales and corynebacteriales and the males gut function. Empty follicles and follicular atresia. Alterations in gene expression levels.
7.	Santos and coworkers, 2021	Cu	Microspheres (Polymer of undisclosed composition)	*Danio rerio* (Zebrafish)	High oxidative stress and mortality in the zebrafish larvae. Inhibition of antioxidant enzymes in the zebrafish larvae. Inhibition of cetylcholinesterase activity. Changes in lactate dehydrogenase activity.
8.	Wang and coworkers, 2022	Cd	PS	*Channa argus* (Northern snakehead)	Gill tissue damage, oxidative stress and altered antioxidant status. Alteration in immune genes on co-exposure.
9.	xin Wang and coworkers, 2022	Cd	PS	*Euplotes vannus* (periphytic marine ciliate)	Decreased population and carbon biomass. Increased bioaccumulation of Cd^{2+} on co exposure. Increased oxidative stress and membrane damage.

triple metal-MPs combination resulting in 70.43% reduction in growth and 64.09% reduction in the chlorophyll content. They disclosed that the adsorption of MPs on the algal cell was the probable cause for the toxic effects.

Yuan and coworkers (2020) explored the adsorption potential of HMs including Pb^{2+}, Cu^{2+}, Cd^{2+}, and Ni^{2+} on two different sizes (10 μm and 50 μm) of PS MPs. They investigated their combined toxicological effect on *Daphnia magna* (water fleas). They concluded that the adsorption of HMs on MPs follow the order as $Pb^{2+} > Cu^{2+} > Cd^{2+} > Ni^{2+}$ with more adsorption occurring on MPs with a smaller size (0.261–0.579 mg/g) with respect to the larger MPs (LMPs) (0.243–0.525 mg/g). They reported the presence of both HMs and MPs on the toxicity towards D. magna increases with increased MPs concentration. However, the toxicity of MPs and HMs in combination depends on the adsorption capacity between them. They further revealed decreased immobilization of D. magna at lower MPs concentration (0.01–10 mg/l), while a rapid increase was observed at higher concentration of MPs (10–1000 mg/l).

Zhou's group (2020) explored the effects of MPs and cadmium (Cd) alone and in combination on the growth and mortality rate of *Eisenia foetida* (earthworm). Based on results they revealed a reduced growth rate and increased mortality of E. foetida after exposure to combination of MPs + Cd (> 3000 mg/kg) for 42 d. Oxidative damage was also observed in E. foetida in the presence of MPs, whereas the presence of Cd enhanced the adverse effects of MPs. Furthermore, MPs reside within the earthworm with 4.3–67.2 particles/g, resulting in a Cd accumulation of 9.7–161.3% in their bodies.

Yan and coworkers (2020) investigated the individual and combined effects of HMs including Cd, Pb, and Zn and MPs on marine medaka's (*Oryzias melastigma*) intestinal bacteria and gonadal development. Alterations in intestinal microbioata were obtained on exposing marine medaka to MPs, while significant enhanced alterations were obtained in the presence of HMs and their combination with MPs. They confirmed this by observing changes in abundance of proteobacteria. Based on KEGG studies they disclosed triggered alterations in specific bacterial species such as betaproteobacteria, burkholderiales and corynebacteriales and the males gut function. HMs, and also the combination of MPs + HMs exposure causes empty follicles and follicular atresia. In addition, alterations in gene expression levels were obtained related to Hypothalamic-Pituitary-Gonadal (HPG) axis.

Santos and coworkers (2021) studied the toxic effects of MPs (2 mg/l), Cu (60 and 125 mg/L), and their co-effect in combinations (Cu60 + MPs, Cu125 + MPs) on exposing zebrafish (*Danio rerio*) from 2-Hr Post-Fertilization (HPF) until 14-D Post-Fertilization (DPF). They revealed that the MPs and Cu exposure either alone and in combination results in high oxidative stress and mortality in the larvae of zebrafish. They further revealed antioxidant enzymes in the zebrafish larvae got inhibited at 6 DPF which kept on increasing then till 14 DPF. Further neurotoxicity in larvae was observed by the inhibition of the cetylcholinesterase activity on exposing the MPs and Cu, single or combined. Further changes in lactate dehydrogenase activity (LDH) in fish larvae suggests that MPs and Cu interference with the metabolic activity during the early stages of fish. This result causes negative effects on physiological performance of larvae and their fitness.

In view of the negative effects of PS MPs with the combination of HMs, Wang's group (2022) studied the toxic effects of MPs and Cd alone and in combination on the *Channa argus* (Northern snakehead). They investigated the tissue damage and immune responses in presence of 200 µg/L concentration of MPs with different sizes (80 nm, 0.5 µm) and 50 µg/L concentration of Cd at an interval of 24 hr, 48 hr and 96 hr. They revealed that the gill tissue was damaged at 96 hr, oxidative stress and altered antioxidant status on exposure to 0.5-µm MPs in Channa argus. Alteration in immune genes were also observed in co-exposure of Cd and MPs.

The studies on carbon biomass and oxidative stress on *Euplotes vannus* (periphytic marine ciliate) due to MPs and HMs alone and in combination were conducted by Wang et al. (2022). They chose PS MPs and Cd^{2+} to carry out their studies. It was revealed that MPs ingestion by Euplotes vannus resulted in decreased population and carbon biomass. Co-exposure to MPs and Cd^{2+} increases the bioaccumulation of Cd^{2+} in ciliated organisms, thereby reducing the increase in ciliated biomass. In addition, increased oxidative stress and membrane damage were observed.

5. Polychlorinated Biphenyls (PCBs) and Polybrominated Diphenyl Ethers (PBDEs) as Emerging Contaminants

PBDEs and PCBs are both classes of chemicals that are considered emerging contaminants because they have been detected in the environment and in various organisms, including humans, and are associated with adverse health effects (Fig. 4.8) (Lavandier et al., 2013). PBDEs are flame-retardant chemicals that have been used in a number of consumer products such as electronics, furniture, and textiles. They are Persistent Organic Pollutants (POPs) that can accumulate in the environment and in the food chain. Studies have linked PBDE exposure to adverse health effects, such as thyroid hormone disruption, neurodevelopmental delays, and reproductive problems (Ramhøj et al., 2022; Sheikh et al., 2023). PCBs are also POPs that were used in various industrial applications such as electrical transformers and capacitors. Like PBDEs, they are persistent in the environment and can accumulate in organisms. PCB exposure has been linked to adverse health effects, such as immune system dysfunction, developmental delays, and cancer (Montano et al., 2022). Both PBDEs and PCBs have been banned or restricted in many countries, but they are still present in the environment and can continue to pose a risk to human and environmental health.

The fate of PBDEs and PCBs when combined with MPs is an area of ongoing research, and the findings suggest that MPs could enhance the transfer and bioavailability of these chemicals in the environment (Chua et al., 2014; Besseling et al., 2013). When PBDEs and PCBs are attached to MPs, they can be transferred over long distances, including through ocean currents and atmospheric transport.

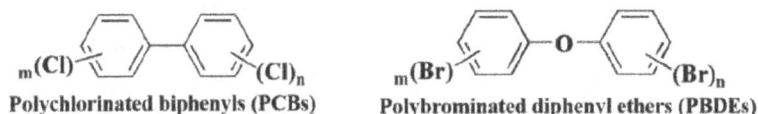

Polychlorinated biphenyls (PCBs)   Polybrominated diphenyl ethers (PBDEs)

Figure 4.8 Chemical structure of PCBs and PBDEs.

Additionally, MPs can act as a sink for these chemicals, increasing their concentrations in areas where MPs accumulate. Studies have also shown that MPs can facilitate the uptake of PBDEs and PCBs by organisms, such as fish, which can accumulate these chemicals in their tissues. This may have potential health effects for the organisms and for humans who consume them. Furthermore, the combination of MPs and PBDEs or PCBs may also result in the release of these chemicals into the environment, as the plastics break down or are ingested by organisms and subsequently excreted. Overall, the interaction of PBDEs and PCBs with MPs has the potential to exacerbate the negative effects of these emerging contaminants on the environment and human health. Current ongoing work to better understand the occurrence and fate of PCBs and PBDEs with MPs in the environment are described here.

5.1 *PCBs Adsorption/Desorption from Pristine and Aged MPs*

PCBs are banned in many countries due to their potential harm to human health and the environment. However, they are still used in some countries where regulations may be less strict or not enforced. The desorption and adsorption process of these chemicals on MPs is an important area of study because MPs are a major source of environmental pollution, and they can act as a vehicle for the transfer of these chemicals into the environment.

Nor and Koelmans (2019) studied the effects of low-density polyethylene (LDPE) and PVC MPs on the 14 PCBs in gut fluid mimic systems after their ingestion. For better understanding they also provided food to the organisms. They revealed 14 to 42% bioavailability of PCBs adsorbed on MPs for lugworm and 45–83% in cod. Furthermore, in a contaminated gut, they revealed the capability of clean MPs in extraction of PCBs from food. They concluded MPs ingestion are responsible for both contamination and cleaning.

Llorca's group (2020) investigated both adsorption and desorption potential of PCBs on the PS, PE and PET MPs of sizes 1–600 μm range in sediment/water system for 21 d. They carry out their studies in realistic environment conditions. The adsorption data fitted better with the Freundlich equation They revealed 20–60% adsorption of PCBs with higher adsorption percentage of PCBs having a lower degree of chlorination. The confirmational impediments in case of PCBs with a higher degree of chlorination results in lower adsorption. They disclosed π–π interactions in the case of PS and PET MPs results in enhanced adsorption.

5.2 *Individual and co-effects of MPs and PCBs on Biota*

PCBs can accumulate in the bodies of organisms and bioaccumulate in the food chain, resulting in high concentrations in top predators such as fish, birds, and mammals. This bioaccumulation of PCBs can further be affected by MPs which has been studied and reported by different research groups (Table 4.4 entry 1–3).

Wang's group (2020) investigated effects of MPs on the bioaccumulation of PCBs in *Eisenia fetida* (earthworm) in soil. They carry out their studies using PE MPs, and PCB52, PCB70, and PCB153 as model PCBs. They set-up two scenarios to expose earthworms in environmental conditions as S1: clean MPs (1%) with PCBs

Table 4.4 Studies towards effects of MPs and PCBs/PBDEs alone and in combination to various species of Biota.

Sr. No.	Research group	PCBs or PBDEs	Microplastics (MPs)	Species (Common name)	Effects (single and in combination)
1.	Wang and Coworkers, 2020	PCB52, PCB70, and PCB153	PE	*Eisenia fetida* (earthworm)	PCBs bioaccumulation was increased by MPs in earthworm.
2.	Li and coworkers, 2022	PCBs	PE	*Danio rerio* (Zebrafish)	MPs has their increased effect on PCB bioaccumulation by 1.8-fold (712.9 ng/g) when taken in above-fugacity PCB concentration.
3.	Wang and Coworkers, 2022	PCB 157	PS	*Paralichthys olivaceus* (juvenile Japanese flounder)	Enhanced effects of porous MPs in combination to PCBs on thyroid hormones level, thyroid tissue disruption, damage to gill morphology and altered gene expression.
4.	Xia and coworkers, 2020	BDE-209	PS	*Chlamys farreri* (Marine scallop)	PS MPs act as both carriers and scavengers for BDE-209 bioaccumulation. Enhanced negative effect on haemocyte phagocytosis by BDE-209.
5.	Chen and coworkers, 2021	BDE-209	PS, PA and PP	*Danio rerio* (Zebrafish)	1.7-fold enhanced triiodothyronine content in zebrafish larvae. Upregulations of thyroid stimulating hormones by 5.9 folds.
6.	Horton and coworkers, 2020	BDE- 47, 99, 100, 153, PBB- 153	PA	*Lymnaea stagnalis* (Pond snail)	Snails exposed to PBDEs in the absence of MPs lost more weight. Increased PBDE concentration in their body. MPs significantly decreases the BDE-47 uptake.
7.	Palacio-Cortés and coworkers, 2022	PBDE-47, 99, 100 and 153	PA	*Chironomus sancticaroli* (Chironomidae)	Faster ingestion of MPs by larvae at 0.5% w/w MPs. MPs reduces PBDE bioavailability and their uptake into the tissues.
8.	Li and Coworkers, 2022	BDE-47 and BDE-209	PE, PS, PHA and PHB	*Epinephelus moara* (Grouper)	Decreased SOD activity and glutathione content with increased catalase activity and malondialdehyde. Inhibition of iL-17 pathway related genes. Upregulation of PPAR pathway genes, GPC-related genes, and CAT gene in the liver. Down-regulation of SOD-regulated genes

contaminated soil and S2: Clean soil with added 1% MPs pre contaminated with PCBs. It was revealed in the S1 scenario PCBs accumulation in earthworm was higher than in S2 scenario due to slower desorption from MPs. However, in S1 scenario PCBs bioaccumulation was decreased by MPs, while PCBs bioaccumulation was increased by MPs in earthworm in the case of S2 scenario. Further based on results of biodynamic model analysis, they suggested 26.1% higher bioaccumulation in S2 due to MPs ingestion as compared to S1 (8.7%). The smaller size MPs results in higher bioaccumulation of PCBs as compared to larger MPs in earthworms. In addition, based on the results of biodynamic model analysis they suggested 26.1% higher bioaccumulation in S2 due to MPs ingestion as compared to S1 (8.7%). Smaller size MPs results in higher bioaccumulation of PCBs as compared to larger MPs in earthworm. They concluded that the diffusion gradient between the organism tissues and MPs affects the bioaccumulation or cleaning effects of PCBs.

Li's group (2022) investigated the effects of PE MPs on the bioaccumulation of PCBs in equi/above-fugacity concentrations on zebrafish (*Danio rerio*) and compared the observations with Food-Borne Particles (FBPs). They revealed no toxic effects on zebrafish in co-exposure of both MPs and FBPs with equi-fugacity and above-fugacity PCB concentrations. Although after exposing zebrafish to equi-fugacity concentration of PCB for 7 d, 177.7–400.5 ng/g PCB concentration was revealed in their body suggesting that the MPs had no enhancing effects on the bioaccumulation of PCB. Still after a 4-d depuration period 58.4–125.1 ng/g PCB concentration remained. They disclosed MPs have their increased effect on of PCB bioaccumulation by 1.8-fold (712.9 ng/g) when taken in above-fugacity PCB concentration. However, they found FBPs results in more PCB bioaccumulation as compared to MPs in zebrafish by 2.8- and 4.2 folds in both equi-fugacity and above-fugacity concentration of PCBs.

Men´endez-Pedriza and coworkers (2022) investigated the toxic effects of PE MPs and two PCBs including 2,2',4,5,5'-pentachlorobiphenyl congener (PCB 101) and 3,3',4,4',5-pentachlorobiphenyl congener (PCB 126) alone and in combination on human hepatoma cell line HepG2. To carry out their investigations they performed an untargeted lipidomic study and the cell viability assessment. They revealed no changes in cell lethality in presence of MPs based on cell lethality evaluations, while the MPs results in increased triglyceride content. However, they observed lipidomic changes on exposure of both PCBs alone with significant alterations in glycerophospholipids and glycerolipids. Based on chemometric analysis after combined exposure (MPs + PCBs) results in more harmful toxic effects due to previously adsorbed pollutants in MPs. They concluded PCBs and MPs induce notable changes to the composition of cell membrane and its permeability.

To gain insights of microbial dichlorination behaviour of PCBs, Li and coworkers (2022) evaluated the effects of PE, PP and PS MPs on anaerobic reductive dechlorination of commercial PCB mixture (Aroclor 1260) using microbes. They revealed the mitigating effects of MPs on microbial reductive dechlorination of Aroclor 1260 as PE (39.43%) > PP (23.97%) > PS (17.53%). They further disclosed primary inhibition of chlorine removal substituted at meta-position of PCBs by MPs. No variability was also observed in the microbial community in both the suspension culture and biofilms on MPs, but Dehalococcoides were detected on the MPs biofilm.

They concluded the synergetic interactions of MPs and microbes could result in the inhibition of PCB microbial dechlorination.

The enhancing effects of MPs in combination with PCBs to the thyroid disruption in *Paralichthys olivaceus* (juvenile Japanese flounder) was investigated by Wang et al (2022). They did an investigation with porous PS MPs and Aroclor 1254 (PCB 157). They exposed the fish to 500 ng/L PCBs concentration alone and in co-exposure with 2, 20, and 200 µg/L concentration of 10-µm porous MPs for 21 d. They revealed the enhancing effects of porous MPs (20, 200 µg/L) in combination to PCBs on the thyroid hormones level, thyroid tissue disruption, damage to gill morphology and altered gene expression in HPT axis. However, decreased concentration of PCBs was observed in seawater with increased MPs concentration confirming PCBs adsorption by MPs in marine environment.

5.3 PBDEs Adsorption/Desorption from Pristine and Aged MPs

Understanding the desorption and adsorption process of PBDEs on MPs is important because it can help to develop strategies for reducing their environmental impact. The ongoing investigations in this field by different researchers has been summarized here.

Sun's group (2019) studied the leaching kinetics of Brominated Flame Retardants (BFRs) including PBDEs and 1,2-bis(2,4,6-tribromophenoxy)ethane (BTBPE) adsorbed on Acrylonitrile Butadiene Styrene (ABS) MPs in water. They reported diffusion in the plastic matrix as the main factor for the leaching rate of BFRs from the ABS MPs. Further they revealed slow leaching of BFRs from the surface of MPs and the half-life to be billions of years if their physical and chemical structure did not change. They concluded that the leaching of BFRs from the surface of MPs is slow but it can be accelerated due to the digestive system and bacteria after their ingestion by organisms. Later they investigated PBDEs leaching kinetics from MPs in fish oil. They carry out their studies using PS, ABS and PP MPs, and their effects on PBDEs bioaccumulation in cod fish by a biodynamic model (Sun et al., 2021). The PBDEs leaching kinetics from MPs follow the order as ABS (1.89×10^{-20}–2.07×10^{-18} m^2 s^{-1}) > PS (1.98×10^{-19}–2.35×10^{-16} m^2 s^{-1}) > PP (4.26×10^{-18}–1.72×10^{-15} m^2 s^{-1}). Further they revealed enhanced bioaccumulation of MPs containing PBDEs by oil in the gastrointestinal tract of the fish. Sun and Zeng, 2022 also explored the chemical diffusion of PBDEs adsorbed on PS or ABS or PP MPs as transfer agents in simulated gut conditions. They explored the role of MP ingestion on the bioaccumulation of PBDEs in lugworm using a biodynamic model. They reported diffusion coefficients of PBDEs from MPs following the order as PS (5.82×10^{-23} to 7.96×10^{-20} m^{2}s^{-1}) > ABS (5.59×10^{-23} to 73.45×10^{-20} m^{2}s^{-1}) > PP (5.58×10^{-21} to 5.79×10^{-17} m^{2}s^{-1}). They further revealed a negligible effect of MPs ingestion with a size larger than 0.5 mm on the bioaccumulation of PBDEs in lugworms. Their contribution towards PBDEs bioaccumulation in lugworm become more significant with their breakdown.

Singla's group (2020) explored the sorption and extraction capabilities of PBDEs from four MPs including PET, PP, LDPE and PS. They concluded that the extraction of PBDEs depends on the polymer composition and the solvent used.

5.4 Individual and Co-effects of MPs and PBDEs on Biota

PBDEs have been associated with thyroid hormone disruption, neurodevelopmental effects, and reproductive problems. To address these issues, scientists are working on developing new methods to detect and measure PBDEs in the environment and biota. Further the presence of MPs can increase their toxicity and recent studies done by various research groups has been included here (Table 4.4 entry 4–8).

Xia et al. 2020 studied the effects of PS MPs on the bioaccumulation of decabromodiphenyl ether (BDE-209) in *Chlamys farreri* (marine scallop) and their co-effects on toxicity at the cellular level with histopathological effects. Based on the experimental results they revealed PS MPs as both carriers and scavengers for BDE-209 bioaccumulation. Furthermore, they revealed an enhanced negative effect on haemocyte phagocytosis by BDE-209. Ultrastructural alterations in the gills and digestive tract of scallops were also observed due to MPs role as carriers for the BDE-209 bioaccumulation.

Chen and coworkers (2021) investigated the effects of degradation products of decabrominated diphenyl ether (BDE-209) in the presence and absence of PS, PA, and PP MPs under only light or in combination of biota conditions on zebrafish (*Danio rerio*). They revealed an increase in BDE-209 concentration by 0.7–2.8 fold due to their desorption from MPs. Further they disclosed a decrease in penta- and hexa-BDE concentrations under light conditions in the presence of PS MPs, while an increase in debrominated products was observed in both light and biota conditions due to high abundance of bacteria including Proteobacteria, Chloroflexi, and Basidiomycotina results in degradation. Furthermore 1.7-fold enhanced triiodothyronine content due to the debrominated products was observed in zebrafish larvae while thyroid stimulating hormones were upregulated by 5.9 folds.

Later the same group in the lead of Horton studied the PA MPs effect alone and coeffect with PDBE mixture containing BDE-47, 99, 100, 153 and PBB-153 on to pond snail *Lymnaea stagnalis* (Horton et al., 2020). They investigated the survival, weight change, microbiome diversity, community composition and accumulation of PBDE in snails. For this they exposed the snails to PBDEs in presence and absence of nylon MPs in quartz sand (1% w/w) for 96 hr. They revealed no mortality under both conditions. Even snails exposed to PBDEs in the absence of MPs lost more weight than those exposed to MPs. With an increase in the concentration of PBDEs in the soil also resulted in increased PBDE concentration in their body however, MPs did not influence the total PBDE uptake. However on further studies they revealed that the MPs significantly decreases the BDE 47 uptake by snails. Furthermore, no change in the diversity of the snail microbiome was observed. They finally concluded a limited effect of MPs individually and in combination of PBDEs on Lymnaea stagnalis.

Palacio-Cortés et al., 2022 investigated the effects of nylon (PA) MPs alone and in combination to PBDEs on the survival and microbiome structure after ingestion by *Chironomus sancticaroli* larvae. They exposed the larvae to PBDEs (PBDE-47, 99, 100 and 153) for 96 hr in presence and absence of 1% w/w MP in the sediment to measure the survival of the PBDE uptake and composition of microbial community. They revealed faster ingestion of MPs by larvae at 0.5% w/w MP, while efficient depuration was observed at 1% w/w MP. After 168 hr of depuration the MP retention still prevailed. Additionally, no significant effect of PBDEs and MP exposure was observed on mortality. Due to the higher adsorption capacity of PBDEs on MPs reduces PBDE bioavailability to C. sancticaroli and their uptake into the tissues. Further they disclosed minimal effects of PBDEs on the microbial community, but MPs show stronger effects on the microbiome even after a short exposure time.

Li's group (2022) explored the combined toxic effects of MPs and PBDEs on oxidative stress and gene expressions of farmed fishes. They chose BDE-47 and BDE-209 and studied their sorption on PE, PS, PHA and PHB MPs followed by their co-effect on *Epinephelus moara* (grouper). They found the sorption capacity follows the order as PHA > PHB > PS > PE. They observed significant alterations in the oxidative system on co-exposure of MPs and PBDEs as compared to their individual effect. Furthermore, decreased SOD activity and glutathione content with increased catalase activity and malondialdehyde were reported influencing the growth of groupers. They also disclosed inhibition of IL-17 pathway related genes, upregulation of Peroxisome Proliferator-Activated Receptors (PPAR) pathway genes, Grain Protein Content (GPC)-related genes and CAT gene in the liver and down-regulation of SOD-regulated genes.

6. Summary and Future Research Directions

MPs, less than 5 mm in size, are present everywhere in the environment. They are found in oceans, freshwater systems, soil, air, and even in the food one eats. Due to their small size, they can be ingested by almost all living organisms, from plankton to humans. This widespread exposure to MPs has raised concerns about their potential health impacts on ecosystems and human health. One of the most concerning aspects of MPs is their ability to adsorb and transfer other toxic contaminants. Emerging contaminants, such as pharmaceuticals, pesticides, and heavy metals, can become attached to the surface of MPs, making them more bioavailable to aquatic organisms. As a result, MPs can amplify the toxic effects of these contaminants when they enter the food chain.

To address these issues, researchers have been studying the interaction between MPs and emerging contaminants in various environmental conditions. They have investigated the effect of physical factors such as weathering, temperature, pH, and salinity on the adsorption and desorption of contaminants from MPs. Further effects of MPs and emerging contaminants alone and in combination on various biota species, including their effects on important enzymes, body muscles, and organs have been explored. Despite these efforts, there is still much to learn about the health impacts of MPs and emerging contaminants. With infinite organisms on Earth, further research is required to fully understand the extent of the problem.

In addition, policymakers should take action to prevent the entry of MPs into the environment and to develop better substitutes for plastic. To solve this problem, the world must come together and take collective action. Governments can implement policies to regulate the use and disposal of plastics. Companies can invest in research to develop safer alternatives to MPs. Individuals can also make a difference by reducing their plastic consumption and correctly disposing of plastic waste. In conclusion, MPs and emerging contaminants are a serious threat to the environment and human health. While researchers have made progress in understanding their impacts, much work remains to be done. With a united effort, one can work towards a cleaner, healthier planet for all living things.

Acknowledgments

Aman Bhalla and Savita Chaudhary gratefully acknowledges support from the Department of Science and Technology (DST), New Delhi. We would like to apologize to those scientists whose work may not have appeared in this chapter either due to the limited scope of the chapter or oversight. Pankaj Kumar acknowledges the financial support from Council of Scientific and Industrial Research (CSIR) New Delhi vides Award No- F.No.09/135/(0823)2018-EMR-I.

Abbreviations

BTBPE	:	1,2-Bis(2,4,6-Tribromophenoxy) Ethane
ABS	:	Acrylonitrile Butadiene Styrene
AML	:	Amlodipine
AMX	:	Amoxicillin
ARGs	:	Antibiotic Resistance Genes
ATV	:	Atorvastatin
ATZ	:	Atrazine
AMWPE	:	Average Molecular Weight Medium Density Polyethylene
NaT	:	Bile Salt
BAF	:	Bioaccumulation Factor
BSA	:	Bovine Serum Albumin
BFRs	:	Brominated Flame Retardants
CAR	:	Carbendazim
CEP-C	:	Cephalosporin C
CPE	:	Chlorinated Polyethylene
CPF	:	Chlorpyrifos
CTC	:	Chlortetracycline
CIP	:	Ciprofloxacin
DPF	:	Days Post-Fertilization
DDTs	:	Dichlorodiphenyltrichloroethanes
DCF	:	Diclofenac
DIFE	:	Difenoconazole
DIF	:	Diflubenzuron
DIP	:	Dipterex
DOM	:	Dissolved Organic Matter
DIR	:	Diuron
FBPs	:	Food-Borne Particles
GPC	:	Grain Protein Content
HMs	:	Heavy Metals
HCHs	:	Hexachlorocyclohexanes
HEX	:	Hexaconazole

HDPE	:	High-Density Polyethylene
HPF	:	Hours Post-Fertilization
HPG	:	Hypothalamic-Pituitary-Gonadal
IMI	:	Imidacloprid
LDH	:	Lactate Dehydrogenase Activity
LEV	:	Levofloxacin
MAL	:	Malathion
MNZ	:	Metronidazole
MPs	:	Microplastics
MGEs	:	Mobile Genetic Elements
MYC	:	Myclobutanil
NSAIDs	:	Nonsteroidal Anti-Inflammatory Drugs
Log D_{ow}	:	Octanol-Water Distribution Coefficients
OFL	:	Ofloxacin
OCPs	:	Organochlorine Pesticides
PAC	:	Paracetamol
PPAR	:	Peroxisome Proliferator-Activated Receptors
POP	:	Persistent Organic Pollutants
PCCPs	:	Personal Care and Cosmetic Products
PHACs	:	Pharmacologically Active Compounds
PHN	:	Phenol
PBAT	:	Poly (Butylene Adipate Co-Terephthalate)
PA	:	Polyamides
PBDEs	:	Polybrominated Diphenyl Ethers
PBS	:	Polybutylene Succinate
PCBs	:	Polychlorinated Biphenyls
PE	:	Polyethylene
PET	:	Polyethylene Terephthalate
PHB	:	Polyhydroxybutyrate
PLA	:	Polylactic Acid
PMMA	:	Polymethyl Methacrylate
PP	:	Polypropylene
PRP	:	Propranolol
PSOM	:	Pseudo-Second-Order Model
ROS	:	Reactive Organic Species
ROX	:	Roxithromycin
SER	:	Sertraline
SBET	:	Specific Surface Areas
SCC	:	Straw Crystalline Cellulose
SMT	:	Sulphamethazine
SMX	:	Sulphamethoxazole
SOD	:	Superoxide Dismutase
TC	:	Tetracycline
TWP	:	Tire Wear Particles
TRI	:	Triadimenol
TCS	:	Triclosan

UHMWPE : Ultra-High Molecular Weight Polyethylene
UPVC : Unplasticized Poly (Vinyl Chloride)
VAC : Vancomycin
VFX : Venlafaxine
WWTP : Waste Water Treatment Plants
WSP : Wheat Straw Powder
WSS : Wheat Straw Segment

References

Alengebawy, A., Abdelkhalek, S.T., Qureshi, S.R. and Wang, M.-Q. 2021. Heavy metals and pesticides toxicity in agricultural soil and plants: ecological risks and human health implications. Toxics. 9: 42.

Allouche, M., Ishak, S., ben Ali, M., Hedfi, A., Almalki, M., Karachle, P.K. et al. 2022. Molecular interactions of polyvinyl chloride microplastics and beta-blockers (Diltiazem and Bisoprolol) and their effects on marine meiofauna: Combined *in vivo* and modeling study. J. Hazard. Mater. 431: 128609.

An, L., Liu, Q., Deng, Y., Wu, W., Gao, Y. and Ling, W. 2020. Sources of microplastic in the environment. pp.143–159. *In*: He, D. and Luo, Y. (eds.). Microplastics in Terrestrial Environment. The Handbook of Environmental Chemistry. Springer, Cham.

Balali-Mood, M., Naseri, K., Tahergorabi, Z., Khazdair, M.R. and Sadeghi, M. 2021. Toxic mechanisms of five heavy metals: Mercury, Lead, Chromium, Cadmium, and Arsenic. Front. Pharmacol. 12: DOI 10.3389/fphar.2021.643972.

Besseling, E., Wegner, A., Foekema, E.M., van den Heuvel-Greve, M.J. and Koelmans, A.A. 2013. Effects of Microplastic on Fitness and PCB Bioaccumulation by the Lugworm *Arenicola marina* (L.). Environ. Sci. Technol. 47: 593.

Campanale, Massarelli, Savino, Locaputo and Uricchio. 2020. A detailed review study on potential effects of microplastics and additives of concern on human health. Int. J. Environ. Res. Public. Health. 17: 1212.

Chen, G., Feng, Q. and Wang, J. 2020. Mini-review of microplastics in the atmosphere and their risks to humans Sci. Total. Environ. 703: 135504.

Chen, Q., Zhang, X., Xie, Q., Lee, Y.H., Lee, J.S. and Shi, H. 2021. Microplastics habituated with biofilm change decabrominated diphenyl ether degradation products and thyroid endocrine toxicity. Ecotoxicol. Environ. Saf. 228: 112991.

Chen, Y., Qian, Y., Shi, Y., Wang, X., Tan, X. and An, D. 2022. Accumulation of chiral pharmaceuticals (ofloxacin or levofloxacin) onto polyethylene microplastics from aqueous solutions. Sci. Total Environ. 823: 153765.

Chua, E.M., Shimeta, J., Nugegoda, D., Morrison, P.D. and Clarke, B.O. 2014. Assimilation of Polybrominated Diphenyl Ethers from Microplastics by the Marine Amphipod, *Allorchestes Compressa*. Environ. Sci. Technol. 48: 8127.

Concha-Graña, E., Moscoso-Pérez, C.M., López-Mahía, P. and Muniategui-Lorenzo, S. 2022. Adsorption of pesticides and personal care products on pristine and weathered microplastics in the marine environment. Comparison between bio-based and conventional plastics. Sci. Total Environ. 848: 157703.

di Mauro, R., Kupchik, M.J. and Benfield, M.C. 2017. Abundant plankton-sized microplastic particles in shelf waters of the northern Gulf of Mexico. Environ. Pollut. 230: 798.

Efimova, I., Bagaeva, M., Bagaev, A., Kileso, A. and Chubarenko, I.P. 2018. Secondary microplastics generation in the sea swash zone with coarse bottom sediments: Laboratory experiments. Front. Mar. Sci. 5: DOI 10.3389/fmars.2018.00313.

Elizalde-Velázquez, A., Subbiah, S., Anderson, T.A., Green, M.J., Zhao, X. and Cañas-Carrell, J.E. 2020. Sorption of three common nonsteroidal anti-inflammatory drugs (NSAIDs) to microplastics. Sci. Total Environ. 715: 136974.

Fan, X., Gan, R., Liu, J., Xie, Y., Xu, D., Xiang, Y. et al. 2021. Adsorption and desorption behaviors of antibiotics by tire wear particles and polyethylene microplastics with or without aging processes. Sci. Total Environ. 771: 145451.

Fang, S., Yu, W., Li, C., Liu, Y., Qiu, J. and Kong, F. 2019. Adsorption behavior of three triazole fungicides on polystyrene microplastics. Sci. Total Environ. 691: 1119.

Francisco, L.F.V., do Amaral Crispim, B., Spósito, J.C.V., Solórzano, J.C.J., Maran, N.H., Kummrow, F. et al. 2019. Metals and emerging contaminants in groundwater and human health risk assessment. Environ. Sci. Pollut. Res. 26: 24581.

Fu, Q., Tan, X., Ye, S., Ma, L., Gu, Y., Zhang, P. et al. 2021. Mechanism analysis of heavy metal lead captured by natural-aged microplastics. Chemosphere 270: 128624.

Godoy, V., Martín-Lara, M.A., Calero, M. and Blázquez, G. 2020. The relevance of interaction of chemicals/pollutants and microplastic samples as route for transporting contaminants. Process Saf. Environ. Prot. 138: 312.

Guo, X., Chen, C. and Wang, J. 2019a. Sorption of sulfamethoxazole onto six types of microplastics. Chemosphere 228: 300.

Guo, X., Liu, Y. and Wang, J. 2019b. Sorption of sulfamethazine onto different types of microplastics: A combined experimental and molecular dynamics simulation study. Mar. Pollut. Bull. 145: 547.

Guo, X. and Wang, J. 2019c. Sorption of antibiotics onto aged microplastics in freshwater and seawater. Mar. Pollut. Bull. 149: 110511.

Hanachi, P., Karbalaei, S. and Yu, S. 2021. Combined polystyrene microplastics and chlorpyrifos decrease levels of nutritional parameters in muscle of rainbow trout (*Oncorhynchus mykiss*). Environ. Sci. Pollut. Res. 28: 64908.

Harding, K., Dennis, J., Vonblottnitz, H. and Harrison, S. 2007. Environmental analysis of plastic production processes: Comparing petroleum-based polypropylene and polyethylene with biologically-based poly-β-hydroxybutyric acid using life cycle analysis. J. Biotechnol. 130: 57.

He, B., Liu, A., Duodu, G.O., Wijesiri, B., Ayoko, G.A. and Goonetilleke, A. 2023. Distribution and variation of metals in urban river sediments in response to microplastics presence, catchment characteristics and sediment properties. Sci. Total Environ. 856: 159139.

Horton A.A., Newbold, L.K., Palacio-Cortés, A.M., Spurgeon, D.J., Pereira, M.G., Carter, H. et al. 2020. Accumulation of polybrominated diphenyl ethers and microbiome response in the great pond snail *Lymnaea stagnalis* with exposure to nylon (polyamide) microplastics. Ecotoxicol. Environ. Saf. 188: 109882.

Huang, Y., Ding, J., Zhang, G., Liu, S., Zou, H., Wang, Z. et al., 2021. Interactive effects of microplastics and selected pharmaceuticals on red tilapia: Role of microplastic aging. Sci. Total Environ. 752: 142256.

Jia, W., Karapetrova, A., Zhang, M., Xu, L., Li, K., Huang, M. et al. 2022. Automated identification and quantification of invisible microplastics in agricultural soils. Sci. Total Environ. 844: 156853.

Jiang, F., Wang, M., Ding, J., Cao, W. and Sun, C. 2022. Occurrence and seasonal variation of microplastics in the effluent from wastewater treatment plants in Qingdao, China. J. Mar. Sci. Eng. 10: 58.

Jiang, M., Hu, L., Lu, A., Liang, G., Lin, Z., Zhang, T. et al. 2020. Strong sorption of two fungicides onto biodegradable microplastics with emphasis on the negligible role of environmental factors. Environ. Pollut. 267: 115496.

Ju, H., Yang, X., Osman, R. and Geissen, V. 2023. Effects of microplastics and chlorpyrifos on earthworms (*Lumbricus terrestris*) and their biogenic transport in sandy soil. Environ. Pollut. 316: 120483.

Juliano, C. and Magrini, G. 2017. Cosmetic ingredients as emerging pollutants of environmental and health concern. A mini-review. Cosmetics 4: 11.

Khan, N., Kalair, E., Abas, N., Kalair, A.R. and Kalair, A. 2019. Energy transition from molecules to atoms and photons. Eng. Sci. Technol. an Int. J. 22: 185.

Lan, T., Cao, F., Cao, L., Wang, T., Yu, C. and Wang, F. 2022. A comparative study on the adsorption behavior and mechanism of pesticides on agricultural film microplastics and straw degradation products. Chemosphere. 303: 135058.

Lan, T., Wang, T., Cao, F., Yu, C., Chu, Q. and Wang, F. 2021. A comparative study on the adsorption behavior of pesticides by pristine and aged microplastics from agricultural polyethylene soil films. Ecotoxicol. Environ. Saf. 209: 111781.

Laskar, N. and Kumar, U. 2019. Plastics and microplastics: A threat to environment. Environ. Technol. Innov. 14: 100352.

Lavandier, R., Quinete, N., Hauser-Davis, R.A., Dias, P.S., Taniguchi, S., Montone, R. and Moreira, I. 2013. Polychlorinated biphenyls (PCBs) and Polybrominated Diphenyl ethers (PBDEs) in three fish species from an estuary in the southeastern coast of Brazil. Chemosphere 90: 2435.

Li J., Huang, X., Hou, Z. and Ding, T. 2022. Sorption of diclofenac by polystyrene microplastics: Kinetics, isotherms and particle size effects. Chemosphere 290: 133311.

Li, H., Wang, F., Li, J., Deng, S. and Zhang, S. 2021. Adsorption of three pesticides on polyethylene microplastics in aqueous solutions: Kinetics, isotherms, thermodynamics, and molecular dynamics simulation. Chemosphere 264: 128556.

Li, H., Li, Y., Maryam, B., Ji, Z., Sun, J. and Liu, X. 2022. Polybrominated diphenyl ethers as hitchhikers on microplastics: Sorption behaviors and combined toxicities to *Epinephelus moara*. Aquat. Toxicol. 252: 106317.

Li, M., Chen, Q., Ma, C., Gao, Z., Yu, H., Xu, L. and Shi, H. 2022. Effects of microplastics and food particles on organic pollutants bioaccumulation in equi-fugacity and above-fugacity scenarios. Sci. Total Environ. 812: 152548.

Li, X., Xu, Q., Cheng, Y., Chen, C., Shen, C., Zhang, C. et al., 2022. Effect of microplastics on microbial dechlorination of a polychlorinated biphenyl mixture (Aroclor 1260). Sci. Total Environ. 831: 154904.

Li, Y., Li, M., Li, Z., Yang, L. and Liu, X. 2019. Effects of particle size and solution chemistry on Triclosan sorption on polystyrene microplastic. Chemosphere 231: 308.

Li, Y., Sun, Y., Li, J., Tang, R., Miu, Y. and Ma, X. 2021. Research on the influence of microplastics on marine life. IOP Conf. Ser. Earth Environ. Sci. 631: 012006.

lin Zhu, Z., Chun Wang, S., Fei Zhao, F., Guang Wang, S., Fei Liu, F. and Zhou Liu, G. 2019. Joint toxicity of microplastics with triclosan to marine microalgae *Skeletonema costatum*. Environ. Pollut. 246: 509.

Liu, G., Dave, P.H., Kwong, R.W.M., Wu, M. and Zhong, H. 2021. Influence of microplastics on the mobility, bioavailability, and toxicity of heavy metals: A review. Bull. Environ. Contam. Toxicol. 107: 710.

Liu, G., Zhu, Z., Yang, Y., Sun, Y., Yu, F. and Ma, J. 2019. Sorption behavior and mechanism of hydrophilic organic chemicals to virgin and aged microplastics in freshwater and seawater. Environ. Pollut. 246: 26.

Liu, P., Lu, K., Li, J., Wu, X., Qian, L., Wang, M. and Gao, S. 2020. Effect of aging on adsorption behavior of polystyrene microplastics for pharmaceuticals: Adsorption mechanism and role of aging intermediates. J. Hazard. Mater. 384: 121193.

Liu, P., Wu, X., Liu, H., Wang, H., Lu, K. and Gao, S. 2020. Desorption of pharmaceuticals from pristine and aged polystyrene microplastics under simulated gastrointestinal conditions. J. Hazard. Mater. 392: 122346.

Liu, P., Wu, X., Shi, H., Wang, H., Huang, H., Shi, Y. et al., 2022. Contribution of aged polystyrene microplastics to the bioaccumulation of pharmaceuticals in marine organisms using experimental and model analysis. Chemosphere 287: 132412.

Liu, W., Pan, T., Liu, H., Jiang, M. and Zhang, T. 2023. Adsorption behavior of imidacloprid pesticide on polar microplastics under environmental conditions: critical role of photo-aging. Front. Environ. Sci. Eng. 17: DOI 10.1007/s11783-023-1641-0.

Llorca, M., Ábalos, M., Vega-Herrera, A., Adrados, M.A., Abad, E. and Farré, M. 2020. Adsorption and desorption behaviour of polychlorinated biphenyls onto microplastics' surfaces in water/sediment systems. Toxics. 8: 59.

López-Martínez, S., Morales-Caselles, C., Kadar, J. and Rivas, M.L. 2021. Overview of global status of plastic presence in marine vertebrates. Glob. Chang. Biol. 27: 728.

Ma, J., Zhao, J., Zhu, Z., Li, L. and Yu, F. 2019. Effect of microplastic size on the adsorption behavior and mechanism of triclosan on polyvinyl chloride. Environ. Pollut. 254: 113104.

Martinho, S.D., Fernandes, V.C., Figueiredo, S.A. and Delerue-Matos, C. 2022. Microplastic pollution focused on sources, distribution, contaminant interactions, analytical methods, and wastewater removal strategies: A review. Int. J. Environ. Res. Public Health. 19: 5610.

McDevitt, J.P., Criddle, C.S., Morse, M., Hale, R.C., Bott, C.B. and Rochman, C.M. 2017. Addressing the issue of microplastics in the wake of the microbead-free waters act-A new standard can facilitate improved policy. Environ. Sci. Technol. 51: 6611.

McDougall, L., Thomson, L., Brand, S., Wagstaff, A., Lawton, L.A. and Petrie, B. 2022. Adsorption of a diverse range of pharmaceuticals to polyethylene microplastics in wastewater and their desorption in environmental matrices. Sci. Total Environ. 808: 152071.

Menéndez-Pedriza, A., Jaumot, J. and Bedia, C. 2022. Lipidomic analysis of single and combined effects of polyethylene microplastics and polychlorinated biphenyls on human hepatoma cells. J. Hazard. Mater. 421: 126777.

Mensah, P.K., C.G. Palmer and W.J. Muller. 2014. Lethal and sublethal effects of pesticides on aquatic organisms: The case of a freshwater shrimp exposure to roundup®. Pesticides - Toxic Aspects, InTech. DOI: 10.5772/57166.

Miranda, M.N., Lado Ribeiro, A.R., Silva, A.M.T. and Pereira, M.F.R. 2022. Can aged microplastics be transport vectors for organic micropollutants? - Sorption and phytotoxicity tests. Sci. Total Environ. 850: 158073.

Mohamed Nor, N.H. and Koelmans, A.A. 2019. Transfer of PCBs from microplastics under simulated gut fluid conditions is biphasic and reversible. Environ. Sci. Technol. 53: 1874.

Montano, L., Pironti, C., Pinto, G., Ricciardi, M., Buono, A., Brogna, C. et al., 2022. Polychlorinated biphenyls (PCBs) in the environment: occupational and exposure events, effects on human health and fertility. Toxics. 10: 365.

Munoz, M., Ortiz, D., Nieto-Sandoval, J., de Pedro, Z.M. and Casas, J.A. 2021. Adsorption of micropollutants onto realistic microplastics: Role of microplastic nature, size, age, and NOM fouling. Chemosphere 283: 131085.

Napper, I.E. and Thompson, R.C. 2020. Plastic debris in the marine environment: history and future challenges. Global Challenges 4: 1900081.

Nawalage, N.S.K. and Bellanthudawa, B.K.A. 2022. Synthetic polymers in personal care and cosmetics products (PCCPs) as a source of microplastic (MP) pollution. Mar. Pollut. Bull. 182: 113927.

Ngo, P.L., Pramanik, B.K., Shah, K. and Roychand, R. 2019. Pathway, classification and removal efficiency of microplastics in wastewater treatment plants. Environ. Pollut. 255: 113326.

Nicolopoulou-Stamati, P., Maipas, S., Kotampasi, C., Stamatis, P. and Hens, L. 2016. Chemical pesticides and human health: the urgent need for a new concept in agriculture. Front. Public Health. 4: DOI 10.3389/fpubh.2016.00148.

Nor, N.H.M and A.A. Koelmans. 2019. Transfer of PCBs from microplastics under simulated gut fluid conditions is biphasic and reversible. Environ. Sci. Technol. 53: 1874.

Nugnes, R., Russo, C., Lavorgna, M., Orlo, E., Kundi, M. and Isidori, M. 2022. Polystyrene microplastic particles in combination with pesticides and antiviral drugs: Toxicity and genotoxicity in *Ceriodaphnia dubia*. Environ. Pollut. 313: 120088.

Palacio-Cortés, A.M., Horton, A.A., Newbold, L., Spurgeon, D., Lahive, E., Pereira, M.G. et al. 2022. Accumulation of nylon microplastics and polybrominated diphenyl ethers and effects on gut microbial community of *Chironomus sancticaroli*. Sci. Total Environ. 832: 155089.

Peña, A., Rodríguez-Liébana, J.A. and Delgado-Moreno, L. 2023. Interactions of microplastics with pesticides in soils and their ecotoxicological implications. Agronomy 13: 701.

Pinto-Poblete, A., Retamal-Salgado, J., López, M.D., Zapata, N., Sierra-Almeida, A. and Schoebitz, M. 2022. Combined effect of microplastics and cd alters the enzymatic activity of soil and the productivity of strawberry plants. Plants 11: 536.

Prata, J.C. 2018. Microplastics in wastewater: State of the knowledge on sources, fate and solutions. Mar. Pollut. Bull. 129: 262.

Prata, J.C., Lavorante, B.R.B.O., da, M., Maria, C. da and Guilhermino, L. 2018. Influence of microplastics on the toxicity of the pharmaceuticals procainamide and doxycycline on the marine microalgae *Tetraselmis chuii*. Aquatic Toxicology 197: 143.

Puckowski, A., Cwięk, W., Mioduszewska, K., Stepnowski, P. and Białk-Bielińska, A. 2021. Sorption of pharmaceuticals on the surface of microplastics. Chemosphere 263: 127976.

Qu, H., Ma, R., Wang, B., Yang, J., Duan, L. and Yu, G. 2019. Effects of microplastics on the uptake, distribution and biotransformation of chiral antidepressant venlafaxine in aquatic ecosystem. J. Hazard. Mater. 370: 203

Qu, H., Ma, R., Wang, B., Zhang, Y., Yin, L., Yu, G. et al., 2018. Enantiospecific toxicity, distribution and bioaccumulation of chiral antidepressant venlafaxine and its metabolite in loach (*Misgurnus anguillicaudatus*) co-exposed to microplastic and the drugs. J. Hazard. Mater. 359: 104.

Ramhøj, L., T. Svingen, K. Mandrup, U. Hass, S.P. Lund, Vinggaard, A.M. et al. 2022. Developmental exposure to the brominated flame-retardant DE-71 reduces serum thyroid hormones in rats without hypothalamic-pituitary-thyroid axis activation or neurobehavioral changes in offspring. PLoS One. 17: e0271614.

Razanajatovo, R.M., Ding, J., Zhang, S., Jiang, H. and Zou, H. 2018. Sorption and desorption of selected pharmaceuticals by polyethylene microplastics. Mar. Pollut. Bull. 136: 516.

Rios-Fuster, B., Alomar, C., Viñas, L., Campillo, J.A., Pérez-Fernández, B., Álvarez, E. et al. 2021. Organochlorine pesticides (OCPs) and polychlorinated biphenyls (PCBs) occurrence in Sparus aurata exposed to microplastic enriched diets in aquaculture facilities. Mar. Pollut. Bull. 173: 113030.

Sajjad, M., Huang, Q., Khan, S., Khan, M.A., Liu, Y., Wang, J. et al., 2022. Microplastics in the soil environment: A critical review. Environ. Technol. Innov. 27: 102408.

Santos, D., Félix, L., Luzio, A., Parra, S., Bellas, J. and Monteiro, S.M. 2021. Single and combined acute and subchronic toxic effects of microplastics and copper in zebrafish (*Danio rerio*) early life stages. Chemosphere 277: 130262.

Santos, L.H.M.L.M., Rodríguez-Mozaz, S. and Barceló, D. 2021. Microplastics as vectors of pharmaceuticals in aquatic organisms—An overview of their environmental implications. Case Studies in Chemical and Environmental Engineering 3: 100079.

Schuhen, K., Toni Sturm, M. and A.F. 2019. Technological approaches for the reduction of microplastic pollution in seawater desalination plants and for sea salt extraction. pp. 1–16. *In*: Plastics in the Environment. IntechOpen. DOI: 10.5772/intechopen.81180.

Sharma, S. and Chatterjee, S. 2017. Microplastic pollution, a threat to marine ecosystem and human health: a short review. Environ. Sci. Pollut. Res. 24: 21530.

Sheikh, I.A., Beg, M.A., Hamoda, T.A.-A.A.-M., Mandourah, H.M.S. and Memili, E. 2023. An analysis of the structural relationship between thyroid hormone-signaling disruption and polybrominated diphenyl ethers: potential implications for male infertility. Int. J. Mol. Sci. 24: 3296.

Shi, W., Han, Y., Sun, S., Tang, Y., Zhou, W., Du, X. et al., 2020. Immunotoxicities of microplastics and sertraline, alone and in combination, to a bivalve species: size-dependent interaction and potential toxication mechanism. J. Hazard. Mater. 396: 122603.

Singh, P. and Sharma, V.P. 2016. Integrated plastic waste management: environmental and improved health approaches. Procedia. Environ. Sci. 35: 692.

Singla, M., Díaz, J., Broto-Puig, F. and Borrós, S. 2020. Sorption and release process of polybrominated diphenyl ethers (PBDEs) from different composition microplastics in aqueous medium: Solubility parameter approach. Environ. Pollut. 262: 114377.

Smital, T., T. Luckenbach, R. Sauerborn, A.M. Hamdoun, R.L. and Vega, D. Epel. 2004. Emerging contaminants-pesticides, PPCPs, microbial degradation products and natural substances as inhibitors of multixenobiotic defense in aquatic organisms. Mutat. Res. - Fundam. Mol. Mech. Mutagen. 552: 101.

Sturm, M.T., Horn, H. and Schuhen, K. 2021. Removal of microplastics from waters through agglomeration-fixation using organosilanes-effects of polymer types, water composition and temperature. Water (Basel) 13: 675.

Sun, B. and Zeng, E.Y. 2022. Leaching of PBDEs from microplastics under simulated gut conditions: Chemical diffusion and bioaccumulation. Environ. Pollut. 292: 118318.

Sun, B., Liu, J., Zhang, Y.Q., Leungb, K.M.Y. and Zeng, E.Y. 2021. Leaching of polybrominated diphenyl ethers from microplastics in fish oil: Kinetics and bioaccumulation. J. Hazard. Mater. 406: 124726.

Sun, B., Hu, Y., Cheng, H. and Tao, S. 2019. Releases of brominated flame retardants (BFRs) from microplastics in aqueous medium: Kinetics and molecular-size dependence of diffusion. Water Res. 151: 215.

Šunta, U., Prosenc, F., Trebše, P., Bulc, T.G. and Kralj, M.B. 2020. Adsorption of acetamiprid, chlorantraniliprole and flubendiamide on different type of microplastics present in alluvial soil. Chemosphere 261: 127762.

Takai, Y., Tokusumi, H., Sato, M., Inoue, D., Chen, K., Takamura, T. et al. 2022. Combined effect of diazepam and polystyrene microplastics on the social behavior of medaka (*Oryzias latipes*). Chemosphere 299: 134403.

Tiller, R. and Nyman, E. 2018. Ocean plastics and the BBNJ treaty-is plastic frightening enough to insert itself into the BBNJ treaty, or do we need to wait for a treaty of its own? J. Environ. Stud. Sci. 8: 411.

Tunali, M., Uzoefuna, E.N., Tunali, M.M. and Yenigun, O. 2020. Effect of microplastics and microplastic-metal combinations on growth and chlorophyll a concentration of *Chlorella vulgaris*. Sci. Total Environ. 743: 140479.

Valdez-Carrillo, M., Abrell, L., Ramírez-Hernández, J., Reyes-López, J.A. and Carreón-Diazconti, C. 2020. Pharmaceuticals as emerging contaminants in the aquatic environment of Latin America: a review. Environ. Sci. Pollut. Res. 27: 44863.

Verla, A.W., Enyoh, C.E., Verla, E.N. and Nwarnorh, K.O. 2019. Microplastic–toxic chemical interaction: a review study on quantified levels, mechanism and implication. SN Appl. Sci. 1: 1400.

Villegas, L., Cabrera, M., Moulatlet, G.M. and Capparelli, M. 2022. The synergistic effect of microplastic and malathion exposure on fiddler crab *Minuca ecuadoriensis* microplastic bioaccumulation and survival. Mar. Pollut. Bull. 175: 113336.

Wagstaff, A. and Petrie, B. 2022. Enhanced desorption of fluoxetine from polyethylene terephthalate microplastics in gastric fluid and sea water. Environ. Chem. Lett. 20: 975.

Wang F., Gao, J., Zhai, W., Liu, D., Zhou, Z. and Wang, P. 2020. The influence of polyethylene microplastics on pesticide residue and degradation in the aquatic environment. J. Hazard. Mater. 394: 122517.

Wang, H., Liu, P., Wang, M., Wu, X., Shi, Y., Huang, H. et al. 2021. Enhanced phototransformation of atorvastatin by polystyrene microplastics: Critical role of aging. J. Hazard. Mater. 408: 124756.

Wang, H.J., Lin, H.H.H., Hsieh, M.C. and Lin, A.Y.C. 2022. Photoaged polystyrene microplastics serve as photosensitizers that enhance cimetidine photolysis in an aqueous environment. Chemosphere 290: 133352.

Wang, J., Li, J., Wang, Q. and Sun, Y. 2020. Microplastics as a Vector for HOC Bioaccumulation in earthworm *Eisenia fetida* in soil: importance of chemical diffusion and particle size. Environ. Sci. Technol. 54: 12154.

Wang, J., Li, X., Li, P., Li, L., Zhao, L. Ru, S. et al., 2022. Porous microplastics enhance polychlorinated biphenyls-induced thyroid disruption in juvenile Japanese flounder (*Paralichthys olivaceus*). Mar. Pollut. Bull. 174: 113289.

Wang, Q., Zhang, Y., Wangjin, X., Wang, Y., Meng, G. and Chen, Y. 2020. The adsorption behavior of metals in aqueous solution by microplastics effected by UV radiation. J. Environ. Sci. (China). 87: 272.

Wang, S., Xie, S., Wang, Z., Zhang, C., Pan, Z., Sun, D. et al., 2022. Single and combined effects of microplastics and cadmium on the cadmium accumulation and biochemical and immunity of *Channa argus*. J. Zou, Biol. Trace Elem. Res. 200: 3377.

Wang, T., Yu, C., Chu, Q., Wang, F., Lan, T. and Wang, J. 2020. Adsorption behavior and mechanism of five pesticides on microplastics from agricultural polyethylene films. Chemosphere. 244: 125491.

Wang, Z., Ding, J., Razanajatovo, R.M., Huang, J., Zheng, L., Zou, H. et al. 2022. Sorption of selected pharmaceutical compounds on polyethylene microplastics: Roles of pH, aging, and competitive sorption. Chemosphere 307: 135561.

Wang, Z., Gao, J., Zhao, Y., Dai, H., Jia, J. and Zhang, D. 2021. Plastisphere enrich antibiotic resistance genes and potential pathogenic bacteria in sewage with pharmaceuticals. Sci. Total Environ. 768: 144663.

Xia, B., Zhang, J., Zhao, X., Feng, J., Teng, Y., Chen, B. et al. 2020. Polystyrene microplastics increase uptake, elimination and cytotoxicity of decabromodiphenyl ether (BDE-209) in the marine scallop *Chlamys farreri*. Environ. Pollut. 258: 113657.

xin Wang, Y., jian Liu, M., Hui Geng, X., Zhang, Y., Qi Jia, R., Ning Zhang, Y. et al. 2022. The combined effects of microplastics and the heavy metal cadmium on the marine periphytic ciliate *Euplotes vannus*. Environ. Pollut. 308: 119663.

Xu, B., Liu, F., Brookes, P.C. and Xu, J. 2018. Microplastics play a minor role in tetracycline sorption in the presence of dissolved organic matter. Environ. Pollut. 240: 87.

Yan, W., Hamid, N., Deng, S., Jia, P.P. and Pei, D.S. 2020. Individual and combined toxicogenetic effects of microplastics and heavy metals (Cd, Pb, and Zn) perturb gut microbiota homeostasis and gonadal development in marine medaka (*Oryzias melastigma*). J. Hazard. Mater. 397: 122795.

Yu, H., Zhang, Z., Zhang, Y., Fan, P., Xi, B. and Tan, W. 2021. Metal type and aggregate microenvironment govern the response sequence of speciation transformation of different heavy metals to microplastics in soil. Sci. Total Environ. 752: 141956.

Yuan, W., Zhou, Y., Chen, Y., Liu, X. and Wang, J. 2020. Toxicological effects of microplastics and heavy metals on the *Daphnia magna*. Sci. Total Environ. 746: 141254.

Zhang, C., Lei, Y., Qian, J., Qiao, Y., Liu, J., Li, S. et al. 2021. Sorption of organochlorine pesticides on polyethylene microplastics in soil suspension. Ecotoxicol. Environ. Saf. 223: 112591.

Zhang, S., Ding, J., Razanajatovo, R.M., Jiang, H., Zou, H. and Zhu, W. 2019. Interactive effects of polystyrene microplastics and roxithromycin on bioaccumulation and biochemical status in the freshwater fish red tilapia (*Oreochromis niloticus*). Sci. Total Environ. 648: 1431.

Zhang, X., Su, H., Gao, P., Li, B., Feng, L., Liu, Y. et al. 2022. Effects and mechanisms of aged polystyrene microplastics on the photodegradation of sulfamethoxazole in water under simulated sunlight. J. Hazard. Mater. 433: 128813.

Zhou, Y., Liu, X. and Wang, J. 2020. Ecotoxicological effects of microplastics and cadmium on the earthworm *Eisenia foetida*. J. Hazard. Mater. 392: 122273.

Zhu, Z.L., S.C. Wang, F.F. Zhao, S.G. Wang, F.F. Liu and G.Z. Liu. 2019. Joint toxicity of microplastics with triclosan to marine microalgae *Skeletonema costatum*. Environ. Pollut. 246: 509.

Zong, X., Zhang, J., Zhu, J., Zhang, L., Jiang, L., Yin, Y. et al. 2021. Effects of polystyrene microplastic on uptake and toxicity of copper and cadmium in hydroponic wheat seedlings (*Triticum aestivum* L.). Ecotoxicol. Environ. Saf. 217: 112217.

Zou, J., Liu, X., Zhang, D. and Yuan, X. 2020. Ecotoxicological effects of microplastics and cadmium on the earthworm *Eisenia foetida*. Chemosphere 248: 126064.

CHAPTER 5

Challenges and Strategies for Degradation of Microplastics in Environment

Harshika Sharma, Divya Sharma, Rutika Sehgal and
*Reena Gupta**

1. Introduction

In the early 1970s, the first scientific report of tiny plastic fragments floating on the ocean surface appeared (Carpenter and Smith, 1972). Subsequent papers detailed the experiments that identified plastic fragments in birds (Harper and Fowler, 1987). The term "microplastic" first appeared in use while referring to marine debris. In the 1990s, it was mentioned in cruise reports of the Sea Education Association about the findings of the survey of South African beaches (Ryan and Moloney, 1990). There was no precise description of the size established at the time, but the phrase generally denoted materials that could only be identified with the use of a microscope. Although the term microplastic was not technically recognized earlier, but now it is extensively used to denote small fragments of plastic in the millimeter to sub-millimeter size range. Even though it has not been publicly suggested for adoption by the international research community, a more rigorous definition of plastics based on science mentions the nano-, micro-, meso-, macro- and mega-size ranges as depicted in Fig. 5.1.

Department of Biotechnology, Himachal Pradesh University, Summerhill, Shimla, 171005, India.
  E-mail: sharshika351@gmail.com; sharmadivyanvs@gmail.com; rutikasehgal@yahoo.in
* Corresponding author: reenagupta_2001@yahoo.com

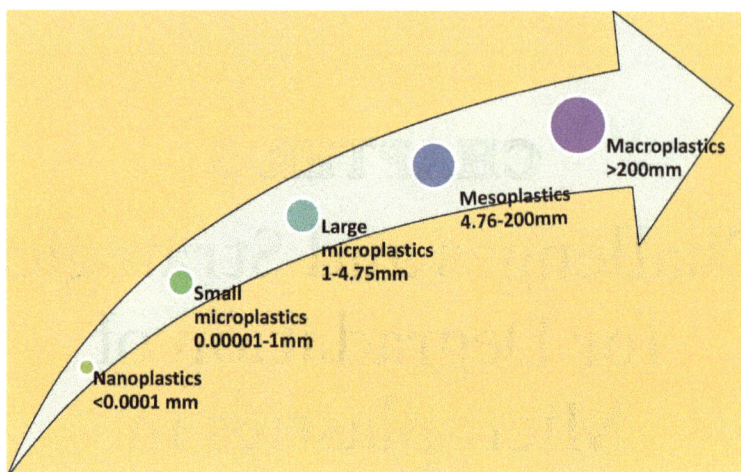

Figure 5.1 Different types of plastics based on their size.

Professor Richard Thompson, a marine biologist at the University of Plymouth in the United Kingdom, coined the term "microplastics" in 2004 (Thompson, 2018). According to the National Oceanic and Atmospheric Administration (NOAA), microplastics are plastic fragments that are less than 5 mm in length (Collignon et al., 2014). The term "macroplastics" is used to distinguish smaller plastic waste from larger plastic waste, such as plastic bottles. They enter the environment through many different channels, such as cosmetics, clothing, food packaging and industrial wastes. Clothing and textile account for 35% of all ocean microplastics, primarily as a result of the deterioration of polyester, acrylic or nylon-based clothing, which occurs during washing.

Microplastics are known to accumulate at higher quantities in the environment, notably in aquatic and marine habitats, where they cause water pollution. On the other hand, microplastics also build up in the troposphere and terrestrial ecosystems. Microplastics are likely to be ingested, incorporated, and accumulated in the bodies and tissues of many species because of the slow decomposition of plastics (typically more than hundreds to thousands of years). Additionally, harmful chemicals from runoff and the ocean can bio magnify their way up the food chain with the microplastics (Chamas et al., 2020). Although research has been conducted on the topic, the processing and transfer of microplastics in nature are not fully understood. A deep layer ocean sediment survey in China showed the presence of plastics in deposition layers that are thousands of years older than the invention of plastics. It suggests that the amount of microplastics in the surface sample ocean survey are probably underestimated (Xue et al., 2020). Away from their source, microplastics have also been found in high mountains. In 2014, it was calculated that there were between 93,000 and 236,000 metric tons or 15 to 51 trillion tiny pieces of microplastics floating around in the oceans (Karbalaei et al., 2018).

2. Classification of Microplastics

Microplastics are generally classified into two types as shown in Fig. 5.2. Plastic pieces or particles that are 5.0 mm in size or smaller before entering the atmosphere are considered primary microplastics. Some examples include plastic pellets, microbeads and clothing-related microfibers (Boucher and Friot, 2017). When bigger plastic products are exposed to natural weathering processes, secondary microplastics are created into the environment. Some examples of secondary microplastics sources include water and soda bottles, plastic bags, fishing nets, microwaveable containers, tea bags and tires (Kovoochich et al., 2021).

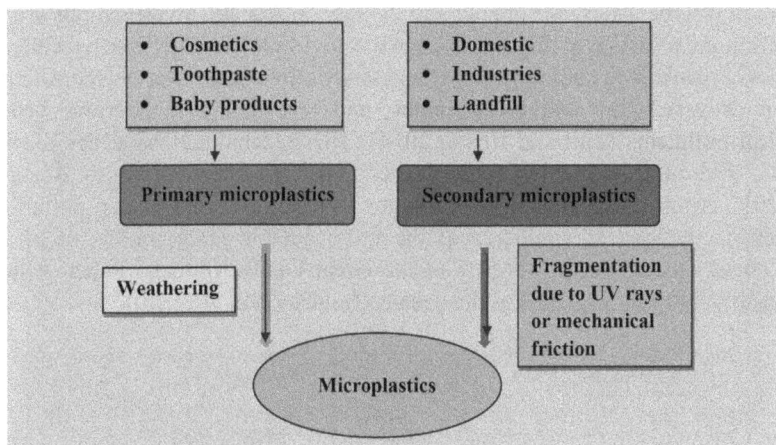

Figure 5.2 Classification of microplastics on the basis of formation.

2.1 Primary Microplastics

Primary microplastics are minute pieces of plastic. They are typically used in air blasting technology, cosmetics and facial cleansers. There have been reports of their use in medicine as a drug delivery system (Patel et al., 2009). Natural ingredients like powdered nut shells, oats and pumice have been replaced by microplastic "scrubbers" which are present in exfoliating hand cleansers and facial scrubs. In order to remove rust and paint from machines, engines, and boat hulls, primary microplastics such as acrylic, melamine and polyester scrubbers are used in the air blasting process. Due to repeated use, heavy metals like cadmium, lead and chromium are deposited on these scrubbers, which cause them to shrink in size and lose their cutting ability (Cole et al., 2011). Even though many businesses have promised to cut down on microbeads production, there are still plenty of bioplastic microbeads that will degrade slowly over time, much like regular plastic.

2.2 Secondary Microplastics

Small pieces of plastic that are produced as a result of the breakdown of larger plastic waste are known as secondary plastics. The structural integrity of plastic trash can eventually become so compromised that it is no longer visible to the naked eye due to a combination of physical, biological and chemical degradation, including photo-oxidation brought on by exposure to sunlight (Masura et al., 2015). Microplastics are thought to deteriorate and become smaller in size, yet the smallest microplastic apparently identified in the oceans at the moment is 1.6 micrometers in diameter (Conkle et al., 2018).

3. Sources of Microplastics

Over 80% of the microplastic in the environment is found in textiles, tires and city dust, which are the main sources of microplastic pollution (Chamas et al., 2020). The sources of microplastics are summarized in Fig. 5.3. Aquatic investigations are used to confirm the presence of microplastics in the environment. They involve collecting samples of plankton, examining muddy and sandy sediments, observing consumption pattern of vertebrates and invertebrates, and researching interactions between chemical pollutants (Ivar and Costa, 2014). These techniques have demonstrated that the environment contains microplastics from a number of sources. According to a 2017 report, microplastics are a major source of marine plastic pollution in developed countries as compared to the more notable larger pieces of plastics. They could account for up to 30% of the Great Pacific Garbage Patch, which is responsible for polluting most of the oceans (Boucher and Friot, 2017).

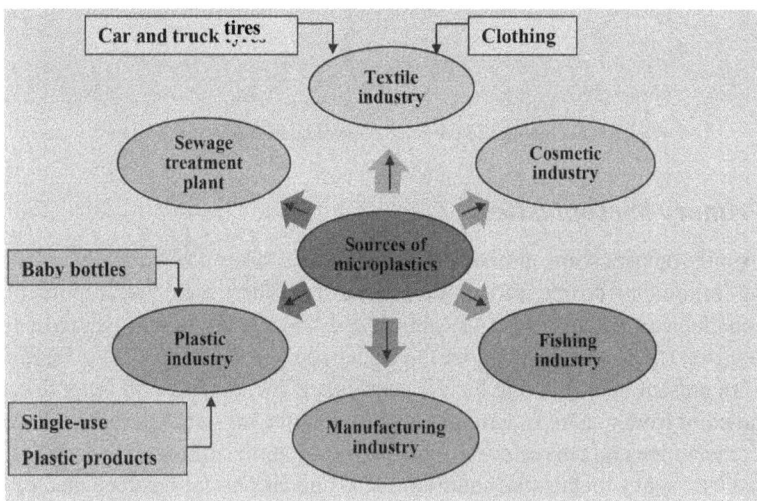

Figure 5.3 Sources of microplastics.

3.1 Textile Industry

An average human exposure to textile fibers has been examined in both indoor and outdoor contexts. Indoor concentrations of microfibers ranged from 1.0 to 60.0 fibers/m^3, while outdoor concentrations ranged from 0.3 to 1.5 fibers/m^3 (Dris et al., 2017). The rate of deposition ranges from 1586 to 11,130 fibers per day/m^3, which added up to about 190 to 670 fibers/mg of dust. The most significant concern regarding these concentrations is that their increased exposure to young children and the elderly people, will negatively impact health. It has been discovered that these microfibers stick to larger animals like whales or even to small creatures like zooplankton (Boucher and Friot, 2017). Polyester, a cheap replacement of cotton that can be made cost effective, is a major fiber that persists throughout the textile industry. In terrestrial birds and marine ecosystems, these fibers play a significant part in the permanence of plastic pollution.

3.1.1 Car and Truck Tires

Tire wear and tear plays a big role in the release of microplastics into the environment. Microplastic emission to the environment is thought to range from 6,100 to 15,400 tons annually. The estimated per capita emission ranges from 0.23 to 4.7 kg/year, with a global average of 0.81 kg/year. When compared to other sources of microplastics, such as tires on aeroplanes (2%), artificial grass (12 to 50%), brakes (8%) and road markings (5%), car tires can degrade to 100%, producing significantly more microplastics in the atmosphere. A recent field study discovered that glass beads were used to shield road markings and their contribution to secondary microplastics emission was only 0.1 to 4.3 g/person/year (Burghardt et al., 2022) or about 0.7% of all emissions of secondary microplastics. It is believed that 5–10% of plastic waste that ends up in the oceans comes from the wear and tear of tires. It is expected to form about 3–7% of particulate matter ($PM_{2.5}$) in the air, which can contribute to the 3 million deaths from air pollution that World Health Organization (WHO) had predicted in 2012 (Wang et al., 2019).

3.1.2 Clothing

Many synthetic fibers such as spandex, nylon, acrylic and polyester have been demonstrated to be capable of shedding off microplastic in the environment (Periyasamy et al., 2022). Each clothing in a cycle of washing can shed more than 1,900 microplastic fibers, with fleeces shedding the most fibers, more than 170% higher than other garments (Katnelson, 2015). Research on whether washing machine filters can reduce the number of microfibers that are processed by sewage treatment plants has also been considered by manufacturers of washing machines (Dris et al., 2017). Over 700,000 fibers can be released per wash with an average washing load of 6 kg (13 pound) (Napper and Thompson, 2016). This has been related to potential health impacts due to the emission of monomers, dispersive dyes, mordants and plasticizers during manufacturing. These fibers have been found to account for 33% of all fibers found in the indoor environment (Browne et al., 2011).

3.2 Cosmetic Industry

Microplastics, which are commonly used as "microbeads" or "micro-exfoliates" in face washes, hand soaps and other personal care products, have replaced natural exfoliating chemicals in the cosmetic business. Although they can also be made of nylon, polypropylene and polyethylene terephthalate (PET), microbeads are generally made of polyethylene, a popular plastic component (Fendall and Sewell, 2009). Despite the fact that many businesses have made a commitment to stop using microbeads in their products, research showed that at least 80 different types of facial scrubs are still available with microbeads as a key ingredient (Anderson et al., 2016). The United Kingdom alone releases 80 metric tons of microbeads every year, which not only harms wildlife and the food chain, but also increases the levels of toxicity. Microbeads have been shown to absorb harmful substances like pesticides and polycyclic aromatic hydrocarbons. According to the European Chemicals Agency (ECHA) and the United Nations Environment Programme (UNEP), there are more than 500 microplastic compounds that are regularly used in cosmetics and personal care products (Rochman et al., 2015). Microbeads are illegal in some countries, such as New Zealand, because even when they are removed from cosmetic products, they are still present in minute quantities which makes the product risky. For instance, when acrylate copolymers were used in cosmetic products, they had toxic effects on human and other organisms.

3.2.1 Fishing industry

Marine industries, recreational and commercial fishing, marine transportation, and marine vessels are direct sources of microplastics that will end up in the ocean and endanger biodiversity. Fishing equipment, which includes nylon netting and plastic monofilament lines, are the types of plastic waste with maritime origin (sometimes called ghost nets). Since they are typically immiscible with water, they can float at various depths in the ocean. Many countries have reported that microplastics from the industry and other sources have been found in various types of seafood. In Indonesia, microplastics have been found in 55% of all fish species, which is comparable to America where 67% fish species have been contaminated with microplastics (Rochman et al., 2015).

Microplastics and toxins bioaccumulate in the food chain. In one investigation, the short-tailed shearwater's stomach was examined for the presence of plastic-derived substances such as polybrominated diphenyl ethers (PBDEs). According to research, one-fourth of the birds had higher brominated congeners that were typically present in their prey (Tanaka et al., 2013).

3.3 Manufacturing Industry

Tiny resin pellets and granules are the primary raw materials used in the production of plastic items. In 2019, the global production of plastic was approximately 368 million tons, with Asia accounting for 51% of the total production. The largest producer in the world, China, created 31% of the global production of plastic (Derraik, 2022). Raw materials of plastic production can enter the aquatic environment by unintentional spills during land or sea transportation, incorrect use as packaging materials and direct outflow from processing plants. Many industrial locations where raw plastics are regularly used, are close to water bodies. These substances have the potential to contaminate rivers if they are spilled during manufacturing (Conkle et al., 2018). Recently, the American Chemistry Council and Society of the plastic industry joined forces to launch Operation Clean sweep, which aims to encourage businesses to commit to zero pellet loss during their processes (Cole et al., 2011).

3.4 Plastic Industry

In one investigative process, microplastic pollutants were discovered in 93% of the samples of bottled water from 11 various brands. Microplastic is twice as prevalent in water from plastic bottles as it is in water from taps. Presumably, the bottling and bundling of the water contributes to some of the microplastics contamination (Mason et al., 2018).

3.4.1 Baby Bottles

Researchers found that in 48 regions, infant feeding bottles made of polypropylene exposed newborns to microplastics at rates ranging from 14,600 to 4,550,000 particles per capita per day. Warmer liquids cause more release of microplastics and products made of polypropylene, such as lunchboxes contribute almost equivalently (Carrington, 2020). Researchers in 2021 discovered that repeated steam sterilization causes silicone rubber baby bottle nipples to deteriorate over time, shedding micro- and nano-sized silicone rubber particles. A child would consume more than 660,000 particles if it consumed food from such heat-degraded nipples for a year (Su et al., 2021).

3.4.2 Single-use Plastic Products

Common single-use plastic items, like paper cups with an interior lining of a thin plastic film, discharge trillions of microplastic particles into food during normal use (Zangmeister et al., 2022). Waterways are contaminated by single-use plastics. Single-use plastics reduction guidelines at the local and state levels have been cited as effective legislative measures that localities can implement to lessen plastic pollution (Rochman et al., 2015).

3.5 Sewage Treatment Plant

Typically, domestic sewage is treated at sewage treatment facilities, which are widely recognized as wastewater treatment plants (WWTPs). They use a number of physical, chemical and biological techniques to remove contaminants (Carr et al., 2016). Most plants in developed regions have both primary and secondary treatment steps. Oils, sand, and other large solids are removed physically during the first stage of treatment using standard filters, clarifiers and settling tanks. Organic matter is broken down by biological processes with the use of bacteria and protozoa in the secondary treatment. Secondary source technologies include trickling filters, constructed wetlands and activated sludge systems (Habib et al., 1998). Disinfection and nutrient removal (nitrogen and phosphorus) are two optional tertiary treatment procedures as shown in Fig. 5.4.

Microplastics have been found in primary and secondary treatment phases of the plants. In a groundbreaking 1998 study, microplastic fibers were predicted to be a long-term indicator of sewage sludge and wastewater treatment plant outfalls. According to a study, microplastics have 99.9% clearance and are released back into the environment at a rate of roughly one particle per liter (Estahbanati and Fahrenfeld, 2016). According to one study, the primary treatment stage, which uses

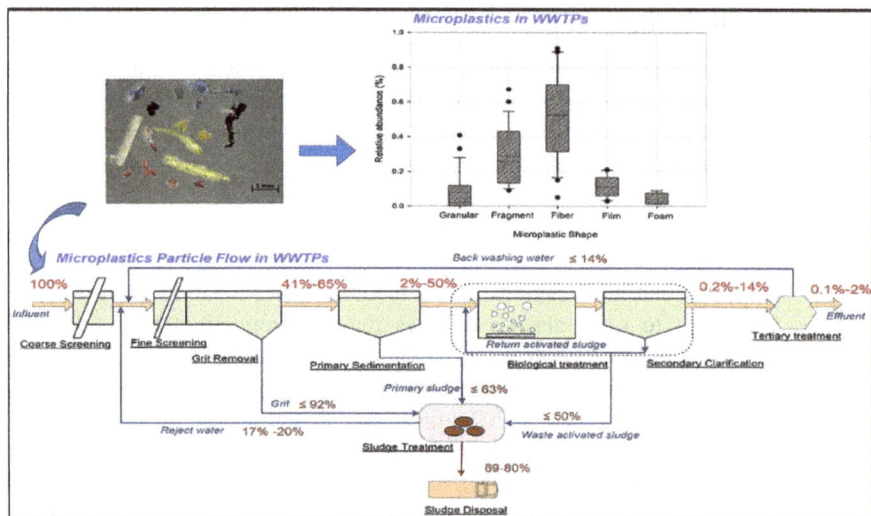

Figure 5.4 Sewage treatment plant (Sun et al., 2019).

solid skimming and sludge settling, actually removes most of the microplastics. When wastewater treatment facilities are running efficiently, they do not contribute to an excessive amount of microplastics to the sea and surface water ecosystems. Samples of sewage sludge disposal sites from six different continents were collected and researchers found an average of one microplastic particle per liter (Murphy et al., 2016).

4. Analysis of Microplastics

Due to the prevalence, bioaccumulation and potential to transport harmful substances in seawater, microplastics have emerged as a major global threat in recent years. The physical and chemical characterization of separated particles forms the basis for the analysis of microplastics. The methods use techniques such as spectroscopy (Fig. 5.5). To decrease the time and effort needed for identification and to find microplastics in environmental samples, new methods should be improved. Strategies to analyze the microplastics in environment are discussed below:

4.1 Sampling of Microplastics

The three main types of microplastic sampling techniques are volume-reduced sampling, bulk sampling and selective sampling (Hidalgo-Ruz et al., 2012a). By using visual recognition, the direct method allows for the selection of microplastics from environmental samples. This method is suitable for gathering plastic samples of surface sediment, especially those with a particle size of 1^{-6} mm (Karlsson et al., 2017). In the concentrated sample approach, a lot of samples are filtered and sorted at the sampling site and the target components are kept for analysis (Redondo-Hasselerharm et al., 2018). The "large sample method" is a type of sampling technique which retains all the samples and is appropriate for situations when there

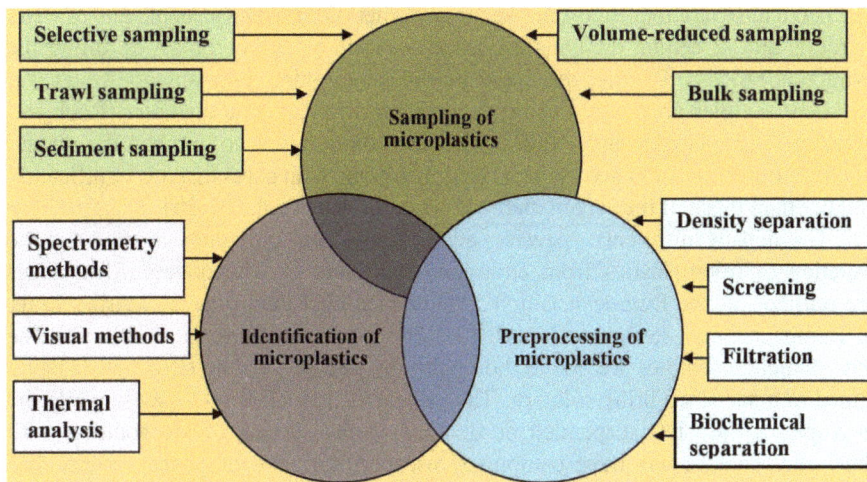

Figure 5.5 Analysis of microplastics.

is a small amount of plastic in the sample that is difficult to see with the naked eyes (Dümichen et al., 2015). Plankton nets with various mesh sizes are typically used to collect the microplastic sample in the aquatic environment. Huge amounts of water samples are taken in order to ensure that the samples collected are representative. Trawl sampling equipment like manta nets and neuston nets etc., are generally used for surface water analysis (Faure et al., 2015).

The manta net is typically used in calm water to prevent damage of the sampling equipment by wave motions or to prevent an incorrect volume of water being sampled due to unstable equipment. The Neuston net is generally used during intense storms. The Bongo net is chosen for intermediate water and benthic trawling is used for deep water at the bottom (do Sul and Costa, 2014). The trawl is positioned towards the windward side of the hull, to prevent sample contact with the hull. Typically, it is pulled for 15 to 30 min at a speed of 2 knots (Auta et al., 2017a).

The collected microplastics come in a number of sizes, according to the various mesh sizes. The sieve pore size determines the size of the particles collected in the trawl and the volume of the filtered water sample. The hole diameter of the sampling screen is approximately 300 mm. A smaller hole diameter increases the likelihood of clogging. Stainless steel shovels, box samplers and other tools are used for sediment sampling (Hidalgo-Ruz et al., 2012a). Beach microplastics are typically studied using a square quadrat sampling. To thoroughly analyze the microplastic interference on the target beach, several samples are combined. After sampling, the preliminary screening can be completed on-site and the samples can also be saved for later laboratory analysis.

4.1.1 Preprocessing of microplastics

The collected environmental samples typically contain interfering contaminants, making it impossible to evaluate the microplastic samples directly; as a result, the microplastics should be extracted and separated from the samples. Pretreatment methods that are normally used include filtering, screening, density separation, biochemical separation (digestion) and many others (Hidalgo-Ruz et al., 2012a). The filtration and screening methods use small pores to catch microplastics during the extraction process. To filter out larger particles and other contaminants, water and sediment samples are filtered using sieves with different pore diameter (Roch and Brinker, 2017). Copper and stainless steel are commonly used to make the screen. The filtration process makes use of a filter membrane with pores that are significantly smaller than those of the screen methods (Hidalgo-Ruz et al., 2012b).

The density difference between microplastics and pollutants is used for the isolation of microplastics from environmental samples. The pretreatment stages are outlined in the European Union's "Guidance on Monitoring of Marine Litter in European Seas" (Schulz et al., 2013). The exact procedure is as follows. The environmental samples are mixed thoroughly by swirling and agitated before being added to a saturated brine solution. The system is then allowed to settle while the microplastics are left suspended or floating on the surface of the solution until the heavy components have completely escaped the aqueous phase system. The microplastics from the upper solution are then gathered. The density of microplastic ranges from 0.8 to 1.4 g cm^{-3}, which includes silica gel, PVC, and others (Hidalgo-Ruz et al., 2012a).

Using high-density saturated salt solutions, microplastic particles can be recovered from more dense sediments (2.65 g cm^{-3}). Saturated NaCl solution (density 1.2 g cm^{-3}) is a commonly used density flotation liquid because of it is inexpensive and is wide accessible (Nuelle et al., 2014). High-density polymers like PVC and PET, however, cannot be completely removed during separation due to their low density. High-density plastic components can be extracted more effectively by saturated ZnCl$_2$ and NaI solutions having densities 1.4–1.6 g cm^{-3} and 1.6–1.8 g cm^{-3} respectively, which is greater than saturated NaCl solution (Van Cauwenberghe et al., 2013).

In order to reduce substrate interference in environmental samples, acid digestion, alkaline digestion and enzymatic digestion are normally used (Enders et al., 2017). The digestion procedure is primarily used to prepare biological samples. A large percentage of the biodegradable components appear to have been digested in some investigations using a 35% H$_2$O$_2$ solution (Nuelle et al., 2014). Treatment of samples with acidic or alkaline solutions are known as acid digestion and alkaline digestion, respectively. Use of a strong acid or a strong base is restricted due to the variable chemical resistance of different types of microplastics. For instance, strong alkali or acid would cause the polyformaldehyde and polycarbonate to react. Styrene and other components can be recovered at a rate of 90 to 98% in some studies that used acid digestion to process samples, but nylon fibers could recover at a rate of almost zero percentage (Desforges et al., 2015). There are numerous uses of the enzyme digestion process. Several studies compared the effects of acid, alkali and enzyme digestion on marine organisms and found that enzyme digestion can break down more than 97% of plankton samples without damaging any microplastic fragments, whereas both acid and alkaline digestion have some influence on the digestion outcomes (Cole et al., 2014).

4.2 Identification of Microplastics

Microplastics come from many sources and vary greatly in terms of their color, form and composition. Analytical identification is therefore a crucial component of microplastic research. There are several methods for identifying microplastics, including visual approaches, spectrometry methods and thermal studies.

4.2.1 Visual Methods

A quick and simple identification technique for microplastic analysis is visual inspection (Hidalgo-Ruz et al., 2012b; Lee et al., 2014). This method is often used to detect microplastics with a particle size of 1 to 15 mm. Microplastics are normally analyzed under a microscope. Based on their color and shape, the microplastics in the biosphere are categorized and identified. However, the use of visual inspection methods has many drawbacks, including the failure to acknowledge many microplastics visually and the influence of experimentalist's visual variations on identification outcomes. When there is interference of minute microplastics, the visual inspection method is less successful (Lee et al., 2014). Studies show that as the size of microplastic particles decreases, the error of the visual examination approach increases (Dekiff et al., 2014). As a result, eye inspection is not recommended as a stand-alone identification method in detection of microplastics. Despite some drawbacks, visual identification of microplastics is the most popular method for their analysis and identification because of its inexpensiveness and simplicity.

4.2.2 Sprectrometry Methods

Two popular methods for figuring out the chemical formation of microplastics are Raman spectroscopy and Fourier Transform Infrared Spectroscopy (FTIR) (Cabernard et al., 2018). FTIR routinely uses Attenuated Total Reflection (ATR), transmission and reflection to detect microplastics (Song et al., 2015b). ATR provides the most reliable surface spectral information. It is widely used to determine microplastic particles when they are larger than or equal to 300 mm. By using this technique, it is possible to complete the analysis quickly and accurately within a minute. Transmission can produce spectra with high resolution, but it needs the material to be penetrated by infrared light. Analysis of thicker, more opaque samples is done in the reflection mode. The benefit of FTIR analysis is that it is unaffected by fluorescence interference and can identify the types of plastic particles quickly and precisely (Tagg et al., 2015).

In addition to providing details on the quality and structure of plastic particles, FTIR can also detect the degree of weathering of microplastics, which is helpful in tracing the origin and path of microplastics (Kappler et al., 2016). While microplastics in the environment with smaller particle size cannot be recognized, FTIR can recognize plastic particles with sizes greater than or equal to 20 mm (Tagg et al., 2015). Furthermore, moisture in the sample affects the result of identification and analysis and it is challenging to identify opaque microplastic particles. Moreover, it takes a lot of time and effort to find and examine each individual microplastic particle beneath the probe (Fischer and Scholz-Bottcher, 2017).

High molecular weight polymers are studied in depth by using Raman Spectroscopy which enables further identification. As opposed to Infrared Spectroscopy, Raman Spectroscopy can identify microplastics with size of particles smaller than 20 mm (Fries et al., 2013). This detection method has the drawback of being slow, which limits its usefulness. Fluorescence-producing materials cannot be found using Raman Spectroscopy. In addition, additives and colors in microplastics can make them difficult to identify (Gü et al., 2015). Furthermore, a significant factor that prevents the identification of the microplastics is that some of the photosensitive materials in the sample produce fluorescence after being activated by the instrument (Song et al., 2015a).

4.2.3 Thermal Analysis

Pyrolysis Gas Chromatography/Mass Spectrometry (Pyr-GC-MS) is another popular microplastic analysis technique. Pyr-GC-MS analyzes the breakdown products of heated samples to pinpoint the chemical elements of microplastics and various types of additives and polymers (Fries et al., 2013). The Pyr-GC-MS method can only ascertain the mass fraction of polymers in comparison to spectroscopy. The quantity and dimensions of microplastics can be determined using spectroscopy, but the process is time-consuming. The Pyr-GC-MS approach also has some drawbacks (Dümichen et al., 2015). Different polymers could produce similar breakdown products, which can cause a misunderstanding. The machine can only hold a tiny amount of the sample, so only one plastic particle can be evaluated at one time. This is completely inadequate for the treatment and analysis of enormous number of samples.

Although the above-mentioned microplastic analysis techniques are commonly used, they have many limitations and shortcomings in practical applications. These techniques, for instance, are limited to the analysis of one or a small number of parameters. When components are complex and samples contain little amounts of microplastics, the reliability of the data that was collected is limited. Therefore, there is more scope in the development of multi-parameter analytical techniques for complex substrates with small sample volumes. It is necessary to enhance the current detection and analytical processes as well as to research and create new techniques and tools. A unified methodology should also be developed to improve quality assurance and control of the entire analytical process in order to enhance the accuracy, reliability, and comparability of monitored data.

5. Environmental Effects of Microplastics

Plastic pollution is a worldwide environmental issue. While most of us are aware of the dangers of plastic bags and other single-use plastics, microplastics are a less visible kind of plastic pollution. Microplastics are found in drinking water, air, and food. Researchers anticipated in 2020 that by 2040, our oceans would contain more than 600 million tons of microplastic garbage (Gutow et al., 2016). Microplastics are just as harmful as conventional plastic debris, posing several damaging consequences on living organisms and the environment. The interaction of microplastics with the soil, disrupt its functioning, degrade its health, and affect the earth's living species. It is also responsible for the extinction of several marine creatures. Humans are not immune to these small plastic particles, as research has proven that microplastics can damage human cells as depicted in Fig. 5.6. Some of the effects of microplastic are discussed as follows:

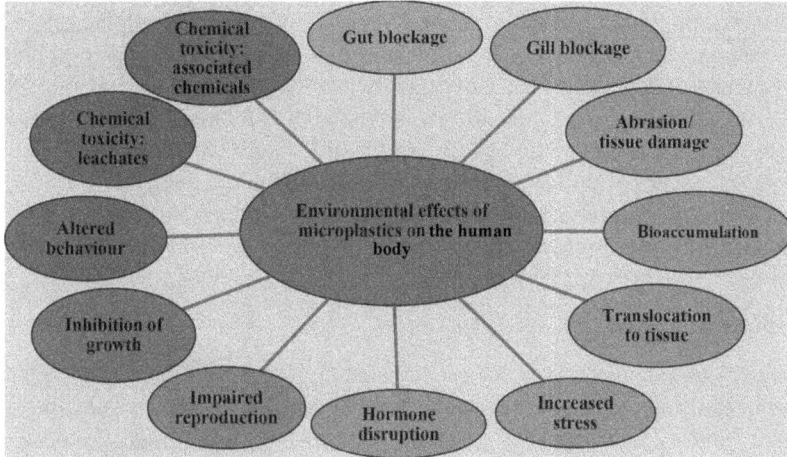

Figure 5.6 Environmental effects of microplastics on human body.

Biological attachment

Microorganisms are frequently found in the form of biofilms. Microplastic particles with microbes on their surfaces can enter the food web and cause damage to higher organisms (Bejgarn et al., 2015). The outer surface of the microplastic can sustain biotic materials like algae, bacteria, and viruses (Gutow et al., 2016). There are two basic reasons why microbes grow and adhere on the microplastic surface. The materials used to manufacture plastic particles, make them ideal for adsorbing contaminants such as inorganic nutrients and organic compounds. These serve as a foundation for bacterial growth on them (Bakir et al., 2014). When compared to a planktonic life in the ocean, microplastics can provide sessile bacteria a more stable home. According to studies, microplastics that enter the water are immediately covered by microorganisms and after about 7 d, a stable biofilm is formed (Holmes et al., 2012).

Microplastics have a complex composition of surface-attached microbes and are difficult to breakdown in the environment (Auta et al., 2017a). The nutritional need that the microbes meet when adhered to the plastic surface are influenced by the water quality, geographic location, and seasonal fluctuations. It is predicted that marine microplastics contain a total of 1000–15,000 trillion of microorganisms (Mincer et al., 2016). The microbial communities attached to microplastics can be identified by using the high-throughput sequencing technology and scanning electron microscopy. These have been shown to contain a number of organisms, including symbiotic and autotrophic ones (Zettler et al., 2013).

In addition to biological material, abiotic contaminants such as perfluorinated chemicals (PFCs) and organic pollutants, can be attached to the surface of microplastics. In comparison to plastics made of polyethylene terephthalate (PET) and polyvinyl chloride (PVC), those composed of polypropylene (PPE) and polyethylene (PE) have greater adsorption ability for polycyclic aromatic hydrocarbons and polychlorinated biphenyls. The organic pollutants' ability to adhere to plastic particles depends on their surface morphology and surface area (Peng et al., 2019). Microplastics become a source of pollution as a result of the adhesion of pollutants (organic pollutants, heavy metals, PFCs, PPCPs etc.), because the particles allow adsorbed pollutants to travel over great distances, which in turn affects the distribution of pollutants in global habitats. Long-distance microplastic migration is brought on by external forces like air flow and water power. The risk of biological invasion is increased by the ability of microplastics to transport attached microorganisms to various biogeographic regions along with their current travel routes (Law and Thompson, 2014).

5.1 Biological Intake

Particles with a smaller size can be accidentally consumed, as plankton feeders frequently mistake microplastics in water for food (Engler, 2012). Thus, ingestion of microplastics has been described in zooplankton, fish, birds, protozoa, and marine mammals. The likelihood that zooplankton will ingest microplastics can vary depending on their abundance, particle size and environmental behavior. A small amount of microplastic will remain in the body after ingestion, but the majority will be excreted after consumption and leave the organism unaltered (Bouwmeester et al., 2015). Microplastics have the capacity to invade a number of tissues and organs and could have toxic effects at the cellular and molecular level (Galloway et al., 2017). The food chain can be traversed by microplastics due to zooplankton's biological ingestion. Multiple dangers can result from ingesting microplastics or microplastic-loaded prey (Farrell and Nelson, 2013). Due to the accumulation and biotransportability of microplastics in living things, research is being done to determine the risks associated with microplastics getting into people's bodies through the food chain.

In addition to causing physical harm, the action of surfactants in living animals' digestive tracts cause the rapid release of several compounds found in microplastics (Wright and Kelly, 2017). These compounds are fat-soluble and are stored in lipid-rich tissues before being concentrated in species further up the food chain. Chemicals adsorbed on microplastics can substantially disrupt the endocrine system of the Japanese Medaka (*Oryzias latipes*), affecting both the male and female reproductive potential (Rochman et al., 2014).

5.2 Toxicological Effect

Plastics are manufactured with a number of additives to increase their heat or corrosion resistance (Lithner et al., 2011). Additives are used to make it softer or to enhance its other qualities. Due to this fact additives are not glued to the synthetic products, over time, microplastics could cause their release into the environment.

Unfortunately, many of these additives are extremely dangerous, such as the antioxidant nonylphenol (Horton et al., 2017). In areas with higher concentrations of microplastics, where the particles are exposed to UV light and at higher temperatures, there is a higher likelihood that plastic additives will be released.

Microplastics recovered from the ocean contain plasticizers and can also have an impact on terrestrial ecosystem (Eriksen et al., 2013). Plastic products have the potential to release plasticizers into the soil environment, endangering terrestrial systems, while effluents may eventually allow contaminants to enter marine systems (Wang et al., 2015). Plasticizers like phthalates have the potential to harm amphipod and crustacean development, impair animal reproduction and result in genetic mutations (Halden, 2010).

Plasticizers harm plants as well as animals. Tetrachlorophenol, for instance, is a thermoplastic additive for heat resistance that is directly toxic to phytoplankton (Rehse et al., 2016). Research shows that microplastics can significantly increase the sedimentation rate of cryptophytes, while significantly reducing the sedimentation rate of diatoms (Long et al., 2015). Microplastics could enter the human body through ingestion of food, skin contact and respiratory tract inhalation. They can also enter the human body through the food chain. Human health is threatened by eating fish that has microplastics in it because it can lead to cell necrosis, cytotoxic effects, inflammation, etc. (Liboiron et al., 2016).

5.3 Combined Pollution

Contaminants adsorbed on the surface of the particles could have a more complex toxicological effect in addition to the environmental effects of endogenous damaging compounds produced by microplastics. Due to their hydrophobic nature and large specific surface area, microplastics quickly bind to hydrophobic persistent organic contaminants (Mason et al., 2016). When such contaminants are found in water or soil, adsorption to microplastics can raise local concentrations of these contaminants (Andrady, 2011). The particle size, plastic type and weathering intensity are variables that affect the ability of hydrophobic contaminants to bind to surfaces. These pollutants provide a major risk to organisms because they are both highly toxic and persistent. Microplastics and the hydrophobic contaminants adsorbed can be directly ingested by organisms and have negative effects. For instance, microplastics can expose mussels (*Mytilus galloprovincialis*) to polycyclic aromatic hydrocarbons, having an adverse effect on many organs, such as the digestive system (Avio et al., 2015).

6. Challenges with Microplastics

Due to their widespread use and possible harm to the ecosystem, microplastics in the environment are a serious and urgent problem that is receiving attention from the public, government, researchers, and funding organizations. These seemingly harmless yet highly dangerous microplastics are likely to be ingested by living beings and cause catastrophic harm to ecological systems. Even worse, the hydrophobicity and enormous specific surface area of microplastics make them suitable substrates for pathogenic bacteria, heavy metal ions and organic pollutants, all of which have detrimental effects on humans and other living beings. However, most of the current solutions used in actual practice are insufficient to address the microplastics problem, leading to the rapid buildup of microplastics in the environment. To create and apply technology that enables the building of an environment free of microplastics, researchers would have to overcome a few obstacles. Some of the challenges are discussed as follow:

1) Although existing classic microplastic removal technologies have shown some efficacy, mineralizing microplastics into innocuous substances such as CO_2 and H_2O are more important. Photocatalysis and Advance Oxidation Processes (AOPs), two powerful organic pollutant degrading mechanisms, will play major roles in the future. According to the chemical stability provided by the structure of most microplastics, as well as the high crystallinity of plastics, the synergy of appropriate external energy sources and reasonable reactor design can overcome intrinsic degradation resistances and accelerate microplastic degradation. Aside from CO_2 and H_2O, it is also possible to establish long-term plans for converting microplastics into valuably fine compounds. Since more and more studies are being published to cover the potential risk associated with these microplastics, it is necessary to create cutting-edge technology to remove microplastics from the air and soil (Wright and Kelly, 2017).

2) The concept of coupling technologies in the application of detection and catalysis, has the potential to boost the development of microplastics removal techniques. Physical-chemical, physical-biological, or chemical-biological coupling techniques could have a synergistic effect on microplastic breakdown efficiency. For example, a functional membrane with photocatalysis and adjustable porosity will provide good microplastics removal efficiency by trapping and *in situ* photocatalytic destruction. At the same time, physical or chemical approaches (such as UV light and AOPs) can be used as a pre-treatment to reduce the molecular weight and increase the hydrophilicity of microplastics through the backbone cleavage and oxygen incorporation, which is required to promote the subsequent biodegradation process (Horton et al., 2017).

3) The deliberate degradation and recycling of bulk plastic wastes should be given priority in addition to the safe treatment methods of environmental microplastics. Reasonable plastic waste recycling measures can effectively minimize a key source of secondary microplastics. In comparison to weathered microplastic particles in the natural environment, bulk plastic wastes that exist for a brief time in nature are easy to collect and relatively pure, making recycling easier and more useful. Furthermore, developing efficient methods for sorting, separation, and purification of plastics from mixed waste streams is critical in practice, because the presence of impurities (including other plastics and additives) can reduce conversion efficiency, generate undesired products and damage equipment (Mason et al., 2016).

4) Catalysts play a critical role in the transformation of plastic into valuable products. In the future, a rational design of dual-functional site catalysts based on interfacial engineering to selectively adsorb specific intermediates followed by

in situ catalytic transformation will be required to alter product selectivity (Avio et al., 2015).

5) The design of the degradable polymer is also essential. Different inorganic or organic motifs can be integrated into the polymer structure as a switch of degradation to start and speed up the deterioration of waste plastics under specific circumstances, presuming that they will not affect the performance. For example, the introduction of inorganic materials with a high expansion coefficient is able to destruct the compact high crystalline structure, such that the decomposition of waste plastics can be expedited by the presence of UV or oxidation reagents. More importantly, biodegradable plastic manufacturing and disposal should be rigorously monitored. Otherwise, it will aggravate short-term microplastics pollution because degradable polymers produce microplastics more quickly (Liborion et al., 2016).

6) With the increasing total volume of plastic waste, it is clearly insufficient to tackle the pollution problem just through the decomposition of plastic particles in the body of individual species, referred to as "*in vivo* degradation." The extraction of critical colonies or enzymes capable of digesting plastic waste from specific species is the only way to efficiently degrade plastic waste. However, it will take a long time and money to accomplish this. Furthermore, biological, inorganic, or other types of catalysts will be used in the micro-biological, enzymatic, and chemical-catalytic degradation processes, which will surely increase the cost (Avio et al., 2015).

7) Biodegradation's great efficiency should be used completely while compensating for their own shortcomings. *In vivo* degradation that occurs in the body of a given species, have a low rate because the degradation capacity is an inherent trait of organisms which is difficult to change. As a result, this is an essential and demanding direction that one should explore and further investigate in the future to achieve a more suitable degrading impact (Horton et al., 2017).

8) Most of the research in microbiological degradation is focused on the extraction and screening of bacteria capable of degrading microplastics or plastics. The screening procedure, on the other hand, is relatively complex and time-consuming. The appropriate application of microbiological deterioration has emerged as a critical concern. The specificity of the enzyme, enzymatic degradation only works with some types of plastics, which is why it is difficult to be used broadly. As a result, hybrid enzymes comprised of several types of enzymes should be investigated further in order to endow the higher practical application value (Eriksen et al., 2013).

9) Many of the catalytic-chemical degradations of microplastic were CO_2 and H_2O and no other high-value chemicals were created as the main or by-product, due to which this process was regarded as very efficient. The type of degradation products is largely influenced by the catalyst's design. More precisely, if the catalyst can attack the weak places on the polymer chain selectively, such as chromophoric groups (Boots et al., 2019), more valuable products such as carboxylic acids and alcohols can be created due to the modification in the reaction path (Galloway et al., 2017).

10) Despite the development of novel technologies in every possible way to reduce the pollution caused by microplastics, another strategy is the synthesis of biopolymers which can potentially replace these synthetic plastics and are capable of complete degradation naturally. These biopolymers can act exactly like synthetic plastics lacking the resistance to biodegradation and accumulation in the environment. Some of the potential biopolymers are polyamides, polyanhydrides, polyoxoesters, and polyphenols which are being used to replace plastics such as polypropylene, polyethylene, and polystyrene. Polyhydroxyalkanoates (PHAs), that belong to the biologically derived polyoxoester family are considered as the best alternatives among these polymers. The challenging issue with these biopolymers is their high production cost. Therefore, use of recombinant strains, genetic and metabolic engineering and novel recovery methods should be promoted for sustainable production of these ecofriendly polymers (Sehgal and Gupta, 2020; Angra et al., 2022).

7. Management Approach of Microplastics

The likelihood of reducing downstream plastic waste is quite low because most of the post-consumer plastic products have low value and poor recovery incentives. To regulate the introduction of plastics into the environment, proper segregation, identification, and quantification are essential, which are currently missing in many nations. Precise annual figures on the generation of plastic trash and information on its classifications are very necessary in order to define the best management techniques regionally. Microplastic management follows three basic ways. These methods include the physical approach, chemical approach and biodegradation as summarized in Fig. 5.7. These approaches have sub levels and methods, which are explained in more detail below:

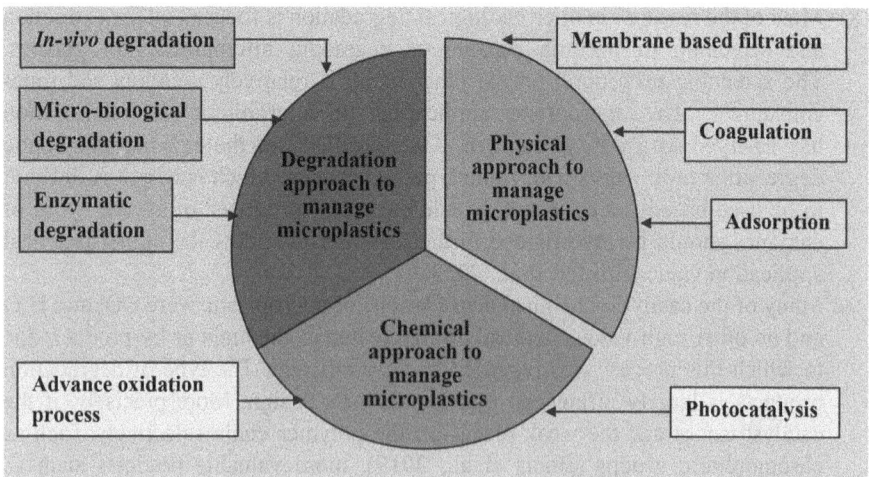

Figure 5.7 Management approach of microplastics.

7.1 Physical Approach to Manage Microplastics

The physical approach usually comprises of physical methods for managing microplastics. Coagulation, membrane-based filtration, and adsorption are three of these approaches. These treatment methods are frequently applied in wastewater management plants. This method makes use of the physicochemical parameters and attachment methods of microplastic as discussed below:

7.1.1 Coagulation

Coagulation is the process of colloidal particles gathering to form a precipitate. In modern water facilities, coagulation is frequently used during advanced treatment to produce high-quality drinking water. To interfere with the charges on suspended particles, colloids or greasy materials, small, highly charged molecules are added to water. By enhancing the performance of the filter and clarifier, the right coagulant can enhance a system's overall performance and in particular, the efficiency of the solids' removal. The coagulants used in this method, such as ferric sulfate or aluminium sulfate, lead to the formation of flocs from the aggregated suspended particulate materials (Sun et al., 2019), which then sediment and can be easily removed from water.

High separation efficiency during this process depends on the coagulants and suspended particles interacting electrostatically and/or through hydrogen bonding. Zeta potential is the measure of a charge's strength in addition to its positive or negative nature. Stronger charges create more stable particle suspensions in water, making the intensity of a charge crucial in wastewater treatment. A stronger negative or positive charge with a more stable suspension in water is indicated by a higher zeta potential value, which is measured on a scale of range −61 to +61 V. At 0, particles will easily come out of suspension, but as they increase significantly beyond 10, coagulation will be required.

The efficacy of microplastic removal via coagulation was checked at the laboratory-scale simulating drinking water treatment using a Fe-based salt ($FeCl_3.6H_2O$) as a coagulant to remove polyethylene (PE) microplastic at pH 7.0 (Ma et al., 2019). At a relatively high Fe salt concentration of 2 mmolL^{-1}, only 13% of microplastics with diameters of 0.5 mm could be eliminated. Due to a poor interaction between the coagulant and the pristine plastic surface, microplastics stuck in unstable flocs were not effectively settled (Carr et al., 2016). Simply adding 15 mgL^{-1} polyacrylamide (PAM) to the $FeCl_3.6H_2O$ resulted in a high microplastic removal efficiency of 90.91%. The researchers postulated that static interactions between cationic Fe-based flocs and anionic PAM led to more stable and dense flocs, which effectively stopped the microplastic escape. The surface characteristics of microplastics have a major effect on the coagulation efficiency of the microplastics removal in addition to the use of additives (Perren et al., 2018).

During weathering processes in the natural environment, changes in the surface chemistry and roughness of the microplastics could affect their affinity for coagulants and flocculants. Unsaturated bonds and oxygen-containing groups (hydroxyl and carboxylic acid groups) can be introduced into weathered microplastics through photo-oxidation action and the attachment of natural organic matter (biofilm or POPs) in the environment (vinyl groups). These newly formed groups could therefore enhance interactions between the coagulant and microplastics by serving as anchoring sites.

7.1.2 Membrane based Filtration

The ability to produce high-quality water from primary or secondary effluents using filtration techniques like microfiltration (MF), Reverse Osmosis (RO), ultrafiltration (UF), Dynamic Membranes (DM) and Membrane-Based Reactor (MBRs) has been demonstrated. These techniques benefit from two important factors: high separation efficiency and a small plant size (Poerio et al., 2019). Impurities such as bacteria, viruses, protozoa, and suspended particles can be successfully eliminated by using asymmetric membranes having micrometer or nanometer-sized pores. Although UF and RO were effective in removing microplastics in waste water treatment plant (WWTP), millions of microplastics remained in the effluent after these treatments (Ziajahromi et al., 2017).

Widespread research has been done on dynamic membrane (DM) technology as a potential competitor in sophisticated water treatment procedures (Salerno et al., 2017). DM technology uses filtration to create a cake layer on a supporting membrane, which acts as a secondary barrier to trap the detritus and impurities, in contrast to standard filtration methods that use membranes with micrometer- or nanometer-sized pores to retain the minute solids. The use of additional chemical agents and the risk of introducing secondary pollutants are eliminated because the DM is entirely composed of solid wastewater contaminants. In many cases, DM filtration can be used in a gravity-driven mode without the use of pumps due to the low filtration resistance and trans-membrane pressure (TMP) during the process. A gravity-driven laboratory-scale system can be used to filter synthetic wastewater made of tap water and diatomite particles (Li et al., 2017). Ninety percent of diatomite particles have diameters between 1 and 90 μm, whereas the pore size of the supplementary membrane is 90 μm. After 20 min of treatment, the turbidity of the synthetic wastewater drops dramatically, from 195 to 1 nephelometric turbidity unit, indicating that the DM technology is effective at removing these tiny microplastic particles.

Recently, the efficiency of removing microplastics from membrane technologies has been improved by combining them with other techniques. Membrane Based Reactor (MBR), a heterogeneous reaction system consisting of a biological reactor and a membrane system, has been created for better wastewater treatment (Gurung et al., 2016). First, the influent enters the bioreactor and begins to degrade. In this instance, the refractory organic contaminants can be effectively removed using biologically activated sludge that has a high biomass concentration and a long solid retention time. A semi-crossflow filtration system is then used to separate the combined suspension. Here, the suspended solids are caught by the membrane and concentrated in the retentate. The synergy of the bio-reaction processes and porous membranes allow the MBR to treat microplastics in contaminated wastewater more effectively.

MBR configuration that included a biodegradation tank with a membrane filtration tank made up of 20 submerged flat-sheet membrane units with an effective membrane area of 8 m^2 and pore size of 0.4 μm (Talvitie et al., 2017). This method resulted in a reduction in the amount of microplastics in primary effluents from WWTPs from 6.9 to 0.005 microplastic particles L^{-1}, indicating a remarkable 99.9% eradication rate.

It is important to note that the biological reactor made it possible for the organic impurities that were stuck to the microplastics to break down, which helps to increase the precision of the qualitative analysis of the microplastics that were separated. It is widely believed that the biodegradation of chemically inert microplastics during the MBR process is constrained, much like the biological process in the secondary treatment. Although the filtering method holds great promise for removing microplastics in the tertiary treatment, its main challenges are membrane fouling and abrasion, which frequently cause the purifying effectiveness to degrade (Enfrin et al., 2019).

7.1.3 Adsorption

Adsorption is the process by which atoms, ions, or molecules from a gas, liquid or dissolved solid attach to a surface. On the surface of the adsorbent, this procedure leaves an adsorbate coating. Adsorption has also been used to remove organic toxins and heavy metals from water. Ion exchange, hydrophobic interactions and hydrogen-bond interactions between absorbents and contaminants affect the adsorption performance. Recently, this strategy has been used to lessen the amount of microplastics in the water treatment process.

Yuan et al. (2020) used 3D Reduced Graphene Oxide (3D RGO) to test the polystyrene (PS) microplastic's ability to adsorb substances. In this study, PS microspheres with an average diameter of 5 μm were used as microplastic samples. The outstanding adsorption capacity of 3D RGO, 617.28 mg PS g^{-3}D RGO1, was produced by the strong chemical interaction between 3D RGO and PS-microplastics. Additionally, this ground-breaking magnetic absorbent creates a new way to deal with the challenge of absorbent separation in real-world water treatment applications by avoiding time-consuming conventional filtration.

Nano Particles (NPs) are much more hazardous to the environment and harder to remove using the current water treatment methods. A type of double-layered Zn-Al hydroxide (Zn-Al LDH) was used to adsorb PS particles with diameters smaller than 1μm from an aqueous solution in order to increase the adsorbing efficiency of NPs (Tiwari et al., 2020). Due to oxidation, the surface characteristics of microplastics found in the natural environment are drastically changed to be more negatively charged. Therefore, PS NPs were pretreated by coating with an anionic surfactant to mimic this effect (Guo et al., 2019). This system has a strong ability to adsorb the anionic contaminants due to its superior anion-exchange capacity (AEC ≈ 3 meq g^{-1}).

Sludge typically contains microplastics that were removed during the multiple stages of the wastewater treatment. Frequently, this microplastic-rich sludge is processed further and used as either landfill or field fertilizer (Zhang and Chen, 2020). Microplastics are not taken into consideration, in spite of the fact that a number of processes are used to remove toxic debris before it is used in agriculture (Mahon et al., 2017). Admittedly, despite many improvements made, the above-mentioned methods based on straightforward physical separation are ineffective at eradicating the pollution of the environment of microplastics because there is no practical way to reuse or completely remove the debris and particles of microplastics that have been physically separated and collected.

7.2 Chemical Approach to Manage Microplastics

The chemical approach is based on chemical bonding and microplastic reaction. The comparison of various physical and chemical strategies are shown in Table 5.1. As discussed below, there are two primary approaches: advance oxidation process and photocatalysis.

Table 5.1 Comparison of various physical and chemical microplastics removal strategies.

Physical strategies		
Strategies	**Advantages**	**Disadvantages**
Coagulation	Highly safe, good flexibility and sustainability.	Expensive, less opportunities with other compound and less conversion efficiency.
Membrane-based filtration	Highly safe, better flexibility and separation efficiency.	Expensive and less conversion efficiency.
Adsorption	High separation efficiency, high opportunities with other compounds and flexible to use.	Less safe, expensive and zero conversion efficiency.
Chemical strategies		
Advance oxidation process	Good conversion efficiency, good separation efficiency and flexible to use.	Expensive, not safe to use and less opportunities with other compound.
Photocatalysis	Highly flexible, sustainable, good separation efficiency and good conversion efficiency.	Expensive and not safe to use.

7.2.1 Advance Oxidation Processes of Microplastics

Advanced oxidation techniques are well-known for their effectiveness in removing organic contaminants by producing reactive oxygen species (ROSs) with high standard reduction potentials, such as sulfate radical and hydroxyl radical (Zhou et al., 2020). Advance Oxidation Processes (AOPs) are powered by external energy sources such as electricity, ultraviolet radiation (UV) or solar light. They are more expensive than traditional biological wastewater treatment. This approach has effectively decomposed or mineralized a wide range of contaminants, including colors, antibiotics and POPs (Ganiyu et al., 2016). Due to their significantly higher Molecular Weights (MWs) than the other low MW organic pollutants mentioned above, microplastics are obviously more difficult to decompose. Advanced oxidation entails numerous processes like the emergence of powerful oxidants (e.g., hydroxyl radicals). These oxidants react with the organic substances in water to produce biodegradable intermediates (Kommineni et al., 2008). Mineralization is the process of oxidizing biodegradable intermediates, i.e., production of water, carbon dioxide and inorganic salts.

The presence of a hydrothermal (HT) condition is crucial to this process because it causes the PE beads to be mechanically sheared, which starts the chain scission of the macromolecules and the degradation process begins. In addition, as peroxymonosulfate (PMS) activation can be heat-driven, the generation of ROSs significantly speeds up. Under HT conditions, the PE backbone's C-C bond is first dissociated into two hydrocarbon radicals. Lower-weight molecules are created when the hydrocarbon radicals are further broken down. These molecules are then converted into new, shorter-chain hydrocarbon radicals via the scission and hydrogen abstraction routes, which are sparked by other hydrocarbons. Finally, the intermediate radicals are attacked and mineralized by the $SO^{4\cdot}$ and $\cdot OH$ produced by the SR-AOPs system. As a result, the extremely persistent PE is transformed into greener intermediates like low hydrocarbon length aldehyde, ketone, and carboxylic acids, which can be used as carbon sources for the growth of algae and are biodegradable. As the SR-AOPs reaction progresses, these intermediates are broken down into low molecular weight organics and further mineralized into CO_2 and H_2O. However the direct application in WWTPs for the removal of microplastics is impractical due to the relatively harsh reaction conditions (high pressure and high temperature). But the chemical degradation process can be strengthened to break down the extremely persistent C-C bond-based macromolecules. As an alternative, Electro-Fenton (EF)-based EAOPs are used to produce ROSs (like $\cdot OH$) and decontaminate Persistent Organic Pollutants (POPs) and they stand out for their adaptability, superior efficiency, excellent environmental compatibility and sustainability (Ye et al., 2020). Hydrogen peroxide (H_2O_2) is produced *in situ* on the cathodes by reducing O_2 via a two-electron oxygen reduction reaction and is then transformed into $\cdot OH$. The cathode material is crucial in influencing the system's degrading efficiency throughout this process (Ding et al., 2020).

7.2.2 Photocatalysis of Microplastics

A well-known, environmently friendly process that uses light energy to fuel the chemical reaction of exciting electron by absorption of high frequency waves is called photocatalysis. Due to its excellent ability to degrade antibiotics, pesticides and dyes, this technology has seen significant use in the water purification industry recently (Duoerkun et al., 2020). The mechanism of photocatalytic degradation involves the interaction of ROSs (for example, hydroxyl ($\cdot OH$), superoxide ($O2\cdot$), generated on the surface of semiconductors with the organic substrate, which break the chemical bonds of organic pollutants and cause their complete mineralization to CO_2 and H_2O. Organic materials can be directly oxidized into CO_2 and H_2O by photo-excited holes (h+VB) formed by an electron transfer from the valance band to the conduction band.

TiO_2 is a well-known photocatalyst capable of efficiently oxidizing organic pollutants. It also has many advantages, such as low toxicity, low cost and great acid and alkali resistance (Sun et al., 2020). TiO_2-based materials have been widely explored in photocatalytic microplastic degradation in recent years. A TiO_2-catalzyed photo process is done under UV radiation because of its comparatively high band gap of 3.2eV. TiO_2 film act as a catalyst for the solid-state degradation of PS microspheres and PE powder under UV radiation (Nabi et al., 2020). The Triton X-100-based

TiO_2 (TiO_2-TXT) layer shows excellent hydrophilicity, specific surface area and the photogenerated electron-hole pair separation and transfer. The remarkable mineralization efficiency of TiO_2-TXT film is attributed mostly to the unique surface hydrophilicity which could promote the interaction between PS microplastics and TiO_2. The increase in the charge carrier production and separation could hasten the generation of $\cdot OH$ and $O_2 \cdot$, which plays an important role in the degradation process.

Photo-excited holes helps to oxidize and mineralize microplastics into CO_2 and H_2O in addition to the oxidative degradation brought on by ROSs. This technique is useful for degrading other microplastic contaminations, as evidenced by the degradation of PE, which revealed a significant change in shape and chemical bonding. this closes a knowledge gap regarding the underlying impact of operating parameters like pH and temperature during the degradation process. In a study, the degradation behavior of HDPE primary microplastics was investigated in the presence of a newly designed C and N-doped TiO_2 catalyst that absorbs light with a wavelength of 428 nm, within the range of visible light. In each experiment, HDPE and photocatalyst were dissolved in water in a batch-style container and placed in a sealed reaction space with an LED light source to produce visible light. The various damaged morphologies of the PE microbeads under various pH and temperature conditions were visible in optical images. At 0°C and pH 3, a maximum mass loss of 72% was attained in 50 hr (Ariza-Tarazona et al., 2020).

7.3 Degradation Approach of Microplastics

Microplastics can be cleaned using ordinary cleaning processes (Schmaltz et al., 2020). The production of secondary microplastics can be decreased, for instance, by gathering up waste plastics from the beach, such as packaging, bags, and mess containers. However, there are no eco-friendly disposal methods for the collected plastics. Incineration and landfilling of plastic waste are not the best solutions for the environmental issue. Although burning plastic waste is a very effective way to deal with it, but it results in the production of greenhouse gases like carbon dioxide (CO_2), methane and carbon monoxide (CO) (Zhu et al., 2009). In addition to using up a lot of land resources, landfills pose a big risk of releasing chemicals like phthalic acid esters (plasticizer) and methyl orange (dye) which causes degradation of the soil (Agarwal, 2020). Several novel approaches have recently been proposed to address the pressing threat posed by microplastics. The following are some approaches for biodegradation of microplastics:

7.3.1 In vivo Degradation of Microplastics

In spite of the fact that microplastics' distinctive structural characteristics allow them to persist in the environment for decades or even hundreds of years without degrading, the ecosystem's capacity for self-regulation must be taken into consideration. Despite being historically undervalued, there are still some species in nature that can partially digest microplastics. At present, *in vivo* degradation of a specific organism has piqued the scientific community's curiosity and it may be a viable option for mitigating the growing issue of microplastic contamination. These organisms have enzyme in their saliva or in their stomach that have the capacity to degrade the plastic under physiological conditions like neutral pH and room temperature. There are mainly two enzymes which belong to phenol oxidase family that can degrade the plastic by oxidation reaction.

Kundungal et al. (2019) discovered that the lesser waxworm (*Achroia grisella*) not only degrades high density polyethylene (HDPE) more effectively, but also goes through the normal stages of its life cycle. More importantly, research asserts that extracellular enzymes from the intestinal flora of waxworms actively take part in PE degradation processes. In contrast to this viewpoint, Peng et al. (2019) discovered that 100 larvae of dark and yellow mealworms decomposed 32.44 mg and 24.30 mg of PS every day, respectively. The intestinal microbial communities of two species linked to PS intake and biodegradation have changed significantly, primarily due to the presence of *Enterobacteriaceae, Treponemaceae,* and *Enterococcus*. This finding proves that bacterial microorganisms within the organism are the main contributors to degradability. In a nutshell, the above data indicate that *in vivo* degradation is a viable solution to microplastic pollution, although the mechanism of degradation requires additional investigation.

7.3.2 Microbiological Degradation of Microplastics

The earth's microbial system is vital in cleaning up "dead" creatures by decomposing all plant and animal components. Biodegradation refers to the destruction of polymer chains by a microbial action. Conventional plastics are so strong that they cannot be digested by current microbes. As a result, in order to expand the use of the microbiological degradation approach, the qualities of the microorganisms should be improved. This concept is novel and it is difficult to implement. A few researches have been published that focus on the use of microorganisms to accomplish this aim. The biodegradation process is separated into four stages: biodeterioration, depolymerization, bio assimilation and mineralization (Haider et al., 2019). Microorganisms first cling to the surface of polymers in a normal biodegradation technique and then the polymers expose themselves to the microbial community. After that, extracellular enzymes cleave or depolymerize the polymer. The produced microscopic compounds are then ingested by microbial cells and transformed into primary and secondary metabolites. Finally, the carbon cycle is restarted by bacteria and the metabolites are converted into CO_2 and H_2O (Tokiwa et al., 2009; Kumar et al., 2022).

In a study, eight bacterial strains were found in the Peninsular Malaysian mangrove sediments. They were tested for their capacity to break down microplastics made of different types of polymers, such as PE, PS, and polyethylene terephthalate (PET) (Auta et al., 2017b). According to the experimental findings of weight loss in 40 d, *Bacillus cereus* degraded PE, PET and PS-based microplastics at the rate of 1.6, 6.6 and 7.4%, respectively, while *Bacillus gottheilii* degraded at the rate of 6.2, 3.0 and 5.8% respectively. However, neither of the bacterial isolate's degradation efficiencies are sufficient for actual use. The microbial biodegradation of microplastic in 200 tons of plastic sewage sludge was seen, where microplastic abundance was 104 items/kg and the majority of the microplastics were made of polypropylene (PP), PE, PS and PET (Chen et al., 2020). The biodegradation conditions in this study were those of "hyperthermophilic composting" (hTC). In comparison to traditional thermophilic composting (cTC) conditions, hTC has a higher level of compost maturity (typically measured by the amount of humus formation in the compost material (Guo and Wang, 2019), higher temperatures (hTC is 20–30°C higher than cTC) and consequently a significantly shorter composting period. Additionally, it was found that, compared to cTC at 40°C, hTC at 70°C degraded 7.3% of the PS microplastics in 56 d.

Furthermore, high-throughput sequencing analysis revealed that *Thermus* (54.2%), *Bacillus* (24.8%) and *Geobacillus* (19.6%) were the primary strains present in this hTC and performed critical roles in the biodegradation of these microplastics under hTC conditions. The abundance, diversity, and activation of most of the bacteria or enzymes rapidly declined at 85°C. In spite of the fact that higher hTC temperatures can introduce oxygen functional (such as C=O/C-O) groups and decrease the hydrophobicity of microplastics, which is advantageous for improving degradation efficiency, composting quality, and compatibility with microorganisms.

7.3.3 Enzymatic Degradation of Microplastics

A class of biomacromolecules called enzymes is routinely produced by living cells. High temperature and harsh pH settings are unfavorable for enzymes because they disrupt their spatial structure and eventually result in enzyme inactivation, in spite of their high efficiency and specificity. Recently, scientists examined using enzymes directly to break down large pieces of plastic waste as well as microplastics. The two stages of enzymatic degradation are enzyme adsorption on the polymer surface and hydro-peroxidation/hydrolysis of the bond. Enzymes for degrading plastic can be found in a number of microbes and in the intestines of some invertebrates. A promising method for depolymerizing waste petro-plastics into polymer monomers and for recycling or for mineralizing waste plastics to produce higher value bioproducts like biodegradable polymers is microbial and enzymatic degradation. According to a recent study, some of the most common enzymes connected with plastic degradation include cutinase, lipase and PETase (an esterase) (Sharma et al., 2021).

Sulaiman et al. (2012) used tributyrate agar plates for screening of LCC cutinase from the leaf branch compost metagenome library. The polymers eventually broke down into terephthalic acid (PTA) and Ethylene Glycol (EG) with a conversion rate of 12 mgh^{-1} enzyme at pH 8 and 50°C temperature. The ester groups of polyethylene glycol terephthalate (PET) were cut off by LCC enzyme. However, this method cannot convert PET at a rate fast enough to keep up with the production of plastic trash and the enzymes are currently too expensive.

The use of garbage classification, which is beneficial for reducing the stringent tendency of microplastics to some extent, results in more effective recycling of plastic waste and limits littering behavior. It is simpler, more thorough, and more economical to eliminate microplastic contamination during the manufacture, use and recycling of plastics. Despite the existence of numerous regulations, a sizable amount of plastic is still produced and society should find a way to deal with it. The creation of microplastics, has led to a serious threat to the daily life of people. One now needs to figure out how to successfully address this environmental problem. In order to address the issue of the increase in microplastics in the environment, more emphasis should be put on creating more efficient degradation mechanisms than biodegradation and catalytic chemical degradation.

8. Conclusion

Microplastics are plentiful, common, and persistent on a global scale. Due to the widespread use of plastic products and their difficulty in degrading, plastic pollution will undoubtedly continue to have a negative impact on the ecological environment for a very long time. They pose a serious threat requiring international action when combined with rising levels of chemical pollutants in the water that are easily adsorbed and concentrated into microplastics. These pollutants can be consumed indiscriminately by aquatic organisms. Chemical pollutants that have been adsorbed to microplastics and chemical additives used in the production of plastic can bioaccumulate in higher trophic levels, including humans, by leaching from microplastics into an aquatic biota tissue. Research on the sources, migration, and transformation of microplastics, their ecotoxicology, enhanced sampling, classification and identification analytical methods and an improved ecological health risk assessment system should be strengthened. The effects of eating seafood containing microplastic on human health are still unknown and research on the toxicity of microplastics to biota is still in its infancy. Further research is required to understand the toxicological mechanism and influencing factors of microplastics. To prevent harm to human health, attention should be paid to the transfer effect of microplastics in the food chain and control measures. For all stakeholders, it is crucial to increase the awareness of the negative effects of microplastics and the incorrect handling of plastic waste. To decrease plastic use and consumption and to provide incentives for preventing plastic pollution and waste reduction, strict policies are needed at the local, national, regional, and international levels.

References

Agarwal, S. 2020. Biodegradable polymers: present opportunities and challenges in providing a microplastic-free environment. Macromol. Chem. Phys. 221: 2000017.

Ahrendt, C., Perez-Venegas, D.J., Urbina, M., Gonzalez, C., Echeveste, P., Aldana, M. et al. 2020. Microplastic ingestion cause intestinal lesions in the intertidal fish *Girella laevifrons*. Mar. Pollut. Bullet. 151: 110–195.

Anderson, A.G., Grose, J., Pahl, S., Thompson, R.C. and Wyles, K.J. 2016. Microplastics in personal care products: Exploring perceptions of environmentalists, beauticians and students. Mar. Pollut. Bullet. 113: 454–460.

Andrady, A.L. 2011. Microplastics in the marine environment. Mar. Pollut. Bull. 62: 1596–1605.

Angra, V., Sehgal, R. and Gupta, R. 2022. Trends in PHA production by microbially diverse and functionally distinct communities. Microb. Ecol. https://doi.org/10.1007/s00248-022-01995-w

Ariza-Tarazona, M.C., Villarreal-Chiu, J.F., Hernández-López, J.M., De la Rosa, J.R., Barbieri, V., Siligardi, C. et al. 2020. Microplastic pollution reduction by a carbon and nitrogen-doped TiO2: Effect of pH and temperature in the photocatalytic degradation process. J. Hazard. Mater. 395: 122–132.

Auta, H.S., Emenike, C.U. and Fauziah, S.H. 2017a. Distribution and importance of microplastics in the marine environment: a review of the sources, fate, effects, and potential solutions. Environ. Int. 102: 165–176.

Auta, H.S., Emenike, C.U. and Fauziah, S.H. 2017b. Screening of *Bacillus* strains isolated from mangrove ecosystems in Peninsular Malaysia for microplastic degradation. Environ. Pollut. 231: 1552–1559.

Avio, C.G., Gorbi, S., Milan, M., Benedetti, M., Fattorini, D., d'Errico, G. et al. 2015. Pollutants bioavailability and toxicological risk from microplastics to marine mussels. Environ. Pollut. 198: 211–222.

Bakir, A., Rowland, S.J. and Thompson, R.C. 2014. Transport of persistent organic pollutants by microplastics in estuarine conditions. Estuar. Coast. Shelf Sci. 140: 14–21.

Bejgarn, S., MacLeod, M., Bogdal, C. and M. Breitholtz. 2015. Toxicity of leachate from weathering plastics: An exploratory screening study with *Nitocra spinipes*. Chem. 132: 114–119.

Boots, B., Russell, C.W. and Green, D.S. 2019. Effects of microplastics in soil ecosystems: above and below ground. Environ. Sci. Technol. 53: 11496–11506.

Boucher, J. and Friot, D. 2017. Primary microplastics in the oceans: a global evaluation of sources. Environ. Manag. 11: 122–188.

Bouwmeester, H., Hollman, P.C. and Peters, R.J. 2015. Potential health impact of environmentally released micro-and nanoplastics in the human food production chain: experiences from nanotoxicology. Environ. Sci. Technol. 49: 8932–8947.

Browne, M.A., Crump, P., Niven, S.J., Teuten, E., Tonkin, A., Galloway, T. et al. 2011. Accumulation of microplastic on shorelines worldwide: sources and sinks. Environ. Sci. Technol. 45: 9175–9179.

Burghardt, T.E., Pashkevich, A., Babić, D., Mosböck, H., Babić, D. and Żakowska, L. 2022. Microplastics and road markings: the role of glass beads and loss estimation. Transp. Res. D Transp. Environ. 102: 103–123.

Cabernard, L., Roscher, L., Lorenz, C., Gerdts, G. and Primpke, S. 2018. Comparison of Raman and Fourier transform infrared spectroscopy for the quantification of microplastics in the aquatic environment. Environ. Sci. Technol. 52: 13279–13288.

Carpenter, E.J. and K.L. Smith. 1972. Plastics on the Sargasso Sea surface. Science. 175: 1240–1241.

Carr, S.A., Liu, J. and Tesoro, A.G. 2016. Transport and fate of microplastic particles in wastewater treatment plants. Water Res. 91: 174–182.

Carrington, D. 2020. Bottle-fed babies swallow millions of microplastics a day, study finds. Guard. Ret. 9: 111–145.

Chamas, A., Moon, H., Zheng, J., Qiu, Y., Tabassum, T., Jang, J.H. et al. 2020. Degradation rates of plastics in the environment. ACS Sustain. Chem. Eng. 8: 3494–3511.

Chen, Z., Zhao, W., Xing, R., Xie, S., Yang, X., Cui, P. et al. 2020. Enhanced *in situ* biodegradation of microplastics in sewage sludge using hyperthermophilic composting technology. J. Hazard. Mater. 384: 121271.

Cole, M., Webb, H., Lindeque, P.K., Fileman, E.S., Halsband, C. and Galloway, T.S. 2014. Isolation of microplastics in biota-rich seawater samples and marine organisms. Sci. Rep. 4: 1–8.

Cole, M., Lindeque, P., Halsband, C. and Galloway, T.S. 2011. Microplastics as contaminants in the marine environment: a review. Mar. Pollut. Bull. 62: 2588–2597.

Collignon, A., Hecq, J.H., Galgani, F., Collard, F. and Goffart, A. 2014. Annual variation in neustonic micro-and meso-plastic particles and zooplankton in the Bay of Calvi (Mediterranean–Corsica). Mar. Pollut. Bullet. 79: 293–298.

Conkle, J.L., Báez Del Valle, C.D. and Turner, J.W. 2018. Are we underestimating microplastic contamination in aquatic environments? Environ. Manag. 61: 1–8.

Dekiff, J.H., Remy, D., Klasmeier, J. and Fries, E. 2014. Occurrence and spatial distribution of microplastics in sediments from Norderney. Environ. Pollut. 186: 248–256.

Derraik, J.G. 2002. The pollution of the marine environment by plastic debris: a review. Mar. Pollut. Bullet. 44: 842–852.

Desforges, J.P.W., Galbraith, M. and Ross, P.S. 2015. Ingestion of microplastics by zooplankton in the Northeast Pacific Ocean. Arch. Environ. Contam. Toxicol. 69: 320–330.

Ding, H., Zhu, Y., Wu, Y., Zhang, J., Deng, H., Zheng, H. et al. 2020. In situ regeneration of phenol-saturated activated carbon fiber by an electro-peroxymonosulfate process. Environ. Sci. Technol. 54: 10944–10953.

do Sul, J.A.I. and Costa, M.F. 2014. The present and future of microplastic pollution in the marine environment. Environ. Pollut. 185: 352–364.

Dris, R., Gasperi, J., Mirande, C., Mandin, C., Guerrouache, M., Langlois, V. et al. 2017. A first overview of textile fibers, including microplastics, in indoor and outdoor environments. Environ. Pollut. 221: 453–458.

Dümichen, E., Barthel, A.K., Braun, U., Bannick, C.G., Brand, K., Jekel, M. et al. 2015. Analysis of polyethylene microplastics in environmental samples, using a thermal decomposition method. Water Res. 85: 451–457.

Duoerkun, G., Zhang, Y., Shi, Z., Shen, X., Cao, W., Liu, T. et al. 2020. Construction of n-TiO$_2$/p-Ag$_2$O junction on carbon fiber cloth with Vis–NIR photoresponse as a filter-membrane-shaped photocatalyst. Adv. Fiber Mater. 2: 13–23.

Enders, K., R. Lenz, S. Beer and C.A. Stedmon. 2017. Extraction of microplastic from biota: recommended acidic digestion destroys common plastic polymers. ICES J. Mar. Sci. 74: 326–331.

Enfrin, M., Dumée, L.F. and Lee, J. 2019. Nano/microplastics in water and wastewater treatment processes–origin, impact and potential solutions. Water Res. 161: 621–638.

Engler, R.E. 2012. The complex interaction between marine debris and toxic chemicals in the ocean. Environ. Sci. Technol. 46: 12302–12315.

Eriksen, M., Maximenko, N., Thiel, M., Cummins, A., Lattin, G., Wilson, S. et al. 2013. Plastic pollution in the South Pacific subtropical gyre. Mar. Pollut. Bull. 68: 71–76.

Estahbanati, S. and Fahrenfeld, N.L. 2016. Influence of wastewater treatment plant discharges on microplastic concentrations in surface water. Chemosphere. 162: 277–284.

Farrell, P. and Nelson, K. 2013. Trophic level transfer of microplastic: Mytilus edulis (L.) to Carcinus maenas (L.). Environ. Pollut. 177: 1–3.

Faure, F., Saini, C., Potter, G., Galgani, F., De Alencastro, L.F. and Hagmann, P. 2015. An evaluation of surface micro-and mesoplastic pollution in pelagic ecosystems of the Western Mediterranean Sea. Environ. Sci. Pollut. Res. 22: 12190–12197.

Fendall, L.S. and Sewell, M.A. 2009. Contributing to marine pollution by washing your face: microplastics in facial cleansers. Mar. Pollut. Bull. 58: 1225–1228.

Fischer, M. and Scholz-Bottcher, B.M. 2017. Simultaneous trace identification and quantification of common types of microplastics in environmental samples by pyrolysis-gas chromatography–mass spectrometry. Environ. Sci. Technol. 51: 5052–5060.

Frias, J.P., Gago, J., Otero, V. and Sobral, P. 2016. Microplastics in coastal sediments from Southern Portuguese shelf waters. Mar. Environ. Res. 114: 24–30.

Fries, E., Dekiff, J.H., Willmeyer, J., Nuelle, M.T., Ebert, M. and Remy, D. 2013. Identification of polymer types and additives in marine microplastic particles using pyrolysis-GC/MS and scanning electron microscopy. Environ. Sci. 15: 1949–956.

Galloway, T.S., Cole, M. and Lewis, C. 2017. Interactions of microplastic debris throughout the marine ecosystem. Nat. Ecol. Evol. 1: 1–8.

Ganiyu, S.O., Oturan, N., Raffy, S., Cretin, M., Esmilaire, R., Van Hullebusch, E. et al. 2016. Sub-stoichiometric titanium oxide (Ti$_4$O$_7$) as a suitable ceramic anode for electrooxidation of organic pollutants: a case study of kinetics, mineralization and toxicity assessment of amoxicillin. Water Res. 106: 171–182.

Gu, Y.G., Lin, Q., Yu, Z.L., Wang, X.N., Ke, C.L. and Ning, J.J. 2015. Speciation and risk of heavy metals in sediments and human health implications of heavy metals in edible nekton in Beibu Gulf, China: a case study of Qinzhou Bay. Mar. Pollut. Bull. 101: 852–859.

Guo, X. and Wang, J. 2019. The chemical behaviors of microplastics in marine environment: A review. Mar. Pollut. Bull. 142: 1–14.

Gurung, K., Ncibi, M.C., Fontmorin, J.M., Särkkä, H. and Sillanpää, M.J.J.M.S.T. 2016. Incorporating submerged MBR in conventional activated sludge process for municipal wastewater treatment: a feasibility and performance assessment. J. Membr. Sci. 6(3): 122–176.

Gutow, L., Eckerlebe, A., Giménez, L. and Saborowski, R. 2016. Experimental evaluation of seaweeds as a vector for microplastics into marine food webs. Environ. Sci. Technol. 50: 915–923.

Habib, D., Locke, D.C. and Cannone, L.J. 1998. Synthetic fibers as indicators of municipal sewage sludge, sludge products, and sewage treatment plant effluents. Wat. Air Soil Pollut. 103: 1–8.

Haider, T.P., Völker, C., Kramm, J., Landfester, K. and Wurm, F.R. 2019. Plastics of the future? The impact of biodegradable polymers on the environment and on society. Angew. Chem. Int. Edit. 58: 50–62.

Halden, R.U. 2010. Plastics and health risks. Annu. Rev. Publ. Health, 31: 179–194.

Harper, P.C. and Fowler, J.C. 1987. Plastic pellets in New Zealand storm-killed prions (*Pachyptila* spp.). Notornis. 34: 65–70.

Hidalgo-Ruz, V., Gutow, L., Thompson, R.C. and Thiel, M. 2012a. Microplastics in the marine environment: a review of the methods used for identification and quantification. Environ. Sci. Technol. 46: 3060–3075.

Hidalgo-Ruz, V., Gutow, L., Thompson, R.C. and Thiel, M. 2012b. Microplastics in the marine environment: a review of the methods used for identification and quantification. Environ. Sci. Technol. 46: 3060–3075.

Holmes, L.A., Turner, A. and Thompson, R.C. 2012. Adsorption of trace metals to plastic resin pellets in the marine environment. Environ. Pollut. 160: 42–48.

Horton, A.A., Walton, A., Spurgeon, D.J., Lahive, E. and Svendsen, C. 2017. Microplastics in freshwater and terrestrial environments: evaluating the current understanding to identify the knowledge gaps and future research priorities. Sci. Total Environ. 586: 127–141.

Ivar do Sul, J.A. and Costa, M.F. 2014. The present and future of microplastic pollution in the marine environment. Environ. Pollut. 185: 352–364.

Käppler, A., Fischer, D., Oberbeckmann, S., Schernewski, G., Labrenz, M., Eichhorn, K.J. et al. 2016. Analysis of environmental microplastics by vibrational microspectroscopy: FTIR, Raman or both? Anal. Bioanal. Chem. 408: 8377–8391.

Karbalaei, S., Hanachi, P., Walker, T.R. and Cole, M. 2018. Occurrence, sources, human health impacts and mitigation of microplastic pollution. Environ. Sci. Pollut. R. 25: 36046–36063.

Karlsson, T.M., Vethaak, A.D., Almroth, B.C., Ariese, F., Van Velzen, M., Hassellöv, M. et al. 2017. Screening for microplastics in sediment, water, marine invertebrates and fish: method development and microplastic accumulation. Mar. Pollut. Bullet. 122: 403–408.

Katsnelson, A. 2015. Microplastics present pollution puzzle. Proc. Natl. Acad. Sci. 112: 5547–5549.

Kommineni, S., Chowdhury, Z., Kavanaugh, M., Mishra, D. and Croue, J.P. 2008. Advanced oxidation of methyl-tertiary butyl ether: pilot study findings and full-scale implications. J. Water Supply: Res. Technol. 57: 403–418.

Kovochich, M., Liong, M., Parker, J.A., Oh, S.C., Lee, J.P., Xi, L. et al. 2021. Chemical mapping of tire and road wear particles for single particle analysis. Sci. Total Environ. 757: 144–185.

Kumar, V., Sehgal, R. and Gupta, R. 2022. Microbes and waste water treatment. pp. 239–255. *In*: Development in Wastewater Treatment Research and Process. Shah, M.P., Rodriguez-Couto, S., Nadda, A.K. and Daverey, A. (eds.). Elsevier.

Kundungal, H., Gangarapu, M., Sarangapani, S., Patchaiyappan, A. and Devipriya, S.P. 2019. Efficient biodegradation of polyethylene (HDPE) waste by the plastic-eating lesser waxworm (*Achroia grisella*). Environ. Sci. Polut. Res. 26: 18509–18519.

Law, K.L. and Thompson, R.C. 2014. Microplastics in the seas. Science 345: 144–145.

Lee, H., Shim, W.J. and Kwon, J.H. 2014. Sorption capacity of plastic debris for hydrophobic organic chemicals. Sci. Total Environ. 470: 1545–1552.

Li, N., He, L., Lu, Y., Zeng, R.J. and Sheng, G. 2017. Robust performance of a novel anaerobic biofilm membrane bioreactor with mesh filter and carbon fiber (ABMBR) for low to high strength wastewater treatment. Chem. Eng. J. 313: 56–64.

Liboiron, M., Liboiron, F., Wells, E., Richárd, N., Zahara, A., Mather, C. et al. 2016. Low plastic ingestion rate in Atlantic cod (*Gadus morhua*) from Newfoundland destined for human consumption collected through citizen science methods. Mar. Pollut. Bull. 113: 428–437.

Lithner, D., Larsson, A. and Dave, G. 2011. Environmental and health hazard ranking and assessment of plastic polymers based on chemical composition. Sci. Total Environ. 409: 3309–3324.

Löder, M.G.J., Kuczera, M., Mintenig, S., Lorenz, C. and Gerdts, G. 2015. Focal plane array detector-based micro-Fourier-transform infrared imaging for the analysis of microplastics in environmental samples. Environ. Chem. 12: 563–581.

Long, M., Moriceau, B., Gallinari, M., Lambert, C., Huvet, A., Raffray, J. et al. 2015. Interactions between microplastics and phytoplankton aggregates: impact on their respective fates. Mar. Chem. 175: 39–46.

Ma, B., Xue, W., Ding, Y., Hu, C., Liu, H. and Qu, J. 2019. Removal characteristics of microplastics by Fe-based coagulants during drinking water treatment. J. Environ. Sci. 78: 267–275.

Mahon, A.M., O'Connell, B., Healy, M.G., O'Connor, I., Officer, R., Nash, R. et al. 2017. Microplastics in sewage sludge: effects of treatment. Environ. Sci.Technol. 51: 810–818.

Mason, S.A., Garneau, D., Sutton, R., Chu, Y., Ehmann, K., Barnes, J. et al. 2016. Microplastic pollution is widely detected in US municipal wastewater treatment plant effluent. Environ. Pollut. 218: 1045–1054.

Mason, S.A., Welch, V.G. and Neratko, J. 2018. Synthetic polymer contamination in bottled water. Front. Chem. 23: 383–407.

Masura, J., Baker, J., Foster, G. and Arthur, C. 2015. Laboratory methods for the analysis of microplastics in the marine environment: Recommendations for quantifying synthetic particles in waters and sediments. Mar. Pollut. Bull. 32: 266–301.

Mincer, T.J., Zettler, E.R. and Amaral-Zettler, L.A. 2016. Biofilms on plastic debris and their influence on marine nutrient cycling, productivity, and hazardous chemical mobility. Hazard. Chem. Asso. Plast. Mar. Environ. 12: 221–233.

Murphy, F., Ewins, C., Carbonnier, F. and Quinn, B. 2016. Wastewater treatment works (WwTW) as a source of microplastics in the aquatic environment. Environ. Sci. Technol. 50: 5800–5808.

Nabi, I., Li, K., Cheng, H., Wang, T., Liu, Y., Ajmal, S. et al. 2020. Complete photocatalytic mineralization of microplastic on TiO2 nanoparticle film. iScience. 23: 101–126.

Napper, I.E. and Thompson, R.C. 2016. Release of synthetic microplastic plastic fibres from domestic washing machines: Effects of fabric type and washing conditions. Mar. Pollut. Bullet. 112: 39–45.

Nuelle, M.T., Dekiff, J.H., Remy, D. and Fries, E. 2014. A new analytical approach for monitoring microplastics in marine sediments. Environ. Pollut. 184: 161–169.

Patel, M.M., Goyal, B.R., Bhadada, S.V., Bhatt, J.S. and Amin, A.F. 2009. Getting into the brain. CNS Drugs. 23: 35–58.

Peng, B.Y., Su, Y., Chen, Z., Chen, J., Zhou, X., Benbow, M.E. et al. 2019. Biodegradation of polystyrene by dark (*Tenebrio obscurus*) and yellow (*Tenebrio molitor*) mealworms (Coleoptera: Tenebrionidae). Environ. Sci. Technol. 53: 5256–5265.

Periyasamy, A.P. and Tehrani-Bagha, A. 2022. A review of microplastic emission from textile materials and its reduction techniques. Polym. Degrad. Stab. 12: 109–151.

Perren, W., Wojtasik, A. and Cai, Q. 2018. Removal of microbeads from wastewater using electrocoagulation. ACS Omega. 3: 3357–3364.

Poerio, T., Piacentini, E. and Mazzei, R. 2019. Membrane processes for microplastic removal. Molecules 24: 4148–4211.

Redondo-Hasselerharm, P.E., Falahudin, D., Peeters, E.T. and Koelmans, A.A. 2018. Microplastic effect thresholds for freshwater benthic macroinvertebrates. Environ. Sci. Technol. 52: 2278–2286.

Rehse, S., Kloas, W. and Zarfl, C. 2016. Short-term exposure with high concentrations of pristine microplastic particles leads to immobilisation of *Daphnia magna*. Chemosphere 153: 91–99.

Roch, S. and Brinker, A. 2017. Rapid and efficient method for the detection of microplastic in the gastrointestinal tract of fishes. Environ. Sci. Technol. 51: 4522–4530.

Rochman, C.M., Tahir, A., Williams, S.L., Baxa, D.V., Lam, R., Miller, J.T. et al. 2015. Anthropogenic debris in seafood: Plastic debris and fibers from textiles in fish and bivalves sold for human consumption. Sci. Rep. 5: 1–10.

Rochman, C.M., Munno, K., Box, C., Cummins, A., Zhu, X. and Sutton, R. 2020. Think global, act local: local knowledge is critical to inform positive change when it comes to microplastics. Environ. Sci. Technol. 55: 4–6.

Rochman, C.M., Kross, S.M., Armstrong, J.B., Bogan, M.T., Darling, E.S., Green, S.J. et al. 2015. Scientific evidence supports a ban on microbeads. Sci. Rep. 112: 132–155.

Rochman, C.M., Kurobe, T., Flores, I. and Teh, S.J. 2014. Early warning signs of endocrine disruption in adult fish from the ingestion of polyethylene with and without sorbed chemical pollutants from the marine environment. Sci. Total Environ. 493: 656–661.

Ryan, P.G. and Moloney, C.L. 1990. Plastic and other artefacts on South African beaches: Temporal trends in abundance and composition. Afr. Tydskr. Wet. 86: 450–452.

Salerno, C., Vergine, P., Berardi, G. and Pollice, A. 2017. Influence of air scouring on the performance of a Self Forming Dynamic Membrane BioReactor (SFD MBR) for municipal wastewater treatment. Bioresour. Technol. 223: 301–306.

Schmaltz, E., Melvin, E.C., Diana, Z., Gunady, E.F., Rittschof, D., Somarelli, J.A. et al. 2020. Plastic pollution solutions: emerging technologies to prevent and collect marine plastic pollution. Environ. Int. 144: 106–167.

Schulz, M., Neumann, D., Fleet, D.M. and Matthies, M. 2013. A multi-criteria evaluation system for marine litter pollution based on statistical analyses of OSPAR beach litter monitoring time series. Mar. Environ. Res. 92: 61–70.

Sehgal, R. and Gupta, R. 2020. Polyhydroxyalkanoate and its efficient production: an ecofriendly approach towards development. 3 Biotech 10: 549–563.

Sharma, D., Bhardwaj, K.K. and Gupta, R. 2021. Immobilization and applications of esterases. Biocatal. Biotransformation. 40: 153–168.

Song, Y.K., Hong, S.H., Jang, M., Han, G.M. and Shim, W.J. 2015. Occurrence and distribution of microplastics in the sea surface microlayer in Jinhae Bay, South Korea. Arch. Environ. Contam. Toxicol. 69: 279–287.

Song, Y.K., Hong, S.H., Jang, M., Han, G.M., Rani, M., Lee, J. et al. 2015. A comparison of microscopic and spectroscopic identification methods for analysis of microplastics in environmental samples. Mar. Pollut. Bullet. 93: 202–209.

Su, Y., Hu, X., Tang, H., Lu, K., Li, H., Liu, S. et al. 2022. Steam disinfection releases micro (nano) plastics from silicone-rubber baby teats as examined by optical photothermal infrared microspectroscopy. Nat. Nanotechnol. 17: 76–85.

Sulaiman, S., Yamato, S., Kanaya, E., Kim, J.J., Koga, Y., Takano, K. et al. 2012. Isolation of a novel cutinase homolog with polyethylene terephthalate-degrading activity from leaf-branch compost by using a metagenomic approach. Appl. Environ. Microbiol. 78: 1556–1562.

Sun, J., Dai, X., Wang, Q., Loosdrecht, M.C.M. and Ni, B.J. 2019. Microplastics in wastewater treatment plants: Detection, occurrence and removal. Water Res. 152: 21–37.

Sun, Y., Mwandeje, J.B., Wangatia, L.M., Zabihi, F., Nedeljković, J. and Yang, S. 2020. Enhanced photocatalytic performance of surface-modified TiO2 nanofibers with rhodizonic acid. Adv. Fiber Mater. 2: 118–122.

Tagg, A.S., Sapp, M., Harrison, J.P. and Ojeda, J.J. 2015. Identification and quantification of microplastics in wastewater using focal plane array-based reflectance micro-FT-IR imaging. Anal. Chem. 87: 6032–6040.

Talvitie, J., Mikola, A., Koistinen, A. and Setälä, O. 2017. Solutions to microplastic pollution–Removal of microplastics from wastewater effluent with advanced wastewater treatment technologies. Water Res. 123: 401–407.

Tanaka, K., Takada, H., Yamashita, R., Mizukawa, K., Fukuwaka, M.A. and Watanuki, Y. 2013. Accumulation of plastic-derived chemicals in tissues of seabirds ingesting marine plastics. Mar. Pollut. Bull. 69: 219–222.

Thompson, A. 2018. Earth has a hidden plastic problem—scientists are hunting it down. Sci. Am. 13: 122–154.

Tiwari, E., Singh, N., Khandelwal, N., Monikh, F.A. and Darbha, G.K. 2020. Application of Zn/Al layered double hydroxides for the removal of nano-scale plastic debris from aqueous systems. J. Hazard. Mater. 397: 122–269.

Tokiwa, Y., Calabia, B.P., Ugwu, C.U. and Aiba, S. 2009. Biodegradability of plastics. Int. J. Mol. Sci. 10: 3722–3742.

Ugolini, A., Ungherese, G., Ciofini, M., Lapucci, A. and Camaiti, M. 2013. Microplastic debris in sandhoppers. Estuar. Coast. Shelf Sci. 129: 19–22.

Van Cauwenberghe, L., Vanreusel, A., Mees, J. and Janssen, C.R. 2013. Microplastic pollution in deep-sea sediments. Environ. Pollut. 182: 495–499.

Wang, J., Chen, G., Christie, P., Zhang, M., Luo, Y. and Teng, Y. 2015. Occurrence and risk assessment of phthalate esters (PAEs) in vegetables and soils of suburban plastic film greenhouses. Sci. Total Environ. 523: 129–137.

Wang, T., Li, B., Zou, X., Wang, Y., Li, Y., Xu, Y. et al. 2019. Emission of primary microplastics in mainland China: invisible but not negligible. Water Res. 162: 214–224.

Wright, S.L. and Kelly, F.J. 2017. Plastic and human health: a micro issue?. Environ. Sci. Technol. 51: 6634–6647.

Xing, L., Zhang, Q., Sun, X., Zhu, H., Zhang, S. and Xu, H. 2018. Occurrence, distribution and risk assessment of organophosphate esters in surface water and sediment from a shallow freshwater Lake, China. Sci. Total Environ. 636: 632–640.

Xue, B., Zhang, L., Li, R., Wang, Y., Guo, J., Yu, K. and Wang, S. 2020. Underestimated microplastic pollution derived from fishery activities and "hidden" in deep sediment. Environ. Sci. Technol. 54: 2210–2217.

Ye, Z., Padilla, J.A., Xuriguera, E., Brillas, E. and Sirés, I. 2020. Magnetic MIL (Fe)-type MOF-derived N-doped nano-ZVI@ C rods as heterogeneous catalyst for the electro-Fenton degradation of gemfibrozil in a complex aqueous matrix. Appl. Catal. B: Environ. 266: 118–174.

Yuan, F., Yue, L., Zhao, H. and Wu, H. 2020. Study on the adsorption of polystyrene microplastics by three-dimensional reduced graphene oxide. Water Sci. Technol. 81: 2163–2175.

Zangmeister, C.D., Radney, J.G., Benkstein, K.D. and Kalanyan, B. 2022. Common single-use consumer plastic products release trillions of sub-100 nm nanoparticles per liter into water during normal use. Environ. Sci. Technol. 56: 5448–5455.

Zettler, E.R., Mincer, T.J. and Amaral-Zettler, L.A. 2013. Life in the "plastisphere": microbial communities on plastic marine debris. Environ. Sci. Technol. 47: 7137–7146.

Zhang, Z. and Chen, Y. 2020. Effects of microplastics on wastewater and sewage sludge treatment and their removal: A review. Chem. Eng. J. 382: 122–155.

Zhou, D., Chen, J., Wu, J., Yang, J. and Wang, H. 2019. Biodegradation and catalytic-chemical degradation strategies to mitigate microplastic pollution. J. Saf. Enivron. 19: 2102–2111.

Zhou, P., Ren, W., Nie, G., Li, X., Duan, X., Zhang, Y. and Wang, S. 2020. Fast and long-lasting iron (III) reduction by boron toward green and accelerated fenton chemistry. Angew. Chem. Int. Ed. 59: 16517–16526.

Zhu, D.P. 2009. Development and recycling for plastics packaging waste. Shanghai Plast. 147: 25–29.

Ziajahromi, S., Neale, P.A., Rintoul, L. and Leusch, F.D. 2017. Wastewater treatment plants as a pathway for microplastics: development of a new approach to sample wastewater-based microplastics. Water Res. 112: 93–99.

CHAPTER 6

Environmental Toxicology, Fate and Risk Assessment of Emerging Contaminants

Vandana Singh,[1,*] *Harsh Sable,*[2] *Vaishali,*[2] *Mohan Ghalley*[3] and *Karma Tenzin*[3]

1. Introduction

With the accelerated growth of modern industry and agriculture, large amounts of Emerging Pollutants (EPs) are continuously released into the environment. Environmental Pollutants (EPs) comprise a diverse range of substances, including pharmaceuticals, personal care products, detergents, and polymers. Despite their potential impact on the environment, there is currently a lack of established regulations governing their environmental safety. The global distribution of these phenomena has been documented across various environmental matrices, including aqueous, terrestrial, and atmospheric systems. The issue of environmental pollution caused by emerging pollutants (EPs) has garnered worldwide attention due to the potential hazards they pose to both the environment and human health. Despite the use of advanced treatment technologies, complete removal of these pollutants from the environment remains a challenge. Thus, environmental policies are gaining prominence in ongoing research on environmental issues (Sahani et al., 2022; Radwan et al., 2022; Morin-Crini et al., 2022a).

In order to effectively prevent and manage the emergence of these compounds, it is necessary to assess their presence, movement, alteration, and removal in various environmental contexts. Identifying the origin and mode of transmission of

[1] Assistant Professor, Department of Microbiology, SSAHS, Sharda University, Greater Noida, (U.P.) India.
[2] Department of Forensic Science, SSAHS, Sharda University, Greater Noida, (U.P.) India.
[3] Department of Medical Lab Sciences, SSAHS, Sharda University, Greater Noida, (U.P.) India.
* Corresponding author: vandana.singh@sharda.ac.in

environmental pollutants (EPs) is crucial for effective environmental management aimed at mitigating their impact on the urbanized community. Despite the discovery of numerous emerging pollutants, only a limited number of them have undergone toxicological assessments. Certain pollutants have been found to be highly toxic to both animals and humans. Therefore, it is imperative to to investigate the toxicity of additional endocrine disruptors that harm humans, plants and animals, as well as the corresponding molecular mechanisms involved. Furthermore, the ecological hazard resulting from the emergence of pollutants is of significant concern, as many of these substances are deemed to possess the potential for deleterious effects on both the fauna and human well-being. Environmental risk assessment is a necessary step before taking the right steps to handle the environment. Most of the time, it is based on large amounts of toxicity data or strong epidemiologic surveys and the exact exposure assessment (Morin-Crini et al., 2022b; Tang et al., 2019; Sahani et al., 2022; Radwan et al., 2022).

Various hazardous materials, including pharmaceuticals for both humans and animals, hormones, personal care products, disinfection by-products, surfactants, household products, flame retardants, industrial compounds, fertilizers, and growth agents, agrochemicals like pesticides, trace metals, microplastics, and nanoparticles, have been identified as pollutants that pose a significant ecological hazard (Fig. 6.1). Studies have demonstrated that these contaminants have detrimental effects on the

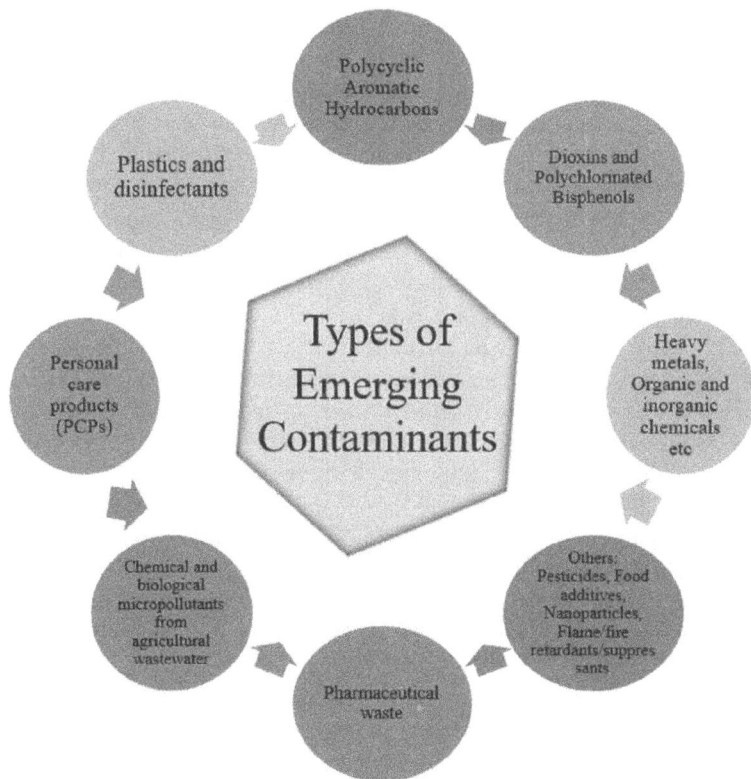

Figure 6.1 Various types of emerging contaminants.

environment's natural resources and biochemical processes. This can result in various harmful outcomes, including endocrine disruption, mutagenesis, carcinogenesis, and congenital disorders in living organisms. As a result, there has been a significant amount of scientific investigation aimed at analyzing the origins, harm, physical and chemical durability, destiny, and the methods of breaking down these substances. This is crucial in order to create successful approaches for reducing their negative effects on both the natural world and living beings (Tang et al., 2019; Cho and Wei 2023; Jatoi et al., 2023).

1.1 Widespread ECs Across the Globe

Emerging Contaminants (ECs) are generally a diverse group of chemical compounds that typically contain carbon, fluorine, nitrogen, oxygen, hydrogen, and sulphur. The occurrence, fate, and transfer of ECs in the environment are influenced by numerous factors, including water characteristics, climatic conditions, contamination sources, socio economic status, available remediation methods, and the physicochemical properties of the compounds such as pharmaceuticals and personal care products are frequently detected among ECs and usually exhibit higher concentrations due to their pervasive consumer use and polar nature.

According to literature reported by Saidulu et al. (2021), the prevalence of significant environmental contaminants (ECs) in several largely populated nations worldwide, including but not limited to China, India, the United States, Germany, the United Kingdom, and Canada. Pharmaceuticals and personal care products are the mostly identified emerging contaminants (ECs) in water systems. Their high concentrations can be attributed to their widespread use by consumers and their polar characteristics. Additionally, with the present situation, more than 50% of the global population will be required to take daily medication. Notably, according to Statista, pharmaceutical product sales are highest in North America, followed by Europe. For instance, by analyzing various analgesics, it has been determined that Ibuprofen concentrations in the influent of wastewater treatment plants (WWTPs) are higher in North America (75.8 g L^{-1}) than in Asia (26.45 g L^{-1}), Australia (10.3 g L^{-1}), and Europe (33.76 g L^{-1}). Similarly, Naproxen, an additional commonly used analgesic, has been detected in concentrations ranging from 0.08 to 25 g L^{-1} in the United States, India, China, Northern America, and several European countries. Another major contaminant are antibiotics, and in the past few decades, there has been a 36% rise in the use of antibiotics, with the majority of this increase being observed in Asia, South America, and European nations. Antibiotics are eliminated from the bodies of humans and animals in diverse forms via urine and faeces and reaches the environment as contaminants. For example, the levels of Ciprofloxacin (an antibiotic) present in the influents of wastewater treatment plants (WWTP) in China are comparatively lower (0.028–0.175 µg L^{-1}) than those found in Spain (0.160–13.6 µg L^{-1}).

Pesticides, which mainly come from agricultural uses, are frequently detected in influent samples from wastewater treatment plants (WWTPs), for example, Tauchnitz et al. (2020) stated that, the presence of pesticides (Glyphosate and AMPA) was found to be responsible for the most significant levels of concentration in both surface water samples (maximum of 58 µg L^{-1}) and soil samples (maximum of 0.19 mg kg^{-1}) of Germany (Tauchnitz et al., 2020). Hexachlorobenzene (HCB) pesticides have been identified in soil samples from different geographical locations. For example, Rajmohan et al. (2020) stated that, soil samples collected from seven metropolitan areas in India were found to have a widespread presence of HCB pesticides. Elevated levels of these pesticides were observed in sites located near automotive production facilities, such as New Delhi (ND-06, 0.72 ng/g). Similarly, unregulated disposal and decomposition of organochlorine pesticides in Tajikistan have led to substantial soil pollution, particularly in the Vakhsh and Konibodam areas, where pesticide concentrations surpass 10 ppm. On the same note. over the last decade, Thailand has also been observed for widespread utilization of insecticides, herbicides, and fungicides as the most frequently used pesticides. According to a study, around 24% of pesticide use in Thailand is linked to the surface application of pesticides on fruits and vegetables (Rajmohan et al., 2020).

The antimicrobial compound known as Triclosan has been found to have elevated concentrations in the influents of wastewater treatment plants in both the United States (3.4 µg L^{-1}) and China (2.08 µg L^{-1}). Nonylphenol and Bisphenol-A, which are among the emerging contaminants, have been identified in wastewater at different concentrations depending on the geographical location. The persistent discharge and prolonged exposure of emerging contaminants have the potential to endanger both human health and aquatic life (Tauchnitz et al., 2020).

Khanverdiluo et al. (2021) reported the detection of Polycyclic Aromatic Hydrocarbons (PAHs) in human breast milk. According to the research results, Mexico showed the maximum concentration of Polycyclic Aromatic Hydrocarbons (PAHs) at 9.830 nanograms per gram (ng/g), whereas Japan demonstrated the minimum recorded concentration at 0.009 ng/g. (Khanverdiluo et al., 2021)

Additionally, other major contaminants are heavy metals, which are metallic chemical elements that possess a high density in comparison to water. Based on the hypothesis that weight and toxicity are correlated, heavy metals encompass metalloids, like arsenic, which can elicit toxic effects at low levels of exposure. Chronic exposure to arsenic is estimated to be faced by millions of individuals globally, especially in countries like Bangladesh, India, Chile, Uruguay, Mexico, and Taiwan. The main route of exposure to arsenic is via groundwater that has been contaminated with high levels of this element. The modes of arsenic exposure encompass oral ingestion, inhalation, dermal contact, and, to a small degree, the parenteral route. Regarding atmospheric contamination, remote regions exhibit arsenic levels ranging from 1 to 3 ng/m3, while urban centres present concentrations between 20 and 100 ng/m3. Arsenic levels in water are usually below 10 µg/L, but they could be higher near natural mineral occurrences or mining sites. Different kinds of food could contain arsenic, with levels ranging from 20 to 140 nanograms per kilogram. Naturally occurring arsenic concentrations in soil typically range from 1 to 40 mg/kg, but pesticide application or incorrect waste disposal practices can lead to elevated levels (Tchounwou et al., 2012).

According to Corrales et al. (2015), Bisphenol A has been widely observed in urban ecosystems in Asia, Europe, and North America. The concentrations of effluents were found to exceed 50% for aquatic life. Nevertheless, the surface water analyses conducted in Asia revealed concentrations reaching up to 80%. Similarly, the maximum concentrations of BPA detected in sediments in Asia exceeded those recorded in Europe. The levels of Bisphenol A (BPA) found in different wildlife species, particularly fish, were seen to vary between 0.2 and 13,000 nanograms per gram (ng/g). According to the National Health and Nutrition Examination Survey conducted by the US Center for Disease Control and Prevention, the median levels were surpassed by 60% in Europe, and 40% in Asia (Corrales et al., 2015). Similarly, some more examples are given in Table 6.1.

Table 6.1 Components of Solid Waste, Proportion of Components, Country, Organic Waste Percentage.

Components of Solid Waste	Proportion (Percentage by Mass)	Country Generating Organic Waste	Organic Waste Percentage (In %)	Reference
Combustible Chips	1.2–7.8	Afghanistan	70	(Karak et al., 2012; Khoshbeen et al., 2020)
Ceramics and Glass	0.5–3.5	Bangladesh	60–75	(Ahmed et al., 2023; Alam and Qiao, 2020)
Inert Components	22–48	China	60	(Ma et al., 2020; Zhu et al., 2021)
Kitchen and Food Waste	40–65	India	48–54	(Patwa et al., 2020; Srivastava et al., 2020)
Metals	0.25–2.4	Nigeria	64	(Adeniyi and Afon, 2022; Mama et al., 2021)
Paper and Plastic Rubbers	1–10	Turkey	50	(Özşeker et al., 2022; Şentürk, Ersoy, and 2020 2020)
Textiles	1.5	Zimbabwe	47	(Kwenda et al., 2022; Muisa Zikali et al., 2022)

2. Sources and Types of Emerging Pollutants

The pollutants referred above will be discussed elaborately further in the chapter. These pollutants mainly contaminate the natural resources that are air, soil, water, which affects and transmits these pollutants to other sources for example, plants absorb and assimilate into themselves that are later consumed by herbivores as well as omnivores (Table 6.2). Environmental pollutants are categorized as;

2.1 Bisphenol A (BPA)

A monomer identified as bisphenol A (BPA) was initially created as a synthesized estrogen in the 1890s and was later shown to exert the same effect as estrone in activating the reproductive system of female rats. Bisphenol A is exogenous which disrupts the normal functioning of endocrine system causing major adverse effects to organisms as well as their leading progeny and which is why it is also known as Endocrine Disruptor (ED). Humans are exposed to this pollutant through air, water, soil, and food. Extensively used daily life substances such as plastics, metallic food cans, toys, cosmetics, detergents, pesticides are the sources of bisphenol A (Cojocaru et al., 2017; Rochester, 2013).

Table 6.2 Contaminants, sources, and toxicity on humans.

Contaminants		Sources	Adverse effect	Refs
Heavy Metals	As	Mining, ore-processing, smelting, groundwater runoff, dyes, soaps, semi-conductor, paints, drugs, pesticides.	Acute poising; nausea, diarrhoea, sensation of pricking, erythrocytes and leukocytes in less amounts, hypertension. Chronic poisoning; Carcinogenic for brain, lungs, liver, skin.	(Ghosh and Sil, 2023; Khan and Flora, 2023)
	Cd	Paint pigment, mining, smelting, batteries, cigarette smoking, alloys, fertilizers, municipal waste.	Acute poisoning; nausea, diarrhoea, hypertension, headache. Chronic poisoning; nephrotoxicity, hepatotoxicity, osteoporosis.	(Yang et al., 2023)
	Pb	Glass production, insecticides, canned food, smoking, paints, drinking water, lead bullets, PVC pipes, toys, batteries, cosmetics.	Acute poisoning; Abdominal pain, severe headaches, renal disorders, arthritis, fatigue, hallucinations, vertigo. Chronic poisoning; renal dysfunctions, neurological disorders, weight loss, dyslexia, psychosis, hyperactivity.	(Balali-Mood et al., 2021; Grant, 2020; Kumar et al., 2020)
	Hg	Industrial and municipal wastewater, mining, agriculture products, fishes.	Acute poisoning; skin rashes, depression, headaches, hair loss, memory loss, nausea, diarrhoea, heart rate and blood pressure increase. Chronic poisoning; Brain, kidney, lungs, liver disorders.	(Basu et al., 2023)
PAHs	Naphthalene Acenaphthylene Pyrene Fluorene Anthracene Acenaphthene	Forest fires, volcanic eruptions, biomass fuel combustion, incomplete combustion of carbon hydrogen containing raw material (crude oil, gas, woods, organic material, coal fuel), combustion of polypropylene, polystyrene, industrial and domestic wastes. Aluminium production, coke manufacturing, power generation, gasoline engine, diesel engine.	Oxidative stress that may lead to cell damage, inflammation, infertility, cardiovascular disorders, poor fetal development, soot exposure leading to skin cancer, lung cancer.	(Soursou et al., 2023; Vijayanand et al., 2023)

Table 6.2 contd. ...

... Table 6.2 contd.

Contaminants	Sources	Adverse effect	Refs
PCBs	Capacitors, transformers, landfills, municipal waste incineration, wastewater effluents, metal recycling scraps, used oil, hydraulic fluids.	Difficulty in breathing, nausea, vomiting, skin, eye, throat irritation, fatigue, weight loss, diarrhoea. Neurotoxicity, endocrine disruption, immunotoxicity, infertility.	(Dreyer and Minkos, 2023; Q. Wang et al., 2023)
PFCs	Electronics, refrigerants, soundproofing windows, aluminium production, fishes.	Breathing problems, skin and eye irritation, nausea, diarrhoea, lung and neurological disorders.	(Kuznetsova and Lyanguzov, 2023)
BPA	Leaching from canned foods, polycarbonate tableware, water bottles, epoxy resins, eyewear, water pipes.	Reproductive disorders, cardiovascular and neurological disorders, insulin resistance, carcinogenic in nature.	(Hahladakis,et al., 2023; Parrado et al., 2023)
Pharmaceutical waste	Dispensed drugs, contaminated and expired products, incorrect disposal by humans, leaching from landfills, veterinary medications and their disposal, drug waste.	Mercury poisoning, breathing problems, neurological disorders, confusion, headaches, diarrhoea, chemical burns, inflicted injuries, severe STD such as HIV-AIDS in case of blood transfusion.	(Li et al., 2014; Maeng et al., 2013a)
Hospital Effluents	Pharmaceutical wastes such as antibiotics, disinfectants, organic matter, microbial wastes.	Headache, nausea, diarrhoea, stomach aches, difficulty in breathing, skin disorders, neurotoxicity.	(Mathur et al., 2023)

Sources: Bisphenol A (BPA) is mostly used as an intermediary material in the production of polycarbonate plastic, epoxy resins, and other specialist resins. Polycarbonate is widely used in a number of industries, including glazing and sheets, electrical and electronic goods, electronic storage media, and domestic items such as bottles, cutlery, and containers. Epoxy resins are commonly used in architectural structures, marine and automotive coatings, container coatings, and printed circuit boards as protective coatings. BPA is also used in the manufacture of phenoplast, phenolic, and unsaturated polyester resins, polyvinyl chloride, and thermal paper (Arnold et al., 2013). Studies have shown that BPA can travel from polycarbonates (PC) both through diffusion and polymer hydrolysis; however, diffusion has far less impact on BPA release from polycarbonates. From plastics, the toxins migrate to the food material present when the plastic is old, brittle, scratched, and heated. Other sources of bisphenol A are dental sealants, polycarbonate products ranging from eyeglass lenses, CDs, kitchenware, etc. Furthermore, meals in cans are largely subjected to BPA, which escapes from the lacquer used to cover tins. These lacquers are created using epoxy resins with BPA diglycidyl ether as their primary ingredient. The build-up of BPA in the atmosphere, the interaction of food with BPA-containing polymers, and the accessibility of livestock and uncooked plants all contribute to the existence of BPA in foodstuffs. Groundwaters close to waste disposal facilities polluted with BPA or waste disposal facilities that contained plastic garbage were found to have noticeably increased BPA contents. BPA is thought to be mostly released into the environment as the outcome of industrial activities. Humans are further exposed to BPA through their respiratory system and skin touch. Due to migration from products created with BPA synthetic polymers, BPA is found in dust (Michałowicz, 2014; Zhang et al., 2023).

Health risks: The process of body maturation, development, and control are all handled by the endocrine system. Different hormones are secreted by several glands, including the pituitary, pancreas, adrenals, and testes. These hormones regulate all bodily processes. The endocrine system's disruption leads to many illnesses and health issues. Endocrine disruptors are substances that interfere with the endocrine system and have negative effects on the immunological, neurological, and reproductive systems as well as on animals (Alharbi et al., 2018). Disorders in how the reproductive system of males and female operate are a part of reproductive issues and illnesses. Birth defects, premature birth, developmental issues, birth weight problems, impotence, decreased fertility, and menstruation abnormalities are a few of these. POP pollution also affects the reproductive systems. Several investigations revealed the well-known consequences of POPs on people. Scientists conducted research and reported on how various testicular cells reacted differently to PCB contaminants. The testicular germ, Leydig, and Sertoli cells were among the afflicted cell types (Alharbi et al., 2018; Hahladakis et al., and 2023; 2022; Kumar et al., 2023).

Toxicity: BPA's estrogenic efficacy is equivalent to that of 17-beta estradiol (E2) for reactions mediated via non-nuclear-estrogen receptors, despite having been reported to possess a lesser affinity for nuclear oestrogen receptors. BPA can also serve as an antiestrogen by contending with endogenous E2 to suppress the estrogenic response.

BPA has the potential to be antiandrogenic by directly binding to androgen receptors and preventing the effects of natural androgens (Rochester, 2013; Cojocaru et al., 2017; Zheng et al., 2013; Mo et al., 2021).

2.2 Poly-aromated Hydrocarbons (PAH)

Constructed with carbon and hydrogen atoms structured as joined rings of benzene in a unidirectional, clustered, or angled orientation, PAH is recognized for its potent carcinogenic, mutagenic, and poisonous effects. Persistent Organic Pollutants (POPs) are a class of hazardous pollutants that include Polycyclic Aromatic Hydrocarbons (PAHs). These are forms of organic pollutants that are chemically repelled to decomposition, can linger in the earth's atmosphere for quite sometime, and possess the possibility of having a negative impact on the ecosystem. One of the main routes for the environmental dissemination and transnational buildup of PAHs is aerial mobility (Bandowe and Meusel, 2017; Sam et al., 2023). Over time, PAHs accumulate in soils and road dust and find their way into aquatic environments. In close proximity to automobiles and other sources of combustion, soils and street debris serve as a direct sink for atmospheric PAHs. Rain and stormwater easily transport PAHs to surrounding aquatic habitats from these environmental compartments. As PAHs are hydrophobic, they preferentially partition in aquatic environments and accumulate in the granular stage in build-up sediment (Hussain et al., 2018).

Although there are hundreds of PAH chemicals in the aerosphere, PAH analysis is typically only used to identify 6 to 16 molecules. PAHs have a wide range of physicochemical characteristics. High molecular weight molecules tend to be more lipophilic in nature, a little volatile, and less soluble in water than reduced molecular weight components. The most well-known composite from this class is the very carcinogenic combination benzo/a/piren (BaP). Both the European Commission and the United States Environmental Protection Agency have identified PAHs as critical contaminants (Dobaradaran et al., 2019).

Environmental PAHs are hazardous to a number of ecosystem-related creatures. Researchers have paid a lot of attention to PAHs since the global mortality toll from cancer is continuously rising. Researchers frequently used ecosystem risk to describe PAH risk to nearby animals and ecology in aquatic habitats by using the Risk Quotient (RQ) of specific PAHs. Risk quotients (RQ) values represented the degree of risk that a particular PAHs posed. Along with benzene, phenols, aldehydes, and other renowned semi-volatile organic pollutants, PAHs are categorized as Hazardous Air Pollutants (HAP) under the category of non-halogenated organic compounds. The physical and chemical features, particularly non-polarity, and hydrophobicity, which cause their persistent nature in various habitats, determine how they act in the environment as a whole. They are highly susceptible to long-distance movement, bioaccumulation, and biomagnification. The ability of PAHs to be adsorbed on soot surfaces and to stay in a gaseous state depends on their volatility and molecular weight (Hussain et al., 2018; Maliszewska-Kordybach, 1999).

Sources: All partial combustion reactions of organic materials result in the production of PAHs. An oxygen-deficient flame, with temperatures ranging between

650 and 900°C, and fuels with low levels of oxidation all assist in their formation. At the moment, the contribution of natural pyrogenic PAH sources like lava eruptions and fires in forests to total PAH exposure is not very substantial. Sources that are human-caused could be divided into two groups: combustion used to minimize waste (such as waste incineration) and combustion used to supply energy (such as coal, oil, gas, and wood). The primary kind consists of static assets like industry (most often, coke and carbon generation, petroleum manufacturing, etc.), house heating systems (furnaces, fireplaces, and burners, gas, and oil burners), power and heat generation (coal, oil, wood, and peat energy plants), and transportation sources like automobiles, trucks, trains, airplanes, and maritime traffic (gasoline and diesel engines). Urban and commercial waste from incineration is included in the second category. Uncontrolled fires like those used for agriculture, recreation, or crematories, as well as vapourization from soils, plants, and other surfaces, are some examples of additional unrelated sources (Kordybach, 1999). Similarly, cigarettes are known and studied as a source of PAH. The low molecular weight and lipophilicity PAH merging with high volatility and water solubility get disposed into the water as well as the environmental air (Dobaradaran et al., 2019).

Toxicity: PAHs can serve as photosensitizers and substantially absorb sunlight. Sunlight significantly increases the toxicity of PAHs to both aquatic and terrestrial vegetation. Plants offer a useful model for researching the toxicity of photoactive pollutants. Two processes—photosensitization and photo-modification reactions— are responsible for the photoinduced toxicity of PAHs. Reactive Oxygen Species (ROS), which are created throughout a living thing and are extremely harmful, are often how photosensitization responses progress. When PAHs are photo-modified, which often involves oxidizing the parent substance, a combination of very hazardous chemicals arises. This is comparable to the elevated PAH risks that follow cytochrome P450 activation. Interestingly, it was discovered that humic acids reduce the toxicity of PAHs to duckweed (Lemna gibba), most probably because they bind to the PAHs. As a result, PAHs serve as an illustration of an environmental toxin in which exposure to light can increase danger, whilst interaction can decrease it.

Health Risks: Cardiac disorders are a result of the interaction of social, behavioural in nature, and metabolic variables, as well as POP pollution. POP-contaminated food is associated with diabetes, obesity, and other behavioural variables. It has also been shown that obesity is associated with PAH. Various other respiratory, reproductive disorders are seen in individuals (Alharbi et al., 2018).

2.3 Heavy Metals

Heavy metals are organically formed elements having an elevated weight of an atom and a density greater than five times that of water. Their widespread dissemination in the natural world is an outcome of their multiple factories, households farming, medical, and technological applications, prompting concerns about their potential effects on the well-being of humans and the natural world. Their toxic effect is governed by a number of variables, which include the amount consumed, the route of being exposed, and the species of chemical, along with the gender, age, genetic makeup, and dietary status of persons exposed. Due to their extreme toxicity, cadmium, arsenic, chromium, mercury, and lead have been named among the foremost metals of health concern to the public. The aforementioned metallic elements have been classed as systemically hazardous chemicals, with modest amounts of exposure causing damage to organs. They are also classified as potential human carcinogens (known or probable) by the US Environmental Protection Agency and the International Agency for Research on Cancer (Ismanto et al., 2023; Molina and Segura, 2021).

Toxicity

A characteristic of toxic metal physiological science is that, despite many being necessary for growth, it is also known that they have very harmful effects on cells, mostly due to their capacity to denature protein molecules. Microbes are able to absorb and accumulate metal cations from their surroundings. Despite the typically small quantities of metal cations required for development, these absorption processes can still function at greater amounts and affect metal toxicity, for both acquiring species individually and to the community of microbes as a whole (Gadd and Griffiths, 1977).

Ionic and oxidative stress processes in living cells are responsible for lead, cadmium, mercury poisoning. Due to a discrepancy between the formation of radicals that are free and the development of antioxidants to detoxify reactive intermediates, living cells experience oxidative stress. Lead's effect results in an increase in reactive oxygen species (ROS) and a reduction in antioxidant levels.

Although these metallic elements have been reported to trigger numerous adverse effects on organs even at extremely low encounter levels, they are tested in systemic contaminants. When heavy metals enter the body by food, drink, air, or skin absorption, they become poisonous and build up in soft tissues if they are not metabolized by the body. The accumulation of heavy metals can lower energy levels and harm the operation of important organs including the nervous system, renal system, respiratory system, hematological system, and red blood cells. Continuous exposure to certain metals could result in degenerative processes that imitate illnesses like multiple sclerosis, Alzheimer's disease, Parkinson's disease, and muscular dystrophy on the physical, muscular, and neurological levels (Bhat et al., 2022; Mashabela et al., 2023).

2.4 Polychlorinated Biphenyls (PCBs)

Polychlorinated biphenyls (PCBs) (Fig. 6.2) are considered the most extensively researched environmental pollutants, with several studies in animals and human populations conducted to determine PCB carcinogenicity. PCBs are combinations of up to 209 distinct chlorinated chemicals (known as congeners or chlorinated hydrocarbons) that are no longer manufactured in the US, but are still present in the environment. PCB exposure causes adult acne-like skin disorders and neurobehavioural and immunological changes in youngsters. PCBs are carcinogenic to animals and have been detected in at least 500 of the US Environmental Protection Agency's (EPA) 1598 National Priorities List locations. PCBs (US trademarked

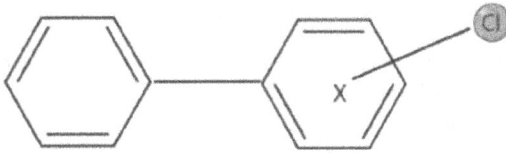

Figure 6.2 Chemical Structure: Polychlorinated biphenyls.

as Aroclor) have no identified natural origins, are uncoloured to light yellow oily liquids or solids, and exist as a vapour in air. PCBs are often odourless and tasteless (Ancona et al., 2017; Dudášová et al., 2016; Lin et al., 2022).

PCBs were discovered in wastewater from companies handling PCB equipment before they were outlawed and before the US Clean Water Act restricted wastewater emissions. These wastewaters were either released directly into surface waterways or transported to municipal wastewater treatment plants. PCB contamination is more frequent in urban industrial locations than in rural regions. While PCBs are not very volatile, they will partition into the air, especially the less chlorinated ones. The most essential mechanism for PCB dispersion is atmospheric transfer. As they are more resistant to metabolism, PCBs with a high degree of chlorination are significantly more persistent in the environment than those with a low degree of chlorination. The metabolism of microbial material is an exceptionally significant process for removing persistent organic contaminants from the environment, such as PCBs. Anaerobic dehalogenation of highly chlorinated PCBs in sediments from aquatics is a primary route for their removal, producing lower chlorinated congeners that can be digested more easily by aerobic enzymes. As a result, PCB levels in the natural environment are gradually reducing over time (Khanverdiluo et al., 2021).

Most exposures are environmental or occupational, with delayed symptoms being the first sign of toxicity. PCBs are resistant to metabolic processes, have a high lipid solubility, and so have substantial longevity in the environment. They bioaccumulate in terrestrial and aquatic creatures and become a significant source of human exposure through the food chain. Other key modes of exposure include dermal contact and through inhalation. Their impact on the environment is gradually diminishing over time (Barroso et al., 2019a; Lodeiro et al., 2019; Saidulu et al., 2021).

Toxicity: Adherence to the aryl hydrocarbon receptor, followed by the generation of reactive oxygen species (ROS) and the development of oxidative stress, is a case that could be similar to all tissues displaying toxicological repercussions as a result of PCB exposure. The aforementioned processes cause oxidative damage to the DNA, organelle and membrane damage, tissue dysfunction, and programmed death of cells, all of which have immunological, neurological, and endocrinological effects. It is estimated that there are approximately 200 stereochemically and structurally distinct PCBs, each having a unique receptor-binding property. As a result, the toxicological impact of the individual PCBs or PCB mixes is given as a toxic equivalent to the most hazardous congener, 2,3,7,8-tetrachlorodibenzop-dioxin (Barroso et al., 2019a; Mohapatra and Kirpalani, 2019; Kadam et al., 2016).

2.5 Hospital Effluents as a Source of Emerging Pollutants

Due to the laboratory as well as research operations or medicine excretion, hospital wastewaters comprise an array of harmful or persistent compounds that consist of pharmaceuticals, radionuclides, solvents, and disinfectants for medical reasons in a wide range of concentrations. Most of these substances are classified as emerging contaminants, which are uncontrolled pollutants that could be contenders for foreseeable regulations based on studies on their possible health impacts along with surveillance of their prevalence. Their key distinguishing feature is that they are not obligated to remain in the environment to create detrimental effects since their high transformation/removal rates could be compensated for by continual introduction into the ecosystem. Some of these substances, most of which are medicines along with personal care items, could also be found in municipal effluent (Fatimazahra et al., 2023; Mathur et al., 2023; Ramírez-Coronel et al., 2023).

2.6 Pharmaceutical Wastes

The prevalence of medications used for both human and veterinarian therapy, as well as their by-products, in aquatic ecosystems, is a growing issue. They infiltrate into the environment as contaminants in a number of ways, including outflow from treatment plants for wastewater or private septic tanks, landfill leachings, and hospital discharges. When extensive animal husbandry and aquaculture are practised on agricultural land, veterinary medications enter the runoff water. This release into the aquatic environment has the potential to harm nontarget aquatic creatures. The overwhelming majority of pharmaceuticals can be found at low amounts (ranging from ng/L to g/L) and in a complex mixture, which could result in unforeseen consequences, particularly with long-term exposure. The main issues of concern have been the spread of antibiotic resistance and the disturbance of endocrine systems, such as male fish feminization (Naghdi et al., 2018; Valdez-Carrillo et al., 2020).

2.7 Perfluorinated Compounds (PFCs)

PFC (Fig. 6.3) contamination of people can occur through a number of routes, which includes eating habits, contact with foodstuffs materials, non-food personal objects, and air both inside and outside. Although an array of writers has been trying to assess the proportion of these specific routes to the overall contamination, the data sets now available are insufficient to draw reliable inferences. Furthermore, numerous researchers have taken a number of approaches to the problem of food contamination. Some believe that eating fish and shellfish is the principal source of PFC contamination in human beings, whilst others believe that eating meat and vegetables is the predominant cause, particularly in countries where only modest amounts of fish are consumed. However, there is widespread agreement that food intake makes the greatest influence. Nonetheless, determining the proportion of specific meals to overall exposure will require comparing data from studies that rigorously identify PFC contamination of a broad spectrum of consumable goods. Aside from the so-called reference chemicals, PFOA and PFOS, toxicokinetic and

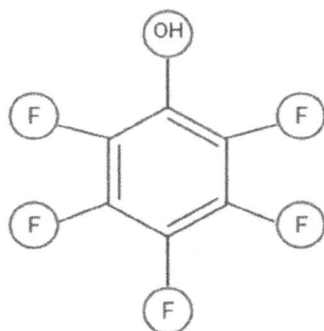

Figure 6.3 Perfluorinated compounds (PFCs).

toxicodynamic data for PFCs are not consistent enough to provide a convincing toxicological evaluation. Numerous articles have appeared in recent years that address the biological features of PFCs; however, these are typically confined to PFOA and PFOS (Costanza et al., 2019; Cordner et al., 2019; L. Liang et al., 2022; Schulz et al., 2020).

2.8 New Emerging Contaminants

A diverse range of emerging contaminants (ECs) have been detected in various environmental matrices such as surface waters, groundwaters, and soils. These ECs can be categorized into different kinds based on their properties(Lodeiro et al., 2019). Some new emerging contaminants are listed below:

2.8.1 Pesticides

These have been subjected to extensive assessments carried out by various regulatory bodies and scientific institutions, which have identified their potential hazards and effects. For centuries, the emphasis was predominantly on the precursor molecules of pesticides. Nevertheless, there is an increasing acknowledgment of the importance of their metabolites, which come from the degradation of the parent compounds in the environment. The biological activity and high toxicity levels of pesticide metabolites have been identified, and their presence in groundwater has been consistently observed, highlighting their significance as emerging contaminants. Multiple scientific investigations have provided evidence of the existence of pesticide metabolites in diverse environmental domains, such as surface waters and groundwaters. The acknowledgement of the potential harmfulness and enduring nature of pesticide by products has prompted a reassessment of the suitability of specific pesticides. Pesticides that were once deemed safe are currently being reclassified as emerging contaminants owing to the existence and potential hazards linked to their metabolites. The US Environmental Protection Agency (US EPA, 2015b) compiled the Contaminant Candidate List 3, which contains several newly discovered pesticide metabolites, including 3-Hydroxycarbofuran, Acephate, Acetochlor Ethane Sulphonic Acid (ESA), Acetochlor Oxanilic Acid (OA), Captan, Clethodim, and others. Including these in this list highlights the increasing acknowledgement of their importance and the requirement for additional research and control measures

(Alengebawy et al., 2021; Rajmohan et al., 2020). The identification and detection of pesticide metabolites in environmental matrices emphasize the significance of comprehensively considering the complete spectrum of transformation products that could emerge from the degradation of pesticides. The metabolites have the potential to display unique toxicological characteristics when compared to their parent compounds, which could pose a threat to both human health and ecosystems. Hence, it is imperative to incorporate both parent pesticides and their metabolites in comprehensive monitoring programs and risk assessments to establish efficient management and mitigation strategies. Further scientific investigations are required to enhance one's comprehension of the destiny, conduct, and probable consequences of pesticide metabolites, with the aim of protecting ecological integrity and human welfare (Lodeiro et al., 2019; Rajmohan et al., 2020; Tauchnitz et al., 2020).

2.8.2 Personal Care Compounds (PCCs)

These are a heterogeneous set of chemical substances that are frequently present in items like cosmetics, toiletries, fragrances, and cleaning agents. Although these compounds have been found to improve personal hygiene and overall health, they have also been identified as environmental pollutants that raise concerns. PCCs can enter the environment in many ways, including as direct releases from wastewater treatment facilities, the flushing of personal care items down toilets, and agricultural practices that use treated wastewater for irrigation (Meyer et al., 2019). PCCs include but are not limited to the following examples:

- Parabens are a class of preservatives that are extensively used in personal care and cosmetic formulations. Some comprise of methylparaben, ethylparaben, and propylparaben.
- Phthalates are a group of chemical compounds that function as plasticizers and are commonly used in a number of consumer goods such as fragrances, lotions, and hair sprays. Frequently encountered instances comprise of diethyl phthalate (DEP) and dibutyl phthalate (DBP).
- UV filters, including oxybenzone and octinoxate, are used in sunscreens and other skincare formulations to safeguard against detrimental UV radiation.
- Triclosan is an antimicrobial agent that exhibits antibacterial properties and is usually present in personal care products such as soaps and toothpastes.

On environmental discharge, PCCs could undergo diverse fate processes. These substances are subject to degradation through both biotic and abiotic mechanisms, including microbial degradation, photolysis, and chemical reactions. Nevertheless, certain polycyclic aromatic hydrocarbons (PCCs) exhibit persistence and can be present in surroundings for prolonged periods. Physicochemical characteristics determine the environmental fate of PCCs (Bell et al., 2011). They can be disseminated through diverse routes, such as the atmosphere, hydrosphere, and lithosphere. PCCs that enter wastewater treatment plants can be partially eliminated during the treatment process, but certain compounds could persist and be released into receiving waterways. In aquatic environments, PCCs can present additional transformations, such as sorption to sediments, bioaccumulation in aquatic organisms, and possible transfer along the food chain (Sahani et al., 2022).

There have been raised concerns regarding the probable toxicity of PCCs to both the environment and human health. Certain polychlorinated compounds (PCCs) have been observed to display characteristics of endocrine disruption, which can interfere with the endocrine systems of various organisms. Phthalates have been linked to reproductive and developmental impacts in aquatic organisms. The association between triclosan and the emergence of antibiotic resistance in bacteria has been established. Furthermore, there is mounting scientific evidence indicating that specific Persistent Chemical Compounds (PCCs), such as ultraviolet (UV) filters, could elicit detrimental impacts on aquatic ecosystems, such as coral bleaching and interference with the physiological processes of aquatic organisms. Efforts are being made to evaluate the environmental fate and toxicity of PCCs, as well as to develop control and management strategies, in order to mitigate potential risks (Chaturvedi et al., 2021; Su et al., 2020). These measures are being implemented in accordance with regulatory guidelines and ongoing research. This involves the advancement of eco-friendly substitutes and enhanced technologies for wastewater treatment to eliminate or decrease the levels of persistent and bio accumulative toxic substances in effluent streams. In addition, enhancing the knowledge of the general public and promoting the conscientious use of personal care products can aid in reducing the discharge of potentially harmful chemicals (PCCs) into the surroundings, thereby safeguarding the well-being of both the biotic community and human population (Priya et al., 2022).

2.8.3 Food Additives

These are compounds that are purposely added to food to improve its flavour, appearance, texture, or shelf life. While most food additives are considered safe to eat, some compounds that could act as pollutants and present health risks are causing increased concern. These pollutants can enter the environment through a number of sources and paths, and depending on their unique properties, their fate and toxicity. Preservatives (such as sulphites, benzoates), artificial sweeteners (such as aspartame, saccharin), colourants (such as tartrazine, caramel colour), flavour enhancers (such as monosodium glutamate, disodium inosinate), and emulsifiers (such as polysorbate 80, lecithins) are a few examples of food additives that can act as contaminants. Processed foods, beverages, and dietary supplements frequently contain these chemicals (Wu et al., 2022).

Environmental pollutants from food additives come from a number of sources. They can result from manufacturing procedures, inappropriate food waste disposal, and the use of food additives in products other than foods, like cosmetics and medications. Additionally, the human body's breakdown and metabolism of food additives can result in the excretion of their metabolites, which can then make their way into the environment and wastewater treatment facilities. The physicochemical characteristics of food additive contaminants determine how they act in the environment. While some additives may be transformed or degrade over time, others could last for a very long time. Water currents in aquatic areas can carry and spread toxins in food additives. They might also bioaccumulate in living things, posing a risk to higher trophic levels, and perhaps infiltrating the food chain.

Depending on a particular additive and its concentration, different food additive contaminants have different toxicities. Many pollutants have been linked to unfavourable health outcomes, including allergic responses, hypersensitivity, and intolerances. For instance, sulphites, a common preservative, could cause respiratory issues in people who are vulnerable. As they could have negative effects on the metabolism and the onset of some diseases, artificial sweeteners like aspartame and saccharin have sparked controversy (Sadighara et al., 2022). Regulatory organizations like the Food and Drug Administration (FDA) set maximum permissible levels and carry out risk analyses to guarantee the safety of food additives. These evaluations take into account things the levels of exposure, toxicology information, and probable health impacts. However, determining their hazards with accuracy continues to be difficult because to the intricacy of chemical interactions and the possibility of additive mixes. Hence, food additives could have an impact on human health by acting as environmental contaminants. To ensure the safe use of food additives and reduce their impact on the environment and consumers, strict monitoring, risk assessments, and regulatory measures are required (Liang, 2022).

2.8.4 Nanoparticles

These are defined as particles with dimensions between 1 and 100 nanometres, have attracted considerable interest due to their unique properties and potential applications in numerous industries. Nanoparticles' increasing production and use have raised concerns about their potential negative effects on the environment and human health, highlighting the need to recognize them as contaminants (Baun and Grieger, 2022).

Nanoparticles can come from natural sources or be synthesized for specific applications. In consumer products, electronics, and biomedical applications, engineered nanoparticles such as silver nanoparticles (AgNPs), titanium dioxide nanoparticles (TiO2NPs), and carbon nanotubes (CNTs) are widely used (Cheriyamundath and Vavilala, 2021). During manufacturing, use, and disposal processes, these nanoparticles can be discharged into the environment. Furthermore, natural nanoparticles, such as naturally occurring metal and metal oxide nanoparticles, can be produced by geological processes, weathering, or combustion. The transmission of nanoparticles in the environment is dependent on many factors, such as their physicochemical properties, surface modifications, and interactions with environmental matrices. Nanoparticles are susceptible to aggregation, sedimentation, and dissolution, all of which influence their transfer and distribution across various environmental compartments. Surface coatings and functionalization can affect a substance's stability and behaviour, thereby altering its environmental fate. In addition, nanoparticles can undergo transformations or modifications as a result of chemical reactions or interactions with organic and inorganic environmental substances (Berekaa, 2016; Khan et al., 2019).

Nanoparticles' potential toxicity is a significant concern due to their small size and high surface area-to-volume ratio, which can increase their reactivity and interactions with biological systems. Nanoparticles can infiltrate organisms via inhalation, ingestion, or skin contact, and have the potential to accumulate in a number of organs and tissues. The precise mechanisms of nanoparticle toxicity are still being studied, but many factors contribute to their negative effects. These include oxidative stress, inflammation, genotoxicity, cytotoxicity, and cellular function disruption. Depending on their composition, size, shape, surface charge, and surface coating, nanoparticle toxicity can vary. Different nanoparticles exhibit variable degrees of toxicity, according to some studies. Silver nanoparticles, for instance, possess antimicrobial properties but can also induce cytotoxicity and genotoxicity in specific cell types. In animal investigations, carbon nanotubes have been linked to lung inflammation and fibrosis (Ng et al., 2023). Commonly used in sunscreens, titanium dioxide nanoparticles can generate reactive oxygen species and induce oxidative stress in cells. It is essential to note, however, that nanoparticle toxicity is complex and can be influenced by factors such as dose, exposure duration, and a particular biological system being studied Baun and Grieger, 2022; Kumari et al., 2022.

2.8.5 Flame/Fire Retardants/Suppressants

Flame retardants play a crucial role in reducing the combustibility of a number of industrial and consumer products, thereby improving fire safety. Concerns have been expressed regarding two kinds of flame retardants, Polybrominated diphenyl ethers (PBDEs) and phosphate-based flame retardants, due to their potential adverse effects on human health and the environment (Lin et al., 2022). As flame retardants, PBDEs have been extensively used in products such as electronics, furniture, and textiles. Studies have demonstrated that PBDEs can migrate from these products over time and contaminate indoor and outdoor environments. These compounds are known for their environmental persistence, which means they are resistant to degradation and can accumulate in living organisms. PBDEs have been detected in air, water, soil, and biota, among other environmental compartments (Pande et al., 2020). Their bio accumulative nature is a cause for concern, as they can accumulate in the lipid tissues of organisms, which could have negative effects on wildlife and human health. Some congeners of PBDEs have been categorized as potential endocrine disruptors due to their association with developmental, reproductive, and neurological effects.

As a substitute for PBDEs, phosphate-based flame retardants have gained popularity due to their lower environmental persistence. New research indicates, however, that certain phosphate-based flame retardants, such as tris (1,3-dichloro-2-propyl) phosphate (TDCPP) and triphenyl phosphate (TPHP), could also create dangers. These compounds have been identified in a number of products, such as foam furniture, electronics, and textiles. They could release phosphate-based flame retardants into the environment (Mohapatra and Kirpalani, 2019). These compounds have been identified in interior dust, air, sediment, and biota. Their potential toxicity and ability to interfere with endocrine systems are primary causes for concern. Some evidence suggests that these flame retardants could have negative effects on neurological development, reproductive health, and thyroid function, although more research is required. In addition to flame retardants, some surfactants including perfluorooctane sulphonate (PFOS) and perfluorooctanoic acid (PFOA) have been widely used in fire-fighting foams. These compounds have been found to contaminate effluents and surface water from wastewater treatment plants. PFOS and PFOA are persistent organic pollutants, which means that they are resistant to degradation and can accumulate in the environment and in organisms. These compounds have been linked to a number of adverse health effects, including liver injury, developmental abnormalities, immune suppression, and the potential to cause cancer. Due to their persistence, bioaccumulation, and potential toxicity, PFOS and PFOA are two emerging contaminants that have garnered considerable attention (Longpré et al., 2020a; 2020b).

To address the concerns surrounding these flame retardants and surfactants, efforts are being made to regulate their use and encourage the development of safer alternatives. For effective risk management and mitigation strategies, the identification, monitoring, and evaluation of their presence in various environmental compartments are essential. In addition, research continues to advance our comprehension of their fate, transfer, and potential health effects, apprising the development of policies and guidelines to reduce their environmental impact and safeguard human health (Naidu et al., 2016).

3. Fate of ECs in Environment

The fate of emerging contaminants (ECs) in environmental waters is a complex process that depends on various factors such as their physicochemical properties, environmental conditions, and the presence of other contaminants. ECs can undergo several transformations in aquatic systems, including biodegradation, photodegradation, hydrolysis, and oxidation. These transformations can result in the formation of metabolites that could have similar or different toxicity profiles compared to parent compounds. ECs can also partition between different phases in water bodies, including the dissolved phase, suspended particles, and sediment. Sorption to particles and sediment can result in the ECs' long-term retention in aquatic systems, where they could continue to exert their effects on the ecosystem. Furthermore, ECs can also be transferred to other water bodies through surface runoff or groundwater flow, leading to contamination of downstream water sources (Fig. 6.4).

Primary treatment, secondary treatment, and tertiary treatment are the three basic steps of traditional wastewater treatment systems. Through procedures like filtration and sedimentation, solid waste materials like settleable particles, plastics, oils and fats, sand, and grit are mechanically removed from wastewater during the initial treatment stage. Techniques used in the secondary treatment stage, which frequently entails the biological breakdown of nutrients and organic materials, can vary, and include fixed bed bioreactors, membrane bioreactors, and moving bed biofilm reactors. The most common method is Conventional Activated Sludge (CAS), which removes organic materials and nitrogen by forming biological floc using dissolved oxygen. Tertiary treatment involves the removal of phosphorus via precipitation and filtering, as well as the disinfection of the effluent using techniques such as UV irradiation or chlorination, prior to its release into the environment. However, some Pharmaceuticals and Personal Care Products (PPCPs) are emerging contaminants (ECs), which current treatments may not be able to eliminate completely. Through biological, chemical, and photochemical degradation and photolysis, a number of Transformation Products (TPs) can be produced, some of which could have higher

Table 6.3 DWEL values of the mentioned countries.

Category		EDCs			PCPs			PhACs		Pesticides	
Emerging Contaminant		Bisphenol a	Estrone-1	Nonylphenol	Diethylhexyl pthalate	Propylparaben	Salicylic acid	Propranolol	Ibuprofen	Triclosan	Atrazine
Dwel Values In North Amercia	Canada	1239	0.02	3717	168	5873	3717188	99125	2726	297	12
	U.S.A.	1319	0.03	3956	179	6251	3956250	105500	2901	317	13
Dwel Values in Europe	Germany	1253	0.03	3759	170	5940	3759375	100250	2757	301	13
	Spain	1180	0.02	3539	160	5592	3539063	94375	2595	283	12
	U.K.	1247	0.02	3741	169	5910	3740625	99750	2743	299	12
Dwel Values in Asia	China	1014	0.02	3042	137	4807	3042188	81125	2231	243	10
	India	856	0.02	2569	116	4059	2568750	68500	1884	206	9
Adi (MG/L)		50	0.001	150	6.77	237	150000	4000	110	12	0.5
Pnec (MG/ Kg/D)		1.5	0.018	6.6	0.01	-	1.28	0.1	0.01	0.07	0.044

References: Hahladakis et al, 2023; Zhang et al., 2023; Majumder et al., 2019; Gogoi et al., 2018a (González et al., 2010; "Water Quality Home" n.d.), (Huang et al., 2021; Impacts and 2021; 2021), Kim et al., 2020 (Dakak et al., 2020)

Afsa et al., 2020; Hajizadeh et al., 2020; Austin et al., 2022; Jan-Roblero and Cruz-Maya, 2023; Reichert et al., 2020

Castro et al., 2021; Fareed et al., 2021

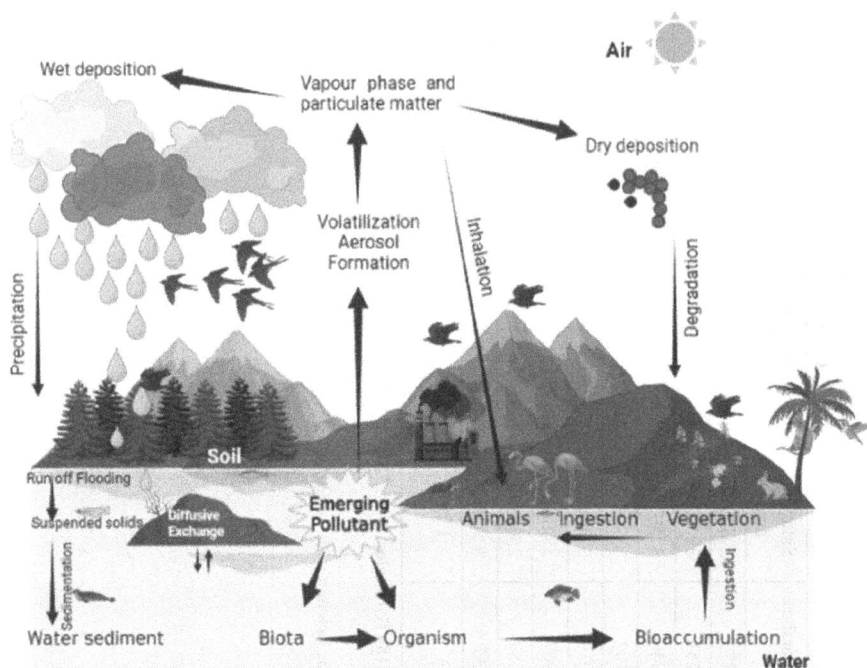

Figure 6.4 EC's fate and the interplay in the surrounding environment.

ecotoxicity than their parent chemicals. A further issue is the hundreds of Disinfection By-Products (DBPs) that have been discovered so far. DBPs are produced when organic matter reacts with disinfectants. Continued study and new ideas are needed to improve the removal of ECs and make sure that treated wastewater released back into the environment is safe. The fate of ECs in environmental waters is an area of ongoing research to better understand the mechanisms underlying their behaviour in aquatic systems. This knowledge is crucial for developing effective strategies to mitigate the impact of ECs on the environment and protect human and ecological health (Gogoi et al., 2018b).

Despite playing a crucial part in the global cycling of volatile and semi-volatile contaminants, the air segment frequently gets overlooked as a result of other environmental compartments. However, on a local as well as a global scale, it acts as a channel for pollution diffusion. Emission sources of environmentally persistent and toxic compounds can have detrimental impacts on the health of humans, animals, and plants, not only in close vicinity to the sources, but also in far-flung places due to long-range transmission of the contaminants. To evaluate the possible dangers to human health and the environmental impact of developing pollutants in the air, it is crucial to comprehend their fate. After being emitted into the atmosphere, these contaminants endure a number of physical and chemical processes that affect their behaviour, transfer, and transformation. Their fate is largely influenced by properties like as volatility, solubility, reactivity, and particle size. Gas-phase pollutants can travel vast distances in the atmosphere due to mechanisms like advection and atmospheric mixing. They could also degrade via reactions with atmospheric oxidants and photolysis induced by solar radiation. Shorter lengths can be used to convey particulate-bound pollutants, which can then wet or dry deposit onto surfaces. Once deposited, contaminants can persist in the environment or endure further transformations, such as leaching into soil or bodies of water. A complete understanding is essential for efficient risk assessment and management strategies since the fate of emerging pollutants in the air is a complicated and dynamic process impacted by a number of environmental factors (Barroso et al., 2019b).

The transfer and transformation of emerging contaminants in soil is a multifaceted phenomenon that is subject to the influence of numerous factors. On introduction into the soil, contaminants undergo various transformations that dictate their fate and potential hazards. Sorption is a key process by which contaminants bond to soil particles through hydrophobic interactions, electrostatic attractions, and van der Waals forces. The sorption's magnitude and capacity are contingent on the physicochemical attributes of the pollutants and the soil's properties. Additionally, the reversibility of sorption is also influenced by these factors. Contrarily, desorption entails the release of pollutants from soil particles, which can be impacted by environmental factors and the contaminants' affinity for the soil matrix. Contaminants can undergo metamorphosis or disintegration as a result of the processes of degradation, which can include both microbial decomposition and abiotic reactions. Transfer of contaminants can also occur via leaching, in which they travel with water through the soil profile, or surface runoff and erosion. The eventual fate of emerging contaminants in soil is notably impacted by various factors, including but not limited to soil pH, moisture level, temperature, and the existence of flora and microorganisms. Comprehending these scientific intricacies is essential in evaluating environmental hazards and executing efficient soil governance and remedial tactics. As an example, a hydrophobic emerging pollutant like Polycyclic Aromatic Hydrocarbons (PAHs). On introduction to the soil, PAHs exhibit a strong tendency to adsorb onto soil particles owing to their hydrophobic properties. The efficacy of this sorption process is contingent on variables such as the soil's mineral composition and organic matter concentration. On sorption, polycyclic aromatic hydrocarbons (PAHs) could experience desorption due to alterations in environmental factors, such as variations in soil moisture or pH levels. Similarly, water-soluble pollutants like nitrate, can be leached through the soil profile when there is an abundance of water in the form of irrigation or rainfall. The degree of leaching is contingent on the soil characteristics, including texture, structure, and the existence of preferential flow channels. The process of nitrate leaching could potentially endanger the quality of groundwater, particularly in regions of agriculture where there is an over-application of fertilizers. Also, some contaminants can get into plants through their roots and build up in plant cells, from where they could enter the food chain. Heavy metals such as lead or cadmium can be taken up by plants, which can potentially lead to adverse health effects in humans on consumption. Comprehending these factors, and the fundamental scientific concepts of sorption, desorption, degradation, and transfer mechanisms is imperative in evaluating the environmental destiny and possible hazards linked with emerging pollutants in soil. This aids in the dissemination of suitable soil management techniques and the formulation of efficient remedial measures (Sarma, 2022).

4. Risk Management of Emerging Contaminants

The scientific evaluation of emerging contaminants involves a rigorous process of risk assessment that comprises multiple interconnected steps aimed at systematically assessing the potential adverse effects and associated risks. The process involves identifying potential hazards, assessing the level of exposure, evaluating the relationship between exposure and the resulting health effects, and characterizing the overall risk associated with emerging contaminants. This approach provides a thorough comprehension of the potential risks posed by these contaminants (Lin et al., 2020).

The process of hazard identification involves a comprehensive examination of the existing toxicological information in order to determine and describe the possible negative impacts of the newly discovered contaminant. This entails evaluating its inherent characteristics, including its chemical composition, physical and chemical properties, and the mechanism of action. Empirical investigations, such as evaluations of acute and chronic toxicity, assessments of reproductive and developmental toxicity, analyses of genotoxicity, and investigations of ecological impacts, are conducted to ascertain the extent and intensity of deleterious outcomes. The assessment of exposure aims to measure the existing or possible exposure to the newly discovered pollutant in different environmental domains and human communities. The process entails the analysis of exposure pathways, quantification of exposure levels, and the recognition of populations or ecological receptors that have been exposed. This procedure is dependent on the acquisition of precise measurement data, the development of accurate environmental models, and the analysis of exposure factors to estimate the extent, frequency, duration, and pathways of exposure (Wang et al., 2021).

The process of dose-response assessment entails determining the correlation between the level of exposure to the newly discovered pollutant and the probability and intensity of unfavourable outcomes. The process incorporates toxicological information to establish correlations between dosage and response, including the identification of levels at which no adverse effects are observed (NOAEL), the benchmark dose (BMD), or the lowest levels at which adverse effects are observed (LOAEL). These associations facilitate the computation of a dosage or level of exposure that is anticipated to have no notable unfavourable impacts or the measurement of hazards linked during any exposure situations (Rathi et al., 2021).

Risk characterization is a process that combines data from earlier stages to evaluate and convey the possible hazards presented by the newly discovered contaminant. This involves the measurement of potential risks through the use of diverse metrics, such as hazard quotients, hazard indices, or cancer risk estimates. Sensitivity analysis and uncertainty analysis are used to obtain a more comprehensive assessment of risks by accounting for uncertainties such as data limitations, variability, and extrapolations. Furthermore, advanced modelling techniques, such as physiologically based pharmacokinetic (PBPK) models or fate and transfer models, can be used to enhance the accuracy of exposure and dose assessments. The outcomes of a thorough evaluation of potential hazards provide insight for determining strategies for managing risks, such as creating guidelines for regulations, executing preventative measures, and determining the order of importance for corrective actions. The implementation of a scientific methodology guarantees the efficient management of newly discovered pollutants with the aim of safeguarding both the well-being of humans and the natural surroundings (Gavrilescu et al., 2015).

Additionally, after the identification of the contaminant, the subsequent step involves the assessment and control of the risk posed to receptors. Currently, the dominant method for handling pollutants is to implement a strategy based on risk assessment. A decision analysis framework, such as that proposed by Khadam and Kaluarachchi, 2003, can assist in evaluating various remediation options by assessing health hazards and identifying the most cost-effective approach. This methodology enables the validation of remedial actions by considering the equilibrium between different types of hazards (i.e., population, individual, and residual) and their corresponding expenses (i.e., reduction of risk and enhancement of efficiency). The incorporation of expenses and potential hazards into this structure provides a potent instrument for the administration of newly discovered pollutants. It is recommended to establish policies that help in guiding for implementing risk-based strategies in the management of sites that have been contaminated. The implementation of policies can surpass the establishment of environmental thresholds, allowing for the development of risk-based alternatives customized to special locations, thereby promoting sustainable administration. Adequate risk management requires appropriate engagement among risk evaluators, interested parties, and other relevant individuals to establish environmental, social, and economic goals for site management and/or remediation at an early stage. Retroactively modifying expectations can create difficulties and incur expenses if the selected remedial measures fail to meet regulatory standards. The activities of risk assessment and risk management are interdependent and necessitate the efficient communication of risk to stakeholders. Risk communication involves the continuous exchange of information regarding health and environmental hazards between all stakeholders, such as risk assessors, risk managers, interested parties, the general public, and the media. The adoption of an inclusive approach results in enhanced efficiency, transparency, and optimized processes, taking into account all viewpoints. In addition, these exchanges of information aid in the formulation of both short- and long-term management and remediation goals for sites, which subsequently impact the identification of suitable remediation alternatives from an initial phase. Due to the considerable uncertainties regarding the destiny and movement of ECs, effective risk communication is of utmost importance in their management. This requires a cautious approach that incorporates a reasonable degree of conservatism. It is important to recognize that environmental exposure cannot be attributed only to a single contaminant, but rather to mixtures consisting of ECs and other chemicals. As a result, it is anticipated that the effects appearing from these combinations will be more significant. Hence, it is crucial to thoroughly evaluate the probable impacts on the ecosystem and the perilous interconnections linked with these amalgamations. Researchers should prioritize research endeavours aimed at comprehending the interplay between ECs and other pollutants, as well as devising novel analytical techniques to detect any potential environmental hazards that could arise from such interactions (Lin et al., 2020; Wang et al., 2021).

5. Conclusion

Emerging contaminants (ECs) are a heterogeneous group of compounds that have been identified globally, encompassing pharmaceuticals, pesticides, industrial by-products, polybrominated flame retardants, nanoparticles, water treatment by-products, food additives, ionic liquids, personal care products etc. The presence of these pollutants presents a substantial ecological hazard and necessitates a thorough and systematic methodology for overseeing and mitigating their effects. An integrative method for monitoring the environment is imperative in order to effectively manage emerging contaminants (ECs). This methodology should involve examining the destiny and consequences of ECs throughout all exposed environmental compartments during their complete life span. The management of ECs poses a multitude of challenges. From a scientific perspective, addressing inadequately studied contaminants and those already known to be of concern requires changes in current practices, including the formulation of ambitious yet feasible policies from a policy and regulation standpoint. Optimal EC regulation involves the amalgamation of scientific and policy-based approaches within frameworks tailored to the prevention, evaluation, and control of polluted areas. From an analytical perspective, the management of environmental contaminants requires the creation of novel techniques for detecting chemicals that could have adverse effects on receptors. In addition, there is a requirement for more extensive evaluations of environmental risks. To accomplish this, it is necessary to integrate chemical and biological evaluations in order to improve the evaluation of the ecological consequences resulting from ECs. The prudent course of action when assessing various remedial options for managing ECs is to adopt a risk-based strategy that takes into account both the reduction of risk and cost-effectiveness. In addition, effective management of environmental contaminants should be accompanied by other strategies, such as enhanced chemical management and used to decrease releases, waste reduction, proper waste disposal, decreased chemical discharge, and upgraded water treatment systems. This encompasses a range of strategies such as all-encompassing monitoring of the environment, alterations to policies, merging of scientific and policy aspects in regulatory structures, creation of novel analytical techniques, thorough evaluation of risks, and the adoption of a risk-oriented methodology to efficiently manage ECs. Moreover, the implementation of enhanced chemical management techniques and state-of-the-art water treatment methodologies are pivotal in mitigating the discharge and consequences of emerging contaminants (ECs) on the ecosystem.

References

Abdelhamid, Ahmed E., Ahmed A. El-Sayed and Ahmed M. Khalil. 2020. Polysulfone Nanofiltration Membranes Enriched with Functionalized Graphene Oxide for Dye Removal from Wastewater. Journal of Polymer Engineering 40(10): 833–41. https://doi.org/10.1515/POLYENG-2020-0141/ HTML.

Adeniyi, Lateef Adeleke and Abel Omoniyi Afon. 2022. Seasonal Quantification and Characterization of Solid Waste Generation in Tertiary Institution: A Case Study. Journal of Material Cycles and Waste Management 24(3): 1172–81. https://doi.org/10.1007/S10163-022-01390-0.

Afsa, Sabrine, Khaled Hamden, Pablo A. Lara Martin and Hedi Ben Mansour. 2020. Occurrence of 40 Pharmaceutically Active Compounds in Hospital and Urban Wastewaters and Their Contribution to Mahdia Coastal Seawater Contamination. Environmental Science and Pollution Research 27(2): 1941–55. https://doi.org/10.1007/S11356-019-06866-5.

Ahmad, Naushad, Ahmed S. Al-Fatesh, Rizwan Wahab, Manawwer Alam and Anis H. Fakeeha. 2020. Synthesis of Silver Nanoparticles Decorated on Reduced Graphene Oxide Nanosheets and Their Electrochemical Sensing towards Hazardous 4-Nitrophenol. Journal of Materials Science: Materials in Electronics 31(14): 11927–37. https://doi.org/10.1007/S10854-020-03747-3.

Ahmed, F., S. Hasan, M.S. Rana and N. Sharmin. 2023. A conceptual framework for zero waste management in Bangladesh. International Journal of Environmental Science and Technology 20(2): 1887–1904. https://doi.org/10.1007/S13762-022-04127-6.

Alam, Ohidul and Xiuchen Qiao. 2020. An In-Depth Review on Municipal Solid Waste Management, Treatment and Disposal in Bangladesh. Sustainable Cities and Society 52(January): 101775. https://doi.org/10.1016/j.scs.2019.101775.

Alengebawy, Ahmed, Sara Taha Abdelkhalek, Sundas Rana Qureshi and Man-Qun Wang. 2021. Heavy Metals and Pesticides Toxicity in Agricultural Soil and Plants: Ecological Risks and Human Health Implications. Toxics 9(3): 42. https://doi.org/10.3390/toxics9030042.

Alharbi, Omar M.L., Al Arsh Basheer, Rafat A. Khattab and Imran Ali. 2018. Health and Environmental Effects of Persistent Organic Pollutants. Journal of Molecular Liquids. Elsevier B.V. https://doi.org/10.1016/j.molliq.2018.05.029.

Anastopoulos, I., I. Pashalidis, A.G. Orfanos, I.D. Manariotis, T. Tatarchuk, L. Sellaoui, A. An Bonilla-Petriciolet, A. Mittal and A. Nú~ Nez-Delgado. 2020. Removal of caffeine, nicotine and amoxicillin from (waste) waters by various adsorbents. A review. Elsevier 261: 110236–110236. https://doi.org/10.1016/j.jenvman.2020.110236.

Ancona, Valeria, Anna Barra Caracciolo, Paola Grenni, Martina Di Lenola, Claudia Campanale, Angelantonio Calabrese et al. 2017. Plant-Assisted Bioremediation of a Historically PCB and Heavy Metal-Contaminated Area in Southern Italy. New Biotechnology 38 (September): 65–73. https://doi.org/10.1016/j.nbt.2016.09.006.

Anfar, Zakaria, Hassan Ait Ahsaine, Mohamed Zbair, Abdallah Amedlous, Abdellah Ait El Fakir, Amane Jada et al. 2020. Recent Trends on Numerical Investigations of Response Surface Methodology for Pollutants Adsorption onto Activated Carbon Materials: A Review. Critical Reviews in Environmental Science and Technology 50(10): 1043–84. https://doi.org/10.1080/10643389.2019.1642835.

Arnold, Scott M., Kathryn E. Clark, Charles A. Staples, Gary M. Klecka, Steve S. Dimond, Norbert Caspers, and Steven G. Hentges. 2013. Relevance of Drinking Water as a Source of Human Exposure to Bisphenol A. Journal of Exposure Science & Environmental Epidemiology 23(2): 137–44. https://doi.org/10.1038/JES.2012.66.

Arola, Kimmo, Bart Van der Bruggen, Mika Mänttäri and Mari Kallioinen. 2019. Treatment Options for Nanofiltration and Reverse Osmosis Concentrates from Municipal Wastewater Treatment: A Review. Critical Reviews in Environmental Science and Technology 49(22): 2049–2116. https://doi.org/10.1080/10643389.2019.1594519.

Austin, Tom, Francesco Bregoli, Dominik Höhne, A. Jan Hendriks and Ad M.J. Ragas. 2022. Ibuprofen Exposure in Europe; EPiE as an Alternative to Costly Environmental Monitoring. Environmental Research 209(June): 112777. https://doi.org/10.1016/j.envres.2022.112777.

Balali-Mood, M., K. Naseri, Z. Tahergorabi, M.R. Khazdair and M. Sadeghi. 2021. Toxic mechanisms of five heavy metals: mercury, lead, chromium, cadmium, and arsenic. Frontiers in Pharmacology 12. https://doi.org/10.3389/FPHAR.2021.643972/FULL.

Bandowe, Benjamin A.Musa and Hannah Meusel. 2017. Nitrated Polycyclic Aromatic Hydrocarbons (Nitro-PAHs) in the Environment—A Review. Science of the Total Environment. Elsevier B.V. https://doi.org/10.1016/j.scitotenv.2016.12.115.

Barroso, Pedro José, Juan Luis Santos, Julia Martín, Irene Aparicio and Esteban Alonso. 2019a. Emerging Contaminants in the Atmosphere: Analysis, Occurrence and Future Challenges. Critical Reviews in Environmental Science and Technology 49(2): 104–71. https://doi.org/10.1080/10643389.2018.1540761.

Barroso, Pedro José, Juan Luis Santos, Julia Martín, Irene Aparicio and Esteban Alonso. 2019b. Emerging Contaminants in the Atmosphere: Analysis, Occurrence and Future Challenges. Https://Doi.Org/10.1080/10643389.2018.1540761 49(2): 104–71. https://doi.org/10.1080/10643389.2018.1540761.

Basu, Niladri, Ashley Bastiansz, José G. Dórea, Masatake Fujimura, Milena Horvat, Emelyn Shroff, Pál Weihe and Irina Zastenskaya. 2023. Our Evolved Understanding of the Human Health Risks of Mercury. Ambio 2023 52:5 52(5): 877–96. https://doi.org/10.1007/S13280-023-01831-6.

Baun, Anders and Khara Grieger. 2022. Environmental Risk Assessment of Emerging Contaminants—The Case of Nanomaterials. Advances in Toxicology and Risk Assessment of Nanomaterials and Emerging Contaminants 349–71. https://doi.org/10.1007/978-981-16-9116-4_15.

Bell, Katherine Y., Martha J.M. Wells, Kathy A. Traexler, Marie-Laure Pellegrin, Audra Morse and Jeff Bandy. 2011. Emerging Pollutants. Wiley Online Library 83(10): 10. https://doi.org/10.2175/1 06143011X13075599870298.

Berekaa, Mahmoud M. 2016. Nanotechnology in Wastewater Treatment; Influence of Nanomaterials on Microbial Systems. International Journal of Current Microbiology and Applied Sciences 5(1): 713–26. https://doi.org/10.20546/ijcmas.2016.501.072.

Bhat, Rouf Ahmad, Dig Vijay Singh, Humaira Qadri, Gowhar Hamid Dar, Moonisa Aslam Dervash, Shakeel Ahmad Bhat, Bengu Turkyilmaz Unal, Munir Ozturk, Khalid Rehman Hakeem and Balal Yousaf. 2022. Vulnerability of Municipal Solid Waste: An Emerging Threat to Aquatic Ecosystems. Chemosphere 287(January). https://doi.org/10.1016/j.chemosphere.2021.132223.

Carvalho, Pedro N. 2021. Constructed Wetlands and Phytoremediation as a Tool for Pharmaceutical Removal. Handbook of Environmental Chemistry 103: 377–413. https://doi.org/10.1007/698_2020_624.

Castro, Muryllo Santos, Fabiana Gonçalves Barbosa, Pablo Santos Guimarães, Camila De Martinez Gaspar Martins and Juliano Zanette. 2021. A Scientometric Analysis of Ecotoxicological Studies with the Herbicide Atrazine and Microalgae and Cyanobacteria as Test Organisms. Environmental Science and Pollution Research 28(20): 25196–206. https://doi.org/10.1007/S11356-020-12213-W.

Chaturvedi, P., P. Shukla, B.S. Giri, P. Chowdhary, R. Chandra, P. Gupta and A. Pandey. 2021. Prevalence and hazardous impact of pharmaceutical and personal care products and antibiotics in environment: A review on emerging contaminants. Environmental Research 194: 110664. https://doi.org/10.1016/j.envres.2020.110664.

Chen, Hang, Yi Meng, Shiyao Jia, Wenqiang Hua, Yi Cheng, Jie Lu and Haisong Wang. 2020. Graphene Oxide Modified Waste Newspaper for Removal of Heavy Metal Ions and Its Application in Industrial Wastewater. Materials Chemistry and Physics 244(April): 122692. https://doi.org/10.1016/j.matchemphys.2020.122692.

Cheriyamundath, Sanith and Sirisha L. Vavilala. 2021. Nanotechnology-Based Wastewater Treatment. Water and Environment Journal. Blackwell Publishing Ltd. https://doi.org/10.1111/wej.12610.

Cho, Seo Won and Haoran Wei. 2023. Surface-Enhanced Raman Spectroscopy for Emerging Contaminant Analysis in Drinking Water. Frontiers of Environmental Science and Engineering 17(5). https://doi.org/10.1007/S11783-023-1657-5.

Cojocaru, Bogdan, Veronica Andrei, Madalina Tudorache, Feng Lin, Chris Cadigan, Ryan Richards and Vasile I. Parvulescu. 2017. Enhanced Photo-Degradation of Bisphenol Pollutants onto Gold-Modified Photocatalysts. Catalysis Today 284: 153–59. https://doi.org/10.1016/j.cattod.2016.11.009.

Cordner, Alissa, Vanessa Y. De La Rosa, Laurel A. Schaider, Ruthann A. Rudel, Lauren Richter and Phil Brown. 2019. Guideline Levels for PFOA and PFOS in Drinking Water: The Role of Scientific Uncertainty, Risk Assessment Decisions, and Social Factors. Journal of Exposure Science & Environmental Epidemiology 29(2): 157–71. https://doi.org/10.1038/s41370-018-0099-9.

Corrales, Jone, Lauren A. Kristofco, W. Baylor Steele, Brian S. Yates, Christopher S. Breed, Spencer Williams, E. and Bryan W. Brooks. 2015. Global Assessment of Bisphenol a in the Environment: Review and Analysis of Its Occurrence and Bioaccumulation. Dose-Response 13(3). https://doi.org/10.1177/1559325815598308/ASSET/IMAGES/LARGE/10.1177_1559325815598308-FIG5.JPEG.

Costanza, Jed, Masoud Arshadi, Linda M. Abriola and Kurt D. Pennell. 2019. Accumulation of PFOA and PFOS at the Air-Water Interface. Environmental Science and Technology Letters 6(8): 487–91. https://doi.org/10.1021/ACS.ESTLETT.9B00355.

Dakak, R.E. and I.H.-E. Processes. 2020, undefined, 2020. The Alleviative Effects of Salicylic Acid on Physiological Indices and Defense Mechanisms of Maize (Zea Mays L. Giza 2) Stressed with Cadmium. Springer 7, 873–884. https://doi.org/10.1007/s40710-020-00448-1.

Dias, Marcela França, Deborah Leroy-Freitas, Elayne Cristina Machado, Letícia da Silva Santos, Cintia Dutra Leal, Gabriel da Rocha Fernandes and Juliana Calábria de Araújo. 2022. Effects of Activated

Sludge and UV Disinfection Processes on the Bacterial Community and Antibiotic Resistance Profile in a Municipal Wastewater Treatment Plant. Environmental Science and Pollution Research 29(24): 36088–99. https://doi.org/10.1007/S11356-022-18749-3.

Dobaradaran, Sina, Torsten C. Schmidt, Nerea Lorenzo-Parodi, Maik A. Jochmann, Iraj Nabipour, Alireza Raeisi, Nenad Stojanović and Marzieh Mahmoodi. 2019. Cigarette Butts: An Overlooked Source of PAHs in the Environment? Environmental Pollution 249(June): 932–39. https://doi.org/10.1016/j.envpol.2019.03.097.

Donga, Cabangani, Shivani B. Mishra, Alaa S. Abd-El-Aziz and Ajay K. Mishra. 2021. Advances in Graphene-Based Magnetic and Graphene-Based/TiO2 Nanoparticles in the Removal of Heavy Metals and Organic Pollutants from Industrial Wastewater. Journal of Inorganic and Organometallic Polymers and Materials 31(2): 463–80. https://doi.org/10.1007/S10904-020-01679-3.

Dreyer, Annekatrin and Andrea Minkos. 2023. Polychlorinated Biphenyls (PCB) and Polychlorinated Dibenzo-Para-Dioxins and Dibenzofurans (PCDD/F) in Ambient Air and Deposition in the German Background. Environmental Pollution 316(January): 120511. https://doi.org/10.1016/J.ENVPOL.2022.120511.

Dudášová, Hana, Katarína Lászlová, Lucia Lukáčová, Marta Balaščáková, Slavomíra Murínová and Katarína Dercová. 2016. Bioremediation of PCB-Contaminated Sediments and Evaluation of Their Pre- and Post-Treatment Ecotoxicity. Chemical Papers 70(8): 1049–58. https://doi.org/10.1515/chempap-2016-0041.

Fareed, Adnan, Abid Hussain, Mohsin Nawaz, Muhammad Imran, Zulfiqar Ali and Sami Ul Haq. 2021. The Impact of Prolonged Use and Oxidative Degradation of Atrazine by Fenton and Photo-Fenton Processes. Environmental Technology & Innovation 24(November): 101840. https://doi.org/10.1016/j.eti.2021.101840.

Fatimazahra, Sayerh, Mouhir Latifa, Saafadi Laila and Khazraji Monsif. 2023. Review of Hospital Effluents: Special Emphasis on Characterization, Impact, and Treatment of Pollutants and Antibiotic Resistance. Environmental Monitoring and Assessment 195(3): 393. https://doi.org/10.1007/S10661-023-11002-5.

Gadd, Geoffrey M. and Alan J. Griffiths. 1977. Microorganisms and Heavy Metal Toxicity. Microbial. Ecology Vol. 4.

Gavrilescu, Maria, Kateřina Demnerová, Jens Aamand, Spiros Agathos and Fabio Fava. 2015. Emerging Pollutants in the Environment: Present and Future Challenges in Biomonitoring, Ecological Risks and Bioremediation. New Biotechnology 32(1): 147–56. https://doi.org/10.1016/j.nbt.2014.01.001.

Ghosh, Jyotirmoy and Parames, C. Sil. 2023. Mechanism for Arsenic-Induced Toxic Effects. Handbook of Arsenic Toxicology. January 223–52. https://doi.org/10.1016/B978-0-323-89847-8.00022-5.

Gogoi, Anindita, Payal Mazumder, Vinay Kumar Tyagi, G.G. Tushara Chaminda, Alicia Kyoungjin An and Manish Kumar. 2018a. Occurrence and Fate of Emerging Contaminants in Water Environment: A Review. Groundwater for Sustainable Development 6 (March): 169–80. https://doi.org/10.1016/J.GSD.2017.12.009.

Gogoi, Anindita, Payal Mazumder, Vinay Kumar Tyagi, G.G. Tushara Chaminda, Alicia Kyoungjin An and Manish Kumar. 2018b. Occurrence and Fate of Emerging Contaminants in Water Environment: A Review. Groundwater for Sustainable Development 6(March): 169–80. https://doi.org/10.1016/J.GSD.2017.12.009.

González, M.M., Martín, J., Santos, J.L., Aparicio, I. and Alonso, E. 2010. Occurrence and Risk Assessment of Nonylphenol and Nonylphenol Ethoxylates in Sewage Sludge from Different Conventional Treatment Processes. Science of The Total Environment 408(3): 563–70. https://doi.org/10.1016/J.SCITOTENV.2009.10.027.

Grant, Lester D. 2020. Lead and Compounds. Environmental Toxicants: Human Exposures and Their Health Effects. January 627–75. https://doi.org/10.1002/9781119438922.CH17.

Hahladakis, J.N. and Iacovidou, E. Integrated, and undefined 2023. 2022. An Overview of the Occurrence, Fate, and Human Risks of the Bisphenol-A Present in Plastic Materials, Components, and Products. Wiley Online Library 19(1): 45–62. https://doi.org/10.1002/ieam.4611.

Hahladakis, John N., Eleni Iacovidou and Spyridoula Gerassimidou. 2023. An Overview of the Occurrence, Fate, and Human Risks of the Bisphenol-A Present in Plastic Materials, Components, and Products. Integrated Environmental Assessment and Management 19(1): 45–62. https://doi.org/10.1002/IEAM.4611.

Hajizadeh, Yaghoub, Ghasem Kiani Feizabadi and Awat Feizi. 2020. Dietary Habits and Personal Care Product Use as Predictors of Urinary Concentrations of Parabens in Iranian Adolescents. Environmental Toxicology and Chemistry 39(12): 2378–88. https://doi.org/10.1002/ETC.4861.

https://www.epa.gov/enforcement/2015-update-1998-us-epa-supplemental-environmental-projects-policy.

Huang, S., Qi, Z, Ma, S., Li, G., Long, C. and Yu, Y. 2021. Environmental Pollution, and undefined 2021. 2021. A Critical Review on Human Internal Exposure of Phthalate Metabolites and the Associated Health Risks. Elsevier. https://doi.org/10.1016/j.envpol.2021.116941.

Hussain, Karishma, Raza R. Hoque, Srinivasan Balachandran, Subhash Medhi, Mohammad Ghaznavi Idris, Mirzanur Rahman, and Farhaz Liaquat Hussain. 2018. Monitoring and Risk Analysis of PAHs in the Environment. pp. 1–35. *In*: Handbook of Environmental Materials Management. Springer International Publishing. https://doi.org/10.1007/978-3-319-58538-3_29-2.

Impacts, B. Prasad. 2021. Environmental Science: Processes and undefined 2021. Phthalate Pollution: Environmental Fate and Cumulative Human Exposure Index Using the Multivariate Analysis Approach. Pubs. Rsc.Org. https://doi.org/10.1039/d0em00396d.

Ismanto, Aris, Tony Hadibarata, Sugeng Widada, Elis Indrayanti, Dwi Haryo Ismunarti, Novia Safinatunnajah et al. 2023. Groundwater Contamination Status in Malaysia: Level of Heavy Metal, Source, Health Impact, and Remediation Technologies. Bioprocess and Biosystems Engineering 46(3): 467–82. https://doi.org/10.1007/S00449-022-02826-5.

Jan-Roblero, Janet and Juan A. Cruz-Maya. 2023. Ibuprofen: Toxicology and Biodegradation of an Emerging Contaminant. Molecules 28(5): 2097. https://doi.org/10.3390/molecules28052097.

Jatoi, Abdul Sattar, Jawad Ahmed, Faheem Akhter, Syed Haseeb Sultan, Ghulam Sever Chandio, Shoaib Ahmed et al. 2023. Recent Advances and Treatment of Emerging Contaminants Through the Bio-Assisted Method: A Comprehensive Review. Water, Air, and Soil Pollution 234(1). https://doi.org/10.1007/S11270-022-06037-2.

Kadam, Atul, Shitalkumar Patil, Sachin Patil and Anil Tumkur. 2016. Pharmaceutical Waste Management An Overview. Indian Journal of Pharmacy Practice 9(1): 2–8. https://doi.org/10.5530/ijopp.9.1.2.

Kadri, Syeda Ulfath Tazeen, Adinath N. Tavanappanavar, Nagesh Babu, R., Muhammad Bilal, Bhaskar Singh, Sanjay Kumar Gupta et al. 2023. Overview of Waste Stabilization Ponds in Developing Countries. Handbook of Environmental Chemistry 117: 153–75. https://doi.org/10.1007/698_2021_790.

Karak, Tanmoy, Bhagat, R.M. and Pradip Bhattacharyya. 2012. Municipal Solid Waste Generation, Composition, and Management: The World Scenario. Critical Reviews in Environmental Science and Technology 42(15): 1509–1630. https://doi.org/10.1080/10643389.2011.569871.

Khan, Ibrahim, Saeed, K. and Khan, Idrees. 2019. Nanoparticles: Properties, applications and toxicities. Arabian Journal of Chemistry 12: 908–931. https://doi.org/10.1016/j.arabjc.2017.05.011.

Khan, Sabiya S. and Swaran Jeet Singh Flora. 2023. Arsenic: Chemistry, Occurrence, and Exposure. Handbook of Arsenic Toxicology, January 1–49. https://doi.org/10.1016/B978-0-323-89847-8.00024-9.

Khadam, I. and J.J. Kaluarachchi. 2003. Applicability of risk-based management and the need for risk-based economic decision analysis at hazardous waste contaminated sites. Environment International 29: 503–519. https://doi.org/10.1016/S0160-4120(03)00009-6.

Khanverdiluo, Shima, Elaheh Talebi-Ghane, Ali Heshmati and Fereshteh Mehri. 2021. The Concentration of Polycyclic Aromatic Hydrocarbons (PAHs) in Mother Milk: A Global Systematic Review, Meta-Analysis and Health Risk Assessment of Infants. Saudi Journal of Biological Sciences 28(12): 6869–75. https://doi.org/10.1016/J.SJBS.2021.07.066.

Khoshbeen, Ahmad Rashid, Mohanakrishnan Logan and Chettiyappan Visvanathan. 2020. Integrated Solid-Waste Management for Kabul City, Afghanistan. Journal of Material Cycles and Waste Management 22(1): 240–53. https://doi.org/10.1007/S10163-019-00936-Z.

Kim, Mi Jin, Chul Hong Kim, Mi Jin An, Ju Hyun Lee, Geun Seup Shin, Jae Yoon Hwang, Jinhong Park, et al. 2020. Propylparaben Induces Apoptotic Cell Death in Human Placental BeWo Cells via Cell Cycle Arrest and Enhanced Caspase-3 Activity. Molecular and Cellular Toxicology 16(1): 83–92. https://doi.org/10.1007/S13273-019-00062-9.

Kumar, Amit, Amit Kumar, Cabral-Pinto M.M.S., Ashish K. Chaturvedi, Aftab A. Shabnam, Gangavarapu Subrahmanyam et al. 2020. Lead Toxicity: Health Hazards, Influence on Food Chain, and Sustainable

Remediation Approaches. International Journal of Environmental Research and Public Health 17(7): 2179. https://doi.org/10.3390/ijerph17072179.

Kumar, P., Aruna Priyanka, R.S., Shalini Priya, P., Gunasree, B., Srivanth, S., Jayasakthi, S. et al. 2023. "Bisphenol A Contamination in Processed Food Samples: An Overview. International Journal of Environmental Science and Technology. https://doi.org/10.1007/S13762-023-04793-0.

Kumar, Rinku and Pradeep Kumar. 2020. Wastewater Stabilisation Ponds: Removal of Emerging Contaminants. Journal of Sustainable Development of Energy, Water and Environment Systems 8(2): 344–59. https://doi.org/10.13044/j.sdewes.d7.0291.

Kumari, Arpna, Vishnu D. Rajput, Dina Nevidomskaya, Saglara S. Mandzhieva, Svetlana Sushkova, Sneh Rajput et al. 2022. Updated Analysis of the Exposure of Plants to the Nanomaterials. pp. 25–45. *In*: Toxicity of Nanoparticles in Plants: An Evaluation of Cyto/Morpho-Physiological, Biochemical and Molecular Responses. Elsevier. https://doi.org/10.1016/B978-0-323-90774-3.00011-8.

Kuznetsova, I.N. and Yu Lyanguzov, A. 2023. The Effect of Perfluorocarbon Nanoparticles on Blood As a Cellular System. Biochemistry (Moscow) Supplement Series A: Membrane and Cell Biology 17(1): 20–27. https://doi.org/10.1134/S1990747823010051/FIGURES/6.

Kwenda, Phyllis Rumbidzai, Gareth Lagerwall, Sibel Eker and Bas Van Ruijven. 2022. A Mini-Review on Household Solid Waste Management Systems in Low-Income Developing Countries: A Case Study of Urban Harare City, Zimbabwe. Waste Management and Research 40(2): 139–53. https://doi.org/10.1177/0734242X21991645.

Li, Y., Zhu, G., Ng, W.J. and Tan, S.K. 2014. Science of the Total Environment, and undefined 2014. A Review on Removing Pharmaceutical Contaminants from Wastewater by Constructed Wetlands: Design, Performance and Mechanism. Elsevier. https://doi.org/10.1016/j.scitotenv.2013.09.018.

Liang, Jiang. 2022. Food Additives. Nutritional Toxicology 167–80. https://doi.org/10.1007/978-981-19-0872-9_6.

Liang, Luyun, Yongling Pan, Lihua Bin, Yu Liu, Wenjun Huang, Rong Li et al. 2022. Immunotoxicity Mechanisms of Perfluorinated Compounds PFOA and PFOS. Chemosphere 291(March): 132892. https://doi.org/10.1016/j.chemosphere.2021.132892.

Lin, Xiaohu, Jingcheng Xu, Arturo A. Keller, Li He, Yunhui Gu, Weiwei Zheng et al. 2020. Occurrence and Risk Assessment of Emerging Contaminants in a Water Reclamation and Ecological Reuse Project. Elsevier 744: 140977. https://doi.org/10.1016/j.scitotenv.2020.140977.

Lin, Yuanjie, Chao Feng, Sunyang Le, Xinlei Qiu, Qian Xu, Shuping Jin et al. 2022. Infant Exposure to PCBs and PBDEs Revealed by Hair and Human Milk Analysis: Evaluation of Hair as an Alternative Biomatrix. Environmental Science and Technology 56(22): 15912–19. https://doi.org/10.1021/ACS.EST.2C04045.

Linge, Kathryn L., Deborah Liew, Yolanta Gruchlik, Francesco Busetti, Una Ryan and Cynthia A. Joll. 2021. Chemical Removal in Waste Stabilisation Pond Systems of Varying Configuration. Environmental Science: Water Research & Technology 7(9): 1587–99. https://doi.org/10.1039/D1EW00129A.

Lodeiro, Carlos, José Luis Capelo, Elisabete Oliveira and Javier Fernández Lodeiro. 2019. New Toxic Emerging Contaminants: Beyond the Toxicological Effects. Environmental Science and Pollution Research 26(1). https://doi.org/10.1007/S11356-018-3003-1.

Longpré, Darcy, Luigi Lorusso, Christine Levicki, Richard Carrier and Philippa Cureton. 2020a. PFOS, PFOA, LC-PFCAS, and Certain Other PFAS: A Focus on Canadian Guidelines and Guidance for Contaminated Sites Management. Environmental Technology & Innovation 18(May): 100752. https://doi.org/10.1016/j.eti.2020.100752.

Longpré, Darcy, Luigi Lorusso, Christine Levicki, Richard Carrier and Philippa Cureton. 2020b. PFOS, PFOA, LC-PFCAS, and Certain Other PFAS: A Focus on Canadian Guidelines and Guidance for Contaminated Sites Management. Environmental Technology & Innovation 18(May): 100752. https://doi.org/10.1016/j.eti.2020.100752.

Ma, S., C. Zhou, C. Chi, Y. Liu and G. Yang. 2020. Estimating physical composition of municipal solid waste in china by applying artificial neural network method. Environ. Sci. Technol. 54: 9609–9617. https://doi.org/10.1021/acs.est.0c01802.

Maeng, Sung Kyu, Byeong Gyu Choi, Kyu Tae Lee and Kyung Guen Song. 2013a. Influences of Solid Retention Time, Nitrification and Microbial Activity on the Attenuation of Pharmaceuticals and

Estrogens in Membrane Bioreactors. Water Research 47(9): 3151–62. https://doi.org/10.1016/j. watres.2013.03.014.

Maeng, Sung Kyu, Byeong Gyu Choi, Kyu Tae Lee and Kyung Guen Song. 2013b. Influences of Solid Retention Time, Nitrification and Microbial Activity on the Attenuation of Pharmaceuticals and Estrogens in Membrane Bioreactors. Water Research 47(9): 3151–62. https://doi.org/10.1016/j. watres.2013.03.014.

Majumder, Abhradeep, Bramha Gupta and Ashok Kumar Gupta. 2019. Pharmaceutically Active Compounds in Aqueous Environment: A Status, Toxicity and Insights of Remediation. Environmental Research 176(September): 108542. https://doi.org/10.1016/J.ENVRES.2019.108542.

Maliszewska-Kordybach, B. 1999. Sources, concentrations, fate and effects of Polycyclic Aromatic Hydrocarbons (PAHs) in the environment. Part A: PAHs in Air, Polish Journal of Environmental Studies.

Mama, Cordelia Nnennaya, Chidozie Charles Nnaji, John P. Nnam and Opata C. Opata. 2021. Environmental Burden of Unprocessed Solid Waste Handling in Enugu State, Nigeria. Environmental Science and Pollution Research 28(15): 19439–57. https://doi.org/10.1007/S11356-020-12265-Y.

Mashabela, Manamele Dannies, Priscilla Masamba and Abidemi Paul Kappo. 2023. Applications of Metabolomics for the Elucidation of Abiotic Stress Tolerance in Plants: A Special Focus on Osmotic Stress and Heavy Metal Toxicity. Plants. MDPI. https://doi.org/10.3390/plants12020269.

Mathur, P., K. Rani and P. Bhatnagar. 2023. Monitoring hospital effluents through physico-chemical characterization and genotoxic testing. Water Environment Research 95. https://doi.org/10.1002/ WER.10843.

Meyer, Michael F., Stephen M. Powers and Stephanie E. Hampton. 2019. An Evidence Synthesis of Pharmaceuticals and Personal Care Products (PPCPs) in the Environment: Imbalances among Compounds, Sewage Treatment Techniques, and Ecosystem Types. Environmental Science and Technology 53(22): 12961–73. https://doi.org/10.1021/ACS.EST.9B02966.

Michałowicz, Jaromir. 2014. Bisphenol A—Sources, Toxicity and Biotransformation. Environmental Toxicology and Pharmacology. Elsevier. https://doi.org/10.1016/j.etap.2014.02.003.

Mo, Limei, Qiaohui Wang and Erping Bi. 2021. Effects of Endogenous and Exogenous Dissolved Organic Matter on Sorption Behaviors of Bisphenol A onto Soils. Journal of Environmental Management 287 (June). https://doi.org/10.1016/j.jenvman.2021.112312.

Mohapatra, Dipti Prakash and Deepak M. Kirpalani. 2019. Advancement in Treatment of Wastewater: Fate of Emerging Contaminants. Canadian Journal of Chemical Engineering 97(10): 2621–31. https://doi.org/10.1002/CJCE.23533.

Molina, Lázaro and Ana Segura. 2021. Biochemical and Metabolic Plant Responses toward Polycyclic Aromatic Hydrocarbons and Heavy Metals Present in Atmospheric Pollution. Plants. MDPI. https:// doi.org/10.3390/plants10112305.

Moosavi, Seyedehmaryam, Chin Wei Lai, Sinyee Gan, Golnoush Zamiri, Omid Akbarzadeh Pivehzhani and Mohd Rafie Johan. 2020. Application of Efficient Magnetic Particles and Activated Carbon for Dye Removal from Wastewater. ACS Omega 5(33): 20684–97. https://doi.org/10.1021/ ACSOMEGA.0C01905.

Morin-Crini, Nadia, Eric Lichtfouse, Guorui Liu, Vysetti Balaram, Ana Rita Lado Ribeiro, Zhijiang Lu et al. 2022a. Worldwide Cases of Water Pollution by Emerging Contaminants: A Review. Environmental Chemistry Letters 20(4): 2311–38. https://doi.org/10.1007/S10311-022-01447-4.

Morin-Crini, Nadia, Eric Lichtfouse, Guorui Liu, Vysetti Balaram, Ana Rita Lado Ribeiro, Zhijiang Lu et al. 2022b. Worldwide Cases of Water Pollution by Emerging Contaminants: A Review. Environmental Chemistry Letters 20(4): 2311–38. https://doi.org/10.1007/S10311-022-01447-4.

Muisa Zikali, Norah, Richman Munyaradzi Chingoto, Beaven Utete and Francisca Kunedzimwe. 2022. Household Solid Waste Handling Practices and Recycling Value for Integrated Solid Waste Management in a Developing City in Zimbabwe. Scientific African 16(July): e01150. https://doi. org/10.1016/j.sciaf.2022.e01150.

Naghdi, Mitra, Mehrdad Taheran, Satinder Kaur Brar, Azadeh Kermanshahi-pour, Mausam Verma and R.Y. Surampalli. 2018. Removal of Pharmaceutical Compounds in Water and Wastewater Using Fungal Oxidoreductase Enzymes. Environmental Pollution. Elsevier Ltd. https://doi.org/10.1016/j. envpol.2017.11.060.

Naidu, Ravi, Victor Andres Arias Espana, Yanju Liu and Joytishna Jit. 2016. Emerging Contaminants in the Environment: Risk-Based Analysis for Better Management. Chemosphere 154(July): 350–57. https://doi.org/10.1016/J.CHEMOSPHERE.2016.03.068.

Ng, Nyuk Ting, Wan Aini Wan Ibrahim, Zetty Azalea Sutirman, Mohd Marsin Sanagi and Aemi Syazwani Abdul Keyon. 2023. Magnetic Nanomaterials for Preconcentration and Removal of Emerging Contaminants in the Water Environment. Nanotechnology for Environmental Engineering 8(1): 297–315. https://doi.org/10.1007/S41204-022-00296-4.

Özşeker, Koray, Yahya Terzi and Coşkun Erüz. 2022. Solid Waste Composition and COVID-19-Induced Changes in an Inland Water Ecosystem in Turkey. Environmental Science and Pollution Research 29(36): 54596–605. https://doi.org/10.1007/S11356-022-19750-6.

Pande, Veni, Satish Chandra Pandey, Diksha Sati, Veena Pande and Mukesh Samant. 2020. Bioremediation: An Emerging Effective Approach towards Environment Restoration. Environmental Sustainability 3(1): 91–103. https://doi.org/10.1007/s42398-020-00099-w.

Parrado, Andrea Cecilia, Luciana S. Salaverry, Rosario Macchi, Marco L. Bessone, Franco M. Mangone, Marisa Castro et al. 2023. Immunomodulatory Effect of Dopamine in Human Keratinocytes and Macrophages under Chronical Bisphenol-A Exposure Conditions. Immunobiology 228(2): 152335. https://doi.org/10.1016/J.IMBIO.2023.152335.

Patwa, Aakash, Divyesh Parde, Devendra Dohare, Ritesh Vijay and Rakesh Kumar. 2020. Solid Waste Characterization and Treatment Technologies in Rural Areas: An Indian and International Review. Environmental Technology & Innovation 20(November): 101066. https://doi.org/10.1016/j.eti.2020.101066.

Priya, A.K., Gnanasekaran, L., Rajendran, S. and Qin, J. 2020. Environmental, and undefined 2022. n.d. Occurrences and Removal of Pharmaceutical and Personal Care Products from Aquatic Systems Using Advanced Treatment—A Review. Elsevier. Accessed June 1, 2023. https://www.sciencedirect.com/science/article/pii/S0013935121015991.

Radwan, Emad K., Hany H. Abdel Ghafar, Ibrahim, M.B.M. and Ahmed S. Moursy. 2022. Recent Trends in Treatment Technologies of Emerging Contaminants. Environmental Quality Management, March. https://doi.org/10.1002/TQEM.21877.

Rajmohan, K.S., Ramya Chandrasekaran and Sunita Varjani. 2020. A Review on Occurrence of Pesticides in Environment and Current Technologies for Their Remediation and Management. Indian Journal of Microbiology 60(2): 125–38. https://doi.org/10.1007/S12088-019-00841-X/TABLES/4.

Ramírez-Coronel, Andrés Alexis, Mohammad Javad Mohammadi, Hasan Sh Majdi, Rahman S. Zabibah, Masoume Taherian et al. 2023. Hospital Wastewater Treatment Methods and Its Impact on Human Health and Environments. Reviews on Environmental Health. https://doi.org/10.1515/REVEH-2022-0216/HTML.

Rathi, B. Senthil, Senthil Kumar, P. and Pau-Loke Show. 2021. A Review on Effective Removal of Emerging Contaminants from Aquatic Systems: Current Trends and Scope for Further Research. Journal of Hazardous Materials 409(May): 124413. https://doi.org/10.1016/j.jhazmat.2020.124413.

Reichert, Gabriela, Alinne Mizukawa, Jhonatas Antonelli, Franciane de Almeida Brehm Goulart, Tais Cristina Filippe and Júlio César Rodrigues de Azevedo. 2020. Determination of Parabens, Triclosan, and Lipid Regulators in a Subtropical Urban River: Effects of Urban Occupation. Water, Air, and Soil Pollution 231(3). https://doi.org/10.1007/S11270-020-04508-Y.

Rochester, Johanna R. 2013. Bisphenol A and Human Health: A Review of the Literature. Reproductive Toxicology. https://doi.org/10.1016/j.reprotox.2013.08.008.

Sadighara, Parisa, Mehdi Safta, Intissar Limam, Kiandokht Ghanati, Zahra Nazari, Marzieh Karami, and Amirhossein Abedini. 2022. Association between Food Additives and Prevalence of Allergic Reactions in Children: A Systematic Review. Reviews on Environmental Health, March. https://doi.org/10.1515/REVEH-2021-0158/HTML.

Sahani, Shalini, Hansa, Yogesh Chandra Sharma and Tae Young Kim. 2022. Emerging Contaminants in Wastewater and Surface Water. Energy, Environment, and Sustainability 9–30. https://doi.org/10.1007/978-981-16-8367-1_2.

Saidulu, Duduku, Bramha Gupta, Ashok Kumar Gupta and Partha Sarathi Ghosal. 2021. A Review on Occurrences, Eco-Toxic Effects, and Remediation of Emerging Contaminants from Wastewater: Special Emphasis on Biological Treatment Based Hybrid Systems. Journal of Environmental Chemical Engineering 9(4). https://doi.org/10.1016/j.jece.2021.105282.

Sam, Kabari, Amarachi P. Onyena, Nenibarini Zabbey, Chuks K. Odoh, Goodluck N. Nwipie, Dumbari K. Nkeeh et al. 2023. Prospects of Emerging PAH Sources and Remediation Technologies: Insights from Africa. Environmental Science and Pollution Research, March. https://doi.org/10.1007/S11356-023-25833-9.

Sarma, Hemen. 2022. Understanding Emerging Contaminants in Soil and Water: Current Perspectives on Integrated Remediation Approaches. Emerging Contaminants in the Environment: Challenges and Sustainable Practices, January 1–38. https://doi.org/10.1016/B978-0-323-85160-2.00002-0.

Schulz, Katarina, Marcia R. Silva and Rebecca Klaper. 2020. Distribution and Effects of Branched versus Linear Isomers of PFOA, PFOS, and PFHxS: A Review of Recent Literature. Science of The Total Environment 733(September): 139186. https://doi.org/10.1016/j.scitotenv.2020.139186.

Şentürk, İ. and Yildirim, B. 2020. Scientific Journal of Mehmet Akif Ersoy and undefined. 2020. A Study on Estimating of the Landfill Gas Potential from Solid Waste Storage Area in Sivas, Turkey. Dergipark. Org. Tr 3 (2): 63–76. https://dergipark.org.tr/en/pub/sjmakeu/issue/56459/765009.

Šereš, Michal, Tereza Hnátková, Petr Maršík, Tomáš Vaněk, Petr Soudek and Jan Vymazal. 2020. Field Study VI: The Effect of Loading Strategies on Removal Efficiencies of a Hybrid Constructed Wetland Treating Mixed Domestic and Agro-Industrial Wastewaters 395–409. https://doi.org/10.1007/978-3-030-29840-1_18.

Simon, Monica, Ajay Kumar, Alok Garg and Manisha. 2021. Biological Treatment of Pharmaceuticals and Personal Care Products (PPCPs) Before Discharging to Environment 259–82. https://doi.org/10.1007/978-981-15-6564-9_14.

Soursou, V., J. Campo and Y. Picó. 2023. Revisiting the analytical determination of PAHs in environmental samples: An update on recent advances. Trends in Environmental Analytical Chemistry 37: e00195–e00195. https://doi.org/10.1016/J.TEAC.2023.E00195.

Srivastava, Vaibhav, Barkha Vaish, Rajeev Pratap Singh and Pooja Singh. 2020. An Insight to Municipal Solid Waste Management of Varanasi City, India, and Appraisal of Vermicomposting as Its Efficient Management Approach. Environmental Monitoring and Assessment 192(3). https://doi.org/10.1007/S10661-020-8135-3.

Su, C., Cui, Y., Liu, D., Zhang, H. and Baninla, Y. 2020. Science of The Total Environment, and undefined 2020. n.d. Endocrine Disrupting Compounds, Pharmaceuticals and Personal Care Products in the Aquatic Environment of China: Which Chemicals Are the Prioritized Ones? Elsevier. Accessed June 1, 2023. https://www.sciencedirect.com/science/article/pii/S0048969720311633.

Su, Chengyuan, Xumeng Lin, Peng Zheng, Yongshen Chen, Lijian Zhao, Yongde Liao et al. 2019a. Effect of Cephalexin after Heterogeneous Fenton-like Pretreatment on the Performance of Anaerobic Granular Sludge and Activated Sludge. Chemosphere 235 (November): 84–95. https://doi.org/10.1016/j.chemosphere.2019.06.136.

Su, Chengyuan, Xumeng Lin, Peng Zheng, Yongshen Chen, Lijian Zhao, Yongde Liao et al.. 2019b. Effect of Cephalexin after Heterogeneous Fenton-like Pretreatment on the Performance of Anaerobic Granular Sludge and Activated Sludge. Chemosphere 235(November): 84–95. https://doi.org/10.1016/j.chemosphere.2019.06.136.

Sundaresan, Ruspika, Vinitha Mariyappan, Tse Wei Chen, Shen Ming Chen, Muthumariappan Akilarasan, Xiaoheng Liu et al. 2023. One-Dimensional Rare-Earth Tungstate Nanostructure Encapsulated Reduced Graphene Oxide Electrocatalyst-Based Electrochemical Sensor for the Detection of Organophosphorus Pesticide. Journal of Nanostructure in Chemistry. https://doi.org/10.1007/S40097-023-00524-6.

Tang, Yankui, Maozhong Yin, Weiwei Yang, Huilan Li, Yaxuan Zhong, Lihong Mo et al. 2019. Emerging Pollutants in Water Environment: Occurrence, Monitoring, Fate, and Risk Assessment. Wiley Online Library 91(10): 984–91. https://doi.org/10.1002/wer.1163.

Tauchnitz, Nadine, Florian Kurzius, Holger Rupp, Gerd Schmidt, Barbara Hauser, Matthias Schrödter et al. 2020. Assessment of Pesticide Inputs into Surface Waters by Agricultural and Urban Sources —A Case Study in the Querne/Weida Catchment, Central Germany. Environmental Pollution 267(December): 115186. https://doi.org/10.1016/J.ENVPOL.2020.115186.

Tchounwou, Paul B., Clement G. Yedjou, Anita K. Patlolla and Dwayne J. Sutton. 2012. Heavy Metal Toxicity and the Environment. EXS 101: 133–64. https://doi.org/10.1007/978-3-7643-8340-4_6/COVER.

Valdez-Carrillo, Melissa, Leif Abrell, Jorge Ramírez-Hernández, Jaime A. Reyes-López and Concepción Carreón-Diazconti. 2020. Pharmaceuticals as Emerging Contaminants in the Aquatic Environment of Latin America: A Review. Environmental Science and Pollution Research 27(36): 44863–91. https://doi.org/10.1007/S11356-020-10842-9.

Vijayanand, Madhumitha, Abiraami Ramakrishnan, Ramakrishnan Subramanian, Praveen Kumar Issac, Mahmoud Nasr, Kuan Shiong Khoo et al. 2023. Polyaromatic Hydrocarbons (PAHs) in the Water Environment: A Review on Toxicity, Microbial Biodegradation, Systematic Biological Advancements, and Environmental Fate. Environmental Research 227(June): 115716. https://doi.org/10.1016/J.ENVRES.2023.115716.

Wang, Liuwei, Wei-Min Wu, Nanthi S. Bolan, Daniel C.W. Tsang, Yang Li, Muhan Qin et al. 2021. Environmental Fate, Toxicity and Risk Management Strategies of Nanoplastics in the Environment: Current Status and Future Perspectives. Journal of Hazardous Materials 401(January): 123415. https://doi.org/10.1016/j.jhazmat.2020.123415.

Wang, Qidi, Shiwei Yan, Chao Chang, Chengkai Qu, Yulu Tian, Jinxi Song et al. 2023. Occurrence, Potential Risk Assessment, and Source Apportionment of Polychlorinated Biphenyls in Water from Beiluo River. Water (Switzerland) 15(3): 459. https://doi.org/10.3390/W15030459/S1.

Water Quality Home. n.d. Accessed June 1, 2023. https://www.waterquality.gov.au/.

Wu, Long, Chenghui Zhang, Yingxi Long, Qi Chen, Weimin Zhang and Guozhen Liu. 2022. "Food Additives: From Functions to Analytical Methods. Critical Reviews in Food Science and Nutrition 62(30): 8497–8517. https://doi.org/10.1080/10408398.2021.1929823.

Yang, Xueke, Lijing Xi, Zhaoyan Guo, Li Liu and Zhiguang Ping. 2023. The Relationship between Cadmium and Cognition in the Elderly: A Systematic Review. Https://Doi.Org/10.1080/03014460.2023.2168755 50(1): 15–25. https://doi.org/10.1080/03014460.2023.2168755.

Zhang, Lei, Jiahuai Zhang, Sai Fan, Yuxin Zhong, Jingguang Li, Yunfeng Zhao et al. 2023. A Case-Control Study of Urinary Concentrations of Bisphenol A, Bisphenol F, and Bisphenol S and the Risk of Papillary Thyroid Cancer. Elsevier 312: 137162. https://doi.org/10.1016/j.chemosphere.2022.137162.

Zheng, Shuilin, Zhiming Sun, Yuri Park, Godwin A. Ayoko and Ray L. Frost. 2013. Removal of Bisphenol A from Wastewater by Ca-Montmorillonite Modified with Selected Surfactants. Chemical Engineering Journal 234(December): 416–22. https://doi.org/10.1016/j.cej.2013.08.115.

Zhu, Yanli, Youxian Zhang, Dongxia Luo, Zhongyi Chong, Erqiang Li and Xuepeng Kong. 2021. A Review of Municipal Solid Waste in China: Characteristics, Compositions, Influential Factors and Treatment Technologies. Environment, Development and Sustainability 23(5): 6603–22. https://doi.org/10.1007/S10668-020-00959-9.

CHAPTER 7

Recent Advances in Modification of Biochar for Removal of Emerging Contaminants from Water Bodies

Jaswinder Kaur, Savita Chaudhary* and Aman Bhalla*

1. Introduction

The escalating population and their burgeoning demands have given rise to a pressing issue of water pollution, which demands urgent attention. Water sources are being directly or indirectly contaminated with domestic and industrial wastewater, making it a formidable task to purify them. The presence of minute quantities of undetectable pollutants or emerging contaminants complicates the treatment process (Ahmaruzzaman, 2021). Emerging Contaminants (ECs) are a group of chemical substances that have garnered attention in recent times due to their potential impacts on the environment and public health. This diverse group of contaminants encompasses various chemical substances, including but not limited to pharmaceuticals, personal care products, pesticides, dyes, microplastics, endocrine disruptors and industrial chemicals (Fig. 7.1) (Shen et al., 2021). They can enter the environment via a range of pathways, including industrial discharges, agricultural runoff, and wastewater treatment plants. They can also persist in the environment for extended periods and could accumulate in the food chain, potentially causing adverse effects on human health and the environment (Cheng et al., 2021).

Department of Chemistry and Centre of Advanced Studies in Chemistry, Panjab University, Chandigarh 160014, India.
* Corresponding authors: amanbhalla@pu.ac.in; schaudhary@pu.ac.in

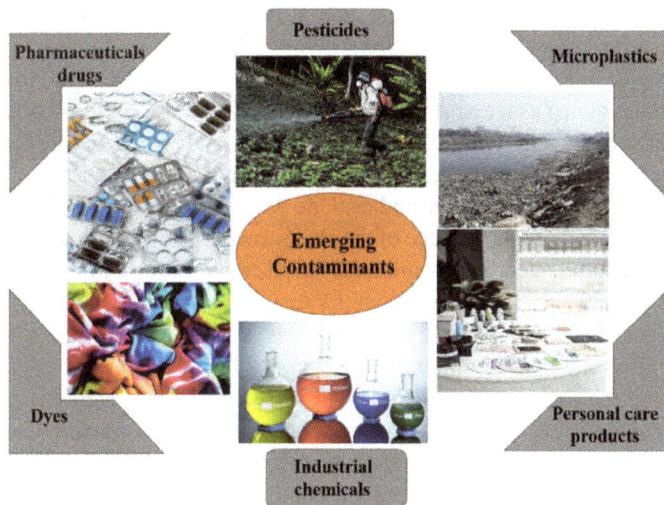

Figure 7.1 Potential source of emerging contaminants.

The issues with ECs are that one has limited knowledge about their potential health impacts, raising concerns about their safety for human health and their presence in the environment are typically in very low concentrations that cannot be easily detected using conventional methods, making it challenging to identify and mitigate their potential risks. The identification, monitoring and reducing the impact of new ECs are essential for protecting human health and the environment. Continued research is aimed at gaining a deeper understanding of the origins, behaviour, and possible health implications of emerging pollutants, while also developing innovative methods to identify and eliminate them (Qiu et al., 2022). ECs are a diverse group of chemicals that possess distinct physical and chemical characteristics, and as a result, exhibit varying degrees of persistence, mobility, and toxicity in the environment. Several factors, such as molecular structure, functional groups, and chemical bonding, can influence the physical and chemical properties of ECs. For instance, the water solubility and mobility of a compound can be increased by the existence of polar functional groups such as hydroxyl (–OH) or amino (–NH$_2$) groups. Similarly, compounds that contain aromatic rings or alkyl groups could exhibit greater resistant to biodegradation, leading to increased persistence within the environment (Bilal et al., 2019).

Fortunately, there are several commercial methods available to remove these pollutants from water bodies. The most common methods include GAC (Granular Activated Carbon) filtration, RO (Reverse Osmosis), AOPs (Advanced Oxidation Process), UV (ultraviolet) disinfection, MBR (membrane bioreactor) filtration, nanofiltration, and ion exchange. These methods have some limitations, such as high energy and maintenance costs, the generation of chemical waste, complex operations, limited removal of dissolved contaminants, and the need for frequent replacement. These factors make it challenging to implement these methods on a large scale. Therefore, it is important to consider these limitations when selecting a water treatment method and to explore alternative or complementary approaches to achieve the desired level of water purity.

Biochar (BC), a porous carbon-rich material that is produced from the pyrolysis of organic matter such as wood chips, crop stalks, animal carcasses and organic waste is used to adsorb ECs making it useful in environmental studies (Qiu et al., 2021). In recent years, researchers have been exploring ways to modify BC to enhance its ability to remove ECs from water bodies. One way to modify BC is through surface functionalization, where the surface of the BC is treated with chemical compounds to increase its ability to adsorb contaminants. Another approach is to incorporate metal oxides into the BC structure to enhance its removal efficiency. Additionally, using a combination of BC and other materials such as zeolites or Activated Carbon (AC) has been shown to improve the removal efficacy of ECs (Rathi and Kumar, 2021). By modifying the surface properties of BC, it is possible to increase its adsorption capacity for specific classes of contaminants. For example, BC that has been modified with oxygen-containing functional groups such as carboxyl (-COOH) or hydroxyl (-OH) groups can effectively adsorb polar compounds such as pharmaceuticals and pesticides (Wang et al., 2020).

Overall, the diverse physical and chemical properties of ECs make it important to tailor treatment approaches to specific compounds or classes of compounds. BC modification is one promising approach for addressing the challenge of removing these contaminants from the environment. This chapter, deals with the recent advances in different modifications of BC for extraction of specific types of ECs.

2. Modification Methods for Tailored/Functionalized Biochar

BC has been gaining attention as a potential low-cost and sustainable material for water treatment due to its high surface area, porosity, and adsorption capacity. However, its ability to remove certain ECs such as PPCPs can be limited. To improve its performance, various modification techniques have been developed (Fig. 7.2). Physical and chemical modifications, including heating at high temperatures in an inert atmosphere and treatment with strong acids or bases, can alter the surface properties of BC, thereby enhancing its ability to remove pollutants from water.

Additionally, BC can be impregnated with various chemicals, metal/metal oxides, or pyrolyzed with additives such as P, N, and S to modify its surface properties. Another approach is to functionalize BC with nanoparticles, such as FeO/TiO or coat it with bio-based polymers such as chitosan or lignin to increase its surface area and improve its adsorption capacity. Functionalized BC has been shown to be effective at removing a wide range of contaminants from water, including PPCPs, heavy metals, and organic compounds (Rajapaksha et al., 2016). The use of modified BC can be a sustainable and cost-effective method for the remediation of contaminated water, contributing to the protection of the environment and human health. In addition to its high adsorption capacity, functionalized BC also has the potential for regeneration and reuse, making it a promising material for sustainable water treatment applications.

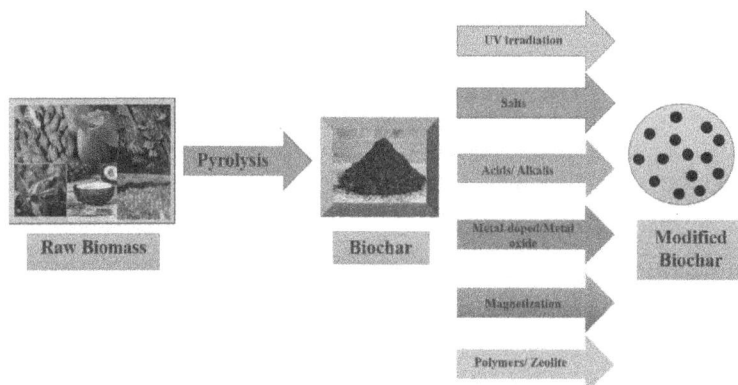

Figure 7.2 Synthesis of biochar and its modification.

2.1 Chemical Modification

To enhance the adsorption capacity and selectivity of BC for ECs, chemical modification is required. One way to achieve this is by functionalizing BC with oxygen and nitrogen-containing groups such as carbonyl, hydroxyl, carboxyl, amino, and amide groups. These functional groups can enhance the interactions between the BC surface and target contaminants, making it more effective at removing them from water.

Chemical modification of BC can also be achieved by treating it with oxidizing agents such as hydrogen peroxide and nitric acid to introduce oxygen-containing groups. Similarly, nitrogen-containing reagents such as ammonia and urea can be used to introduce nitrogen-containing groups.

2.1.1 Alkali/Acid Modification of Biochar

Alkali/acid modification involves treating BC with a base, such as sodium hydroxide or potassium hydroxide or with an acid, such as hydrochloric acid or sulphuric acid, which can increase the surface area and porosity of the material, as well as introduce functional groups that can enhance adsorption and remove mineral impurities.

Both alkali and acid modification as a low-cost and sustainable adsorbent have been shown to improve the adsorption capacity of BC for a number of ECs, including PPCPs (Xue et al., 2022). However, the choice of the modification method depends on the specific properties of the BC and the contaminants of interest, as well as the intended application. It is also important to consider the potential environmental impact of the modification process and the disposal of the resulting waste materials.

2.1.1.1 Removal of Nitrates and Phosphates

Water eutrophication, caused by the enrichment of nutrients such as nitrate and phosphate in waterbodies due to human activities is one of the significant concerns globally. Their accumulation in water bodies can lead to degradation of the quality of water and the ecological environment. Therefore, recovery of phosphate is necessary to reduce the concentration of phosphates in agricultural and industrial wastewaters and prevent their discharge into water bodies. To find the solution to this problem,

Takaya et al. (2016) explored the potential of modified BC for the recovery of phosphate (PO_4-P), which could have significant environmental benefits. To address the low nutrient adsorption capacity, chemical modifications using metal salts, alkali and acids were performed. The results showed that impregnation with magnesium could increase phosphate uptake from 2.1–3.6 (untreated BC) to 66.4–70.3% (treated oak and greenhouse waste BC's). The study suggested that the mineral composition of BC was a crucial factor affecting its phosphate uptake, while the surface area had a less significant influence on sorption. Chemical activation at low temperatures was found to enhance the functionality of BC to some extent. Overall, they concluded that magnesium salt treatment had a significant impact on the recovery of PO_4-P, while other activation methods had a marginal effect.

Yin et al. (2018) described the preparation and characterization of BC, a green and low-cost adsorbent, in the field of environmental remediation, to assess their effectiveness in the removal of nutrients from eutrophic waters in 2018. The study found that Al-modified BC has excellent multifunctional and surface charge properties, leading to enhanced NO_3^- and PO_4^{3-} adsorption capacities. The amount of Al content was critical in determining the optimal adsorption capacity for each nutrient, with the 15 and 20% Al-modified BC being ideal for NO_3^- and PO_4^{3-} adsorption. The maximum adsorption capacities for NO_3^- and PO_4^{3-} was found at 89.58 mg/g and 57.49 mg/g, respectively. They also found that the adsorption property was dependent on pH of the adsorption solution (i.e., optimal pH = 6 for NO_3^- adsorption and pH < 6 was advantageous for PO_4^{3-} adsorption). They concluded that chemical adsorption was the primary mechanism for NO_3^- and PO_4^{3-} adsorption, and the Langmuir-Freundlich model accurately described the adsorption isotherms. The use of BC as an adsorbent has emerged as a promising solution for addressing both the challenges of waste biomass disposal and climate change mitigation. This approach provides an economically viable way to use waste biomass.

2.1.1.2 Removal of Pharmaceutical Compounds

PPCPs are widely used in daily lives and are a growing concern due to their persistence, difficulty in breaking down and potential harm to living organisms when they are introduced into the aquatic ecosystem from various sources such as hospitals, households, and wastewater treatment plants (Sayin et al., 2021). These high concentrations of antibiotics can cause toxicity to microorganisms and disrupt the ecological balance, while even lower concentrations can lead to the development of antibiotic-resistant bacteria. Therefore, the use and disposal of PPCPs require careful consideration to minimize their harmful effects on the environment and human health. It is necessary to explore new treatment technologies to mitigate the adverse impacts of pharmaceutical products in the environment.

Tang et al. (2018) demonstrated a novel method of modifying Municipal Sewage Sludge (MS) BC using a combination of alkali and acid treatments to increase its adsorption capacity for removing organic contaminants like tetracycline (TC). The impact of pyrolysis temperature was also investigated, and the best results were obtained with a BC named SNMS-800. This BC showed an excellent performance for TC removal, with a maximum adsorption capacity of 286.913 mg/g. The enhanced adsorption properties were attributed to its increased porosity and strong interaction between the adsorbent's pore size and adsorbate size. The adsorption mechanism involves both π–π stacking interaction and pore filling. The low-cost and magnetic separation ability, along with its good reusability in natural water samples, makes SNMS-800 a promising option for large-scale removal of aqueous contaminants. Additionally, they provide a cost-effective solution for disposing of MS.

Furthermore, Tang et al. (2022) investigated the use of alkali-modified BC made from straw pyrolysis to remove ECs such as antibiotics (ofloxacin (OFL) and TC) and bisphenol A (BPA) from a complex KW (Kitchen Wastewater) system. The results showed that the BC treated with sodium hydroxide (NBC) was the most effective in removing contaminants, with over 95% of them removed within 9 hr. This was attributed to the higher specific surface area and stronger hydrophobicity of the NBC. The adsorption mechanism involved electrostatic attractions, π–π and hydrophobic interactions and pore filling between NBC and the contaminants. The optimal pH for the highest equilibrium adsorption capacity was found to be around 8, with little effect observed from other environmental factors such as COD (Chemical Oxygen Demand), NH_4^+ and PO_4^{3-} on adsorption. They suggested that NBC can effectively remove environmental contaminants from complex KW systems and could have practical applications for controlling contaminants in real-world settings.

Zhang et al. (2019) investigated the impact of different microwave-assisted chemical modification methods on BC's adsorption capacity and its potential for removing 17β-estradiol (E2). Rice husk biomass waste was used to produce BC, which was then modified using different chemical agents, including alkali (sodium hydroxide and sodium bicarbonate), acid (sulphuric acid and phosphoric acid) and the oxidizing agent (hydrogen peroxide). The results revealed that the alkali-modified BC had the highest adsorption capacity for E2, the maximum being 44.93 mg/g for NBC at 298K, due to its higher surface area, pore volume, more hydrophobic and nonpolar surface. The adsorption mechanism was mainly dominated by the π–π interaction and hydrogen bonding, and the process was spontaneous, endothermic, and involved both physisorption and chemisorption. Overall, this provides valuable insights into the use of modified BC for removing E2 from water, as well as the potential for repurposing biomass waste.

Similarly, to address the problem of water pollution, Sayin et al. (2021) came up with a solution for removing pharmaceutical pollutants. They successfully synthesized a new water-based BC MPCWSB500 at 500°C, from a sour cherry stalk that was modified with H_3PO_4 to remove ciprofloxacin (CFX), a fluoroquinolone based micropollutant from water. The modified feedstock significantly improved the BC's CFX adsorption capacity from 17 to 39 times. The BC is suitable for both batch and continuous treatment systems, and its surface characteristics and possible interactions with CFX were investigated. MPCWSB500 exhibited a specific surface area 93 times larger than that of pristine BC (949.0 m²/g compared to 10.16 m²/g). Researchers found that BC had a short operation time, high sorption capacity (410.06) mg/g) and nearly 100% removal efficiency at optimum experimental conditions (pH: 6.3, contact time: 40 min, MPCWSB500 dose: 15 mg). Furthermore, the presence of Cl⁻, K⁺, Na⁺ and NO³⁻ ions in the adsorption medium had little impact on the sorption performance and it could be regenerated and recycled for up to five sorption-desorption cycles. Ultimately, MPCWSB500 demonstrated effectiveness at removing CFX from polluted water sources including stimulated hospital wastewater and synthetic urine samples. Therefore, providing a viable and practical solution for addressing the issue of CFX contamination in aquatic environments.

Xue et al. (2022) successfully synthesized and used pinewood BC modified with KOH (K-BC) to absorb and degrade model pharmaceutical compounds, including CFX and carbamazepine (CBZ), by loading them with CuO from polluted water in 2022. The K-BC showed a larger surface area (23 m^2/g > 2 m^2/g) and more surface functional groups than pristine BC, resulting in a higher adsorption capacity for CFX. K-BC-CuO exhibited the highest degradation performance for CFX due to its larger surface area, which provides more active sites for CFX removal and the efficient activation of persulphate by the functional groups of K-BC and Cu-O. Additionally, they used Response Surface Methodology (RSM) to investigate the effects of various operating parameters on the degradation of CBZ using the K-BC CuO/PS system, confirming the order of the effect of each factor on the CBZ degradation rate. Scavenging experiments showed that singlet oxygen plays a key role in the degradation of both CFX and CBZ using this system. Furthermore, K-BC-CuO demonstrated an excellent performance for the simultaneous degradation of CFX and CBZ and relatively high Total Organic Carbon (TOC) concentration removal rate. They suggested that modifying BC and integrating it with CuO provides a sustainable approach for enhancing the adsorption and degradation of various environmental pollutants, including emerging contaminants.

2.1.1.3 Removal of Mercury

Mercury stands out among other heavy metals due to its unique characteristics, which make it very mobile and capable of persisting in the atmosphere for a long time. As a result, exposure to mercury can lead to various neurological and renal issues such as memory loss, anxiety, loss of speech and hearing abilities, and kidney and gastrointestinal problems (Faheem et al., 2018). The release of toxic mercury into the environment is caused by both natural and human activities, including volcanic eruptions, rock weathering, biogenic emissions, metal processing, electroplating, and wood pulping. One of the distinct characteristics of mercury is its ability to accumulate in organisms through the food chain, posing a significant threat to human health and the ecosystem.

Tan et al. (2016) investigated the sorption capacity of Na_2S modified biochar (BS), K-BC and AC, which were produced using corn straw biochar (BC) as a precursor, for aqueous Hg (II) and atrazine, both individually and as a mixture. The study found that BS significantly has higher adsorption capacity (76.95%) to remove Hg (II) from water than that of BC. Similarly, the sorption capacity for atrazine increased by 38.66, 46.39% and 47 times for BS, K-BC, and AC, respectively. However, competitive sorption was observed when Hg (II) and atrazine were present together on all sorbents. The results suggested that impregnating BC with sulphur is an effective way to enhance Hg (II) removal due to formation of HgS precipitate and oxygen-containing functional groups. Additionally, they indicated that AC had high BET surface area and was the most effective sorbent for atrazine removal.

New low-cost, amino-grafted modified BC (AMBC) has been successfully created by activating carboxylic groups on pre-oxidized BC (BC-COOH) for removing Hg

(II) from an aqueous solution, with significantly improved adsorption performance (Faheem et al., 2018). Optimized amino-BC2 has a significantly higher capacity (14.1 mg/g) compared to pristine BC (7.1 mg/g). The presence of chemically grafted amino groups and residual -COOH groups on the BC surface was confirmed using the FTIR, SEM and XPS techniques. The success of this modification is attributed to the presence of grafted amino and residual –COOH groups on the adsorbent surface, which facilitate chemisorption. These research findings not only provide a suitable adsorbent for decontaminating Hg (II) from an aqueous solution but also provide a new route for multi-functionalizing BC, making it an environmentally friendly and inexpensive adsorbent.

2.1.1.4 Removal of Lead

Metal pollution, especially lead contamination, is a serious threat to human health due to its persistence and toxicity even at low concentrations. Anthropogenic sources such as mining, industrial, and recycling activities contribute significantly to Pb contamination. The toxic effects of lead on plants, animals, and humans are well-established and can result in severe health problems, including brain and nervous system damage. The impact on young children is of particular concern, as they are more vulnerable to Pb toxicity (Shahib et al., 2022). Therefore, it is crucial to take proactive measures to reduce lead pollution to protect public health and the environment.

Wu et al. (2016) focused on the use of modified BC (MCFB), derived from coconut fibre (CFB), by subjecting them to ammonia, nitric acid, and hydrogen peroxide treatments as an effective and eco-friendly material for immobilizing toxic heavy metals in aqueous environments. Three types of BC were obtained by pyrolyzing coconut fibre at different temperatures (300, 500, and 700°C). The BC treated with these chemicals showed an increased ability to remove lead (Pb) from aqueous environments when compared to the untreated control. However, at higher temperatures (500 or 700°C), chemical modifications did not enhance adsorption of Pb, indicating that the resistance of BC to chemical treatment increased with pyrolysis temperature. The study also revealed that the removal rate of Pb by the CFB pyrolyzed at 300°C and modified with ammonia increased from 71.8 to 99.6% compared to the untreated BC in aqueous solutions containing 100 mgL^{-1} Pb. These findings indicate that the CFBs and MCFBs possess promising physico-chemical properties and further investigations are required to quantify their adsorption capacities for various pollutants.

In continuation, the same group also identified the five most common Pb species (Pb-montmorillonite, $Pb(C_2H_3O_2)_2$, $PbSO_4$, $Pb-Al_2O_3$ and $Pb_3(PO_4)_2$) that were removed by the BC. Specifically, BC modified with ammonia or nitric acid showed the highest capacity for adsorbing Pb, increasing the capacity by 113.3 and 86.85% respectively, while H_2O_2 had no effect. As well as unmodified BC produced at 700°C, hold promise for remediating Pb-contaminated water. This indicates a temperature-dependent response of chemically modified BC to lead sorption (Wu et al., 2017).

Further, Wongrod et al. (2018) investigated the effectiveness of raw digested sludge BC and chemically-modified BC in removing lead from water through adsorption kinetics and isotherm studies. Two types of BC were used, one derived from sewage sludge and the other from the organic fraction of municipal solid waste digestate. The results showed that the Langmuir isotherm model was able to accurately describe the lead sorption data, which followed the pseudo-second-order kinetic model. The study also found that KOH treatment was more efficient than H_2O_2 treatment in enhancing lead sorption, improving both sorption kinetics and capacity. For instance, in the case of sewage sludge BC, the sorption capacity increased by about 2-fold and 10-fold after H_2O_2 and KOH treatment, respectively.

Shahib et al. (2022) conducted chemical modification using chemical reagents like acid (H_3PO_4 and H_2O_2) and alkali (NaOH and K_2CO_3) modifiers to modify the surface of the sludge derived BC (SDBC). The alkali modifier can change the neutral pH of the BC to an alkali pH by adding –OH functional groups, while the H_3PO_4 activation can promote the acidity of the BC by adding carboxylic functional groups. The effectiveness of the SDBC in removing pollutants depends on the type of chemical activation used. For example, the H_2O_2, K_2CO_3 and NaOH activations significantly enhance the removal of positively charged pollutants, while the H_3PO_4 activation enhances the removal of Cr (VI) by creating opposite charge attractions on the surface of the BC. The adsorption process of the NBC was well-described by the Elovich model Electrostatic attraction accounted for the initial fast stage of Pb (II) and MB adsorption, with the rate-determining steps being the chemical adsorption through the heterogeneous surface of the NBC. They found that NaOH was the optimal chemical activation candidate under identical experimental conditions. However, different modifier concentrations could affect the physiochemical properties of the BC and hence they concluded that reuse of sewage sludge through activated BC preparation is a promising technique for sustainable sludge treatment.

2.1.1.5 Removal of Dye

Synthetic dyes are widely used in industries such as textiles, paper, plastic, leather, cosmetics, food processing, wool, and printing, but they have become a significant source of pollution (Liu et al., 2019). Leftover dyes in wastewater are very visible and difficult to break down, creating a serious environmental hazard. They can have adverse effects on both aquatic life and human health.

To solve this issue, Xu et al. (2016) generated a new adsorbent called Citric Acid (CA)-modified BC (CAWB) from water hyacinth biomass by pyrolysis at 300°C in a nitrogen environment. The purpose of modifying BC with citric acid was to increase its ability to adsorb contaminants. Results showed that CAWB was very effective in removing Methylene Blue (MB), a common dye used in textile production with a maximum capacity of 395 mg/g. The properties of CAWB were analyzed using the FTIR, XPS, SEM, and BET analyses, which showed that the introduction of carbonyl groups via esterification with citric acid played a crucial role in MB adsorption. The adsorption process was influenced by factors such as the initial MB concentration, solution pH, background ionic strength, and temperature. The adsorption behaviour was described by the pseudo-second-order kinetic model and Langmuir isotherm. The thermodynamic analysis showed that the adsorption of MB onto CAWB was a process that required heat and was naturally occurring. Moreover, CAWB was found to retain its excellent regeneration and adsorption performance after multiple cycles of adsorption. They suggested that CAWB has great potential for environmental applications in treating actual dye wastewater.

For the removal of dye from polluted water, Liu et al. (2019) used corn stalk to create BC (CSBC), and modified it using KOH and H_3PO_4. They characterized the BC using various techniques and tested their ability to adsorb MB. The modification significantly increased the surface area, pore volume and functional groups of the BC and demonstrated an enhanced increase in MB adsorption capacity (up to 10 times) compared to unmodified BC. The adsorption mechanism involved physical and electrostatic interactions, π–π interactions and hydrogen bonding. Overall, they concluded that modified BC, particularly KOH-CSBC, has potential as an effective adsorbent for removing dye from textile wastewater.

Mian and Liu (2020) reported the use of chemically modified sludge-derived BC as catalysts for peroxymonosulphate (PMS) activation and subsequent degradation of organic pollutants (rhodamine B, methyl orange, acid orange 7, MB and their combination). The researchers investigated the effects of the pyrolysis temperature and chemical modification on the efficiency of PMS activation by the sludge BC. They aimed to find a cost-effective way to produce efficient PMS activators. The results showed that NH_3-S600-KOH (both pre- and post-treated BC produced at pyrolysis temperature of 600°C) was the most effective catalyst for PMS activation and organic pollutants degradation. The best catalyst showed an excellent performance in oxidative degradation of organic pollutants in PMS/acidic media and outperformed many previously reported carbocatalysts. The dominant catalytic mechanism was a non-radical process performed by pyridinic N, while pyrrolic N, activated C (+) and surface area supported the adsorption of pollutants. They provided a facile method for efficient sludge valourization and new scientific insights into the specific role of N species and other functional groups of carbocatalysts in PMS-based AOPs.

2.1.1.6 Removal of Ammonium

Long-term fertilization with nitrogen fertilizers has led to nitrogen discharge into underground water and aquatic systems, causing eutrophication. Research has mainly focused on nitrate adsorption, but little attention has been given to ammonium nitrogen adsorption (Wang et al., 2020). The development of materials to remove ammonium from deteriorated water is crucial for restoring the water quality, as ammonium is a major pollutant affecting both ground and surface water. Although ammonium levels in nature are typically low, certain regions have experienced elevated concentrations of ammonium due to intensive livestock farming and excessive use of nitrogen-based fertilizers. Therefore, studying the adsorption mechanism of ammonium nitrogen is necessary to improve its utilization capacity and reduce loss.

Vu et al. (2017) investigated the modification of BC made from corncob to enhance its efficiency in adsorbing ammonium from synthetic water. The BC was treated with HNO_3 and NaOH under specific conditions, including an acid concentration of 6 M and an impregnation ratio of 1:5 (w/v). The resulting material, MBCC2, showed a maximum adsorption capacity of 22.6 mg/g, as determined by the Langmuir model. The optimum pH for adsorption was between 8 and 9, and the kinetics of adsorption followed the pseudo-second order model. They suggested that MBCC2 has the potential to remove ammonium from water, including groundwater and surface water, as well as in wastewater treatment. They recommended further investigation into desorption and adsorption interference before considering the application of this material in water and wastewater treatment systems.

The impact of modifying wheat straw BC using $FeCl_3$ and HCl, both individually and in combination, on ammonium adsorption onto BC surfaces involving both boundary and intraparticle diffusion, with chemisorption being the rate-limiting step was revealed by Wang et al. (2020). They also analyzed the surface properties of BCs before and after ammonium adsorption to understand the adsorption mechanisms. The findings showed that these modification techniques improved the ammonium adsorption capacity by 14% or more. The improvement was due to increased –OH and O–C=O functional groups, a larger specific surface area, as well as Fe^{3+}/Fe^{2+} redox coupling that acted as an electron shuttle. The mechanism behind ammonium adsorption by Fe-H-WB (wheat straw modified by $FeCl_3$ and HCl) involves the formation of N-(C=O) functional group during the adsorption process.

Wang et al. (2020) investigated the effect of acid and oxidation treatments to stimulate BC ageing conditions and on the adsorption ability of BC towards NH_4^+-N of peanut shell BC (PBC). The study used untreated fresh BC as a control group and compared it with four different types of treated BC. The results showed that acid-aged BC had a significantly increased maximum NH_4^+-N adsorption capacity (from 24.58 to 123.28 mg/g) after H_2O_2 modification, mainly due to physical changes such as increased surface area and porosity. The study suggested that the acid and oxidation ageing treatments did not decrease the adsorption ability of BC in this simulation experiment and aged BC under a weakly acidic oxidant (H_2O_2) could greatly improve its adsorption ability towards NH_4^+-N.

2.1.1.7 Removal of Chromium

Water pollution has become a major public concern (Fig. 7.3), with heavy metals such as chromium posing a significant threat due to their toxicity, bioaccumulation, and persistence in water. Hexavalent chromium is highly toxic and can cause serious health problems, including cancer (Shi et al., 2020). This metal is discharged into water from various industrial processes, such as leather tanning, steel fabrication, and paint manufacturing. Due to incorrect disposal and storage, chromium ions have entered the environment, leading to incidents of water contamination. It is essential to implement proper disposal and treatment methods to mitigate the adverse effects of chromium contamination on the environment and human health.

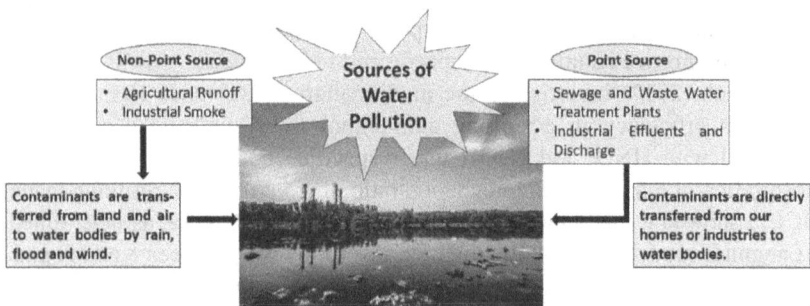

Figure 7.3 Sources of water pollution.

Dong et al. (2017) explored the use of nZVI (nano Zero-Valent Iron) supported on different modified BCs (KOH-BC, HCl-BC, and H_2O_2-BC) for the removal of Cr (VI) from water. The results showed that HCl-BC was the most effective BC for enhancing the removal of Cr (VI) by nZVI@BC due to its larger surface area and lower surface negative charge. The mass ratio of nZVI to HCl-BC (1:1) was found to be crucial in determining the particle dispersion and the Cr (VI) removal. The removal of Cr (VI) was found to be pH dependent, with lower pH favouring the reduction of Cr (VI). The study also found that nZVI@HCl-BC had a unit removal capacity for Cr (VI) and this was influenced by the initial concentration of Cr (VI) in water. The use of BC as a support for nZVI was considered cost-effective and the modifying methods were simple, making nZVI@HCl-BC a favourable remediation material for removing Cr (VI) from waste water. Additionally, they found that the reaction products of Cr (III) hydroxides or Cr (III)/ Fe (III) hydroxides on the surface of HCl-BC made it easier to remove chromium and nZVI from water and eradicated potential secondary pollutants caused by the reaction products.

Asadullah et al. (2019) investigated the modification of BC using KOH at an optimum hydrochar to KOH ratio 3:1 obtained from *Lepironia Articulata* (LA) through hydrothermal carbonization (HTC) for removing heavy metal (Cr (VI)) from wastewater. The physical and chemical properties of both hydrochar and modified BC were characterized. The modification resulted in a porous structure and a significant increase in specific surface area from 69.23 m^2/g to 820.54 m^2/g for the modified and unmodified BC, respectively. The maximum sorption capacity and removal efficiency of the modified BC were 28.75 mg/g and 98.9%, respectively, at 313 K and pH 2.0, which was 30% higher compared to the unmodified BC. The pseudo-second-order model best described the sorption kinetics data, and the Langmuir adsorption model was used to simulate the sorption isotherm. The thermodynamic parameters (negative $\Delta G°$ and positive $\Delta H°$) confirmed that the adsorption process was spontaneous and endothermic. The overall findings suggested that LABC could be used as an efficient, eco and budget friendly adsorbent for Cr (VI) removal from polluted water.

Shi et al. (2020) developed a cost effective and highly efficient adsorbent for removing Cr (VI) from aqueous solutions using glue residue, modified by using various modifiers such as HCl, KOH and $ZnCl_2$. The study found that the BC (Zn2GT700), derived from glue residue modified by a 2:1 ratio of $ZnCl_2$ and pyrolyzed at 700°C, had the highest specific surface area and surface functional groups, making it an efficient and recyclable adsorbent for Cr (VI) adsorption. At a pH of 2, the maximum adsorption capacity of Zn2GT700 was 325.54 mg/g for Cr (VI), with a removal efficiency of 98%, higher than other adsorbents tested. The adsorption data fitted well by the pseudo second order kinetic and Freundlich isotherm models and the thermodynamic study showed that Cr (VI) adsorption on Zn2GT700 was spontaneous. The adsorption mechanism analysis showed that ZnO promoted Cr (VI) removal via electrostatic interaction and ion exchange, while surface functional groups such as -OH and COOH reduced Cr (VI) to Cr (III). Co-precipitation and physical adsorption also played important roles in Cr (VI) removal. Zn2GT700 demonstrated good regeneration, with Cr (VI) removal efficiency remaining at

90% after six cycles. Therefore, they suggested that modified glue BC has potential applications in the treatment of Cr (VI) polluted wastewater.

El-Nemr et al. (2021) achieved a goal related to the production of modified BC and their effectiveness in adsorbing Cr (VI) ions. The study explored novel synthesis methods for producing modified BC from selective cheap biomass of watermelon peel residues. The physicochemical properties of BC surfaces were evaluated and the effect of the modification on the adsorption of Cr (VI) ions was studied. It was found that a small amount of BC can effectively adsorb Cr (VI) ions, which can reduce the cost of wastewater treatment. The modified Melon-BO-NH$_2$ and Melon-BO-TETA BCs were successfully functionalized with amino groups (ammonium hydroxide and Triethylenetetramine (TETA)), which significantly increased the amount of Cr (VI) ions adsorption. The BCs were characterized using various techniques, including SEM, BJH, FT-IR, BET, DSC, TGA and EDAX analyses. The results showed that the chemical modification of BC significantly increased its ability to adsorb Cr (VI) ions, with the highest removal percentage being 69% for Melon-B, 98% for Melon-BO-NH$_2$ and 99% for Melon-BO-TETA BCs. Also, the pore size of the modified BC reduced by amination of BC, but its surface area got expanded. The kinetic study of Cr (VI) ions uptake revealed that the pseudo-second order kinetic model was the most appropriate to fit the adsorption kinetic data. Overall, they highlighted the importance of functional group interactions in controlling the adsorption mechanisms of BC. Acid-base and hydrogen bonding interactions were identified as the primary mechanisms responsible for the adsorption of Cr (VI) ions.

2.1.1.8 Removal of Cadmium

Cadmium is a non-essential element for the human body, but a major environmental pollutant with teratogenic, carcinogenic, and toxic effects. It is widely used in metal plating and cannot degrade naturally, leading to its accumulation in organisms through the food chain. Cadmium affects the renal tubular function, causing renal tubular disease and kidney failure. It exists mainly in ionic and compound forms, easily forming complexes with organic matter and making it hard to degrade naturally. Therefore, effective measures should be taken to control cadmium pollution to protect human health and the environment.

Chen et al. (2019) aimed to develop a method of modifying the surface of BC (derived from rice husks with amino groups and disulphide bonds) with cystamine dihydrochloride (ligand) and crosslinking it with glutaraldehyde (crosslinker), for the efficient removal of cadmium (Cd) from contaminated water. The resulting BC showed a higher affinity for Cd (II) than other heavy metals such as Zn, As, Co, Ni and Cr. The modification of BC enhanced the electrostatic attraction and surface complexation, resulting in a theoretical maximum adsorption capacity of almost 10 times greater than that of raw BC for Cd (II). These results suggest that the modified BC can potentially serve as a cost-effective and eco-friendly alternative for removing Cd (II) from water.

Mo et al. (2021) focused on the use of modified BC from waste eucalyptus wood (KBC) for removing Cd (II) from contaminated water. The modification with potassium permanganate (KMnO$_4$) improved its adsorption properties and removal rate of Cd (II) compared to unmodified BC. The optimal conditions for removal of Cd (II) were found to be pH 5, dosage of 80 mg, contact time of 6 hr, 25°C temperature, and initial concentration of 50 mg/L. The adsorption process followed pseudo-second-order kinetics and the Langmuir isothermal model. The samples were characterized by various techniques (SEM, EDAX, BET, XRD, and XPS) to understand the adsorption mechanism of heavy metals. The mechanism of Cd (II) removal was explored and found to involve complexation, oxidation, and cation-π electron interaction with oxygen- and manganese-containing groups on the surface of KBC. The treatment of potassium permanganate increased the pore size, introduced polar functional groups, and generated manganese oxides, leading to improving heavy metal adsorption performance. Thus, it was shown that KBC modification is an effective and economical method for treating heavy metal-contaminated water.

2.1.1.9 Removal of Fungicide

The use of fungicides has increased with the rise in demand for crop yields due to the rapid growth in global population. However, this increase in fungicide use can lead to a non-point source pollution in surface water through runoff. To address this issue, researchers have identified adsorption as the most efficient treatment option for removing fungicides from water, which comes with low operating costs in water-treatment plants. In conclusion, adsorption can be a viable solution to tackle the problem of fungicide pollution in water bodies.

Lee et al. (2021) investigated the impact of successive $KMnO_4$ and KOH modifications on peanut shell BC (PSB) in removing fungicides (carbendazim, pyrimethanil, tebuconazole) from contaminated water. The physical and chemical properties of PSB and the chemically-modified PSB (PSB_{OX-A}) were analyzed using various techniques, and batch experiments were also conducted to evaluate the effects of temperature, ionic strength, and humic acids on fungicide adsorption. The results showed that the PSB_{OX-A} had a higher specific surface area and total pore volume than PSB alone. The chemisorption process and Langmuir isotherms model played a significant role in the adsorption of the fungicides. Furthermore, the adsorption efficiency of the fungicides increased with temperature and NaCl concentration. They recommended that in water-treatment plants, modified PSB_{OX-A} could be a promising option for removing fungicides (Table 7.1).

2.1.2 Salt/Metal-doped/Metal Oxide Modified Biochar

BC can also be modified using different techniques such as salt modified, metal-doping, and metal oxide-modification. Salt modified BC involves adding salts such as sodium chloride or potassium chloride to the BC during its production. While metal-doped BC consists of incorporating metals such as iron, copper, or silver into the BC and metal oxide-modified BC involves coating the BC with metal oxides such as titanium dioxide or iron oxide. These modifications aim to improve the BC's adsorption capacity and chemical reactivity by increasing its surface area and creating new adsorption sites. As a result, BC becomes more efficient in removing various types of contaminants from wastewater (Fig. 7.4), including heavy metals, organic compounds, and nutrients.

Mahmoud et al. (2016) found that the modified switchgrass BC treated with tetradecyltrimethyl ammonium bromide (SB-TTAB) adsorbent is highly effective in

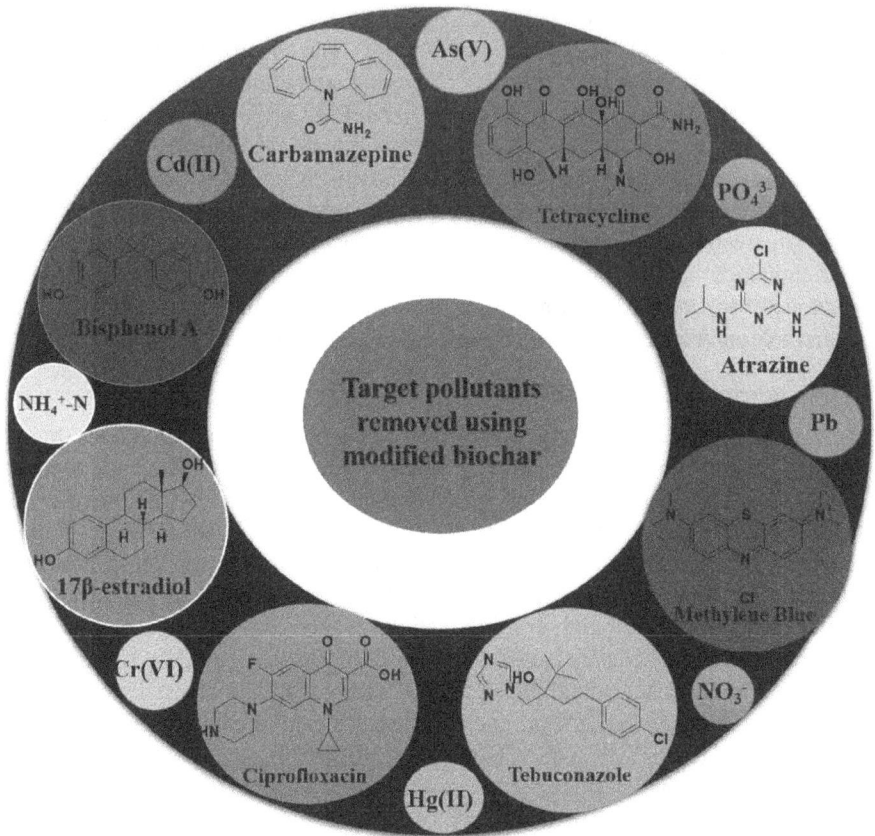

Figure 7.4 Target pollutants removed using modified biochar.

removing colour from water samples containing the reactive red 195 A (RR-195A) dye. The performance of this modified BC adsorbent was compared with other earlier reported BC and it was found to be superior. The dye was extracted from aqueous solutions in a pH range of 1.0–12.0 and the optimal condition was identified as pH 5, with an equilibrium time of 20–30 min. The effect of temperature on the colour removal of RR-195A dye by the SB-TTAB adsorbent was also explored and the thermodynamic parameters confirmed that the adsorption process was exothermic and spontaneous. The Langmuir model was found to fit best for the experimental data, while the pseudo-second-order kinetic model and the intraparticle diffusion process were both effective in describing the adsorptive colour removal of the dye. The SB-TTAB adsorbent was also found to be effective in decolourizing real water samples, including tap water, raw water, industrial water, and sea water. In conclusion, they confirmed the potential application of the SB-TTAB adsorbent for decolourization of RR-195A dye from various water resources.

Liu et al. (2021) focused on the use of lignin, a byproduct of biorefinery processes, to create a promising adsorbent for the removal of MB, a prevalent cationic dye used in the textile industry. The researchers, chemically modified lignin-derived BC with different oxidation number manganese compounds and found that

Table 7.1 Chemical modification using acid/alkali treatment of biochar.

Modifier	Raw biomass	Target pollutants	Adsorption capacity (Q_{max}) (mg g^{-1})	Refs.
Metal salts, acids and alkali followed by impregnation with Mg	Oak wood and paprika waste (Greenhouse waste)	Phosphate PO_4-P	Oak wood (3.6–70.3%) Greenhouse waste (2.1–66.4%)	Takaya et al., 2016
Na_2S, KOH, AC	Corn straw	Hg (II) and atrazine	Hg (II) [BC= 3.23 BS= 5.71 BK= 4.46 AC= 4.57], Atrazine [BC= 1.94 BS= 2.69 BK= 2.84 AC= 92.23]	Tan et al., 2016
Ammonia, hydrogen peroxide, and nitric acid pyrolyzed at 300, 500, and 700°C	Coconut fibre	Pb	increased from 71.8 to 99.6 % compared to the untreated BC (pyrolyzed at 300°C and modified with ammonia)	Wu et al., 2016
Citric acid	Water hyacinth (*Eichornia crassipes*)	Methylene Blue	395 mg g^{-1}	Xu et al., 2016
Ammonia, hydrogen peroxide and nitric acid	Coconut fibre	Lead (Pb-montmorillonite and Pb($C_2H_3O_2)_2$)	Modified with ammonia and nitric acid increased from 49.5 to 105.5 and 85.2 mg g^{-1}, respectively (BC pyrolyzed at 300 °C)	Wu et al., 2017
HNO_3 and NaOH	Corncob	Ammonium (NH_4^+-N)	22.6 mg g^{-1}	Vu et al., 2017
Acid (HCl), base (KOH) and oxidation (H_2O_2) treatment	Cornstalk	Cr (VI)	-	Dong et al., 2017
Alkali-acid	Sewage sludge	Tetracycline	286.913 mg g^{-1}	Tang et al., 2018
Al-modified (AlCl$_3$)	Poplar chips	Nitarte (NO_3^-) and phosphate (PO_4^{3-})	NO_3^-= 89.58 mg g^{-1}, PO_4^{3-}=57.49 mg g^{-1}	Yin et al., 2018

Table 7.1 contd. ...

... Table 7.1 contd.

Modifier	Raw biomass	Target pollutants	Adsorption capacity (Q_{max}) (mg g^{-1})	Refs.
Potassium hydroxide (KOH) and hydrogen peroxide (H_2O_2))	Sewage sludge digestate	Pb	$H_2O_2 = 25$ mg g^{-1}	Wongrod et al., 2018
Amino-grafted modified	Raw corncobs	Hg (II)	14.1 mg g^{-1}	Faheem et al., 2018
Amino groups and disulphide bonds (cystamine dihydrochloride as a modification ligand and glutaraldehyde as a crosslinker)	Rice husk	Cd (II)	81.02 mg g^{-1}	Chen et al., 2019
Alkali-modified (KOH)	*Lepironia articulata*	Cr (VI)	28.75 mg g^{-1}	Asadullah et al., 2019
Microwave-assisted acid (H_2SO_4 and H_3PO_4), alkali (NaOH and NaHCO$_3$) and oxidizing agent (H_2O_2) modification	Rice husk	Estrogen (17β-estradiol (E2))	44.9 mg g^{-1}	Zhang et al., 2019
KOH and H_3PO_4	Corn stalk	Methylene Blue	KOH-CSBC = 406.43 mg g^{-1}, H_3PO_4-CSBC = 230.39 mg g^{-1}	Liu et al., 2019
NH_4OH, KOH, or HCl	Sludge	organic pollutants (acid orange 7, rhodamine B, methylene blue, methyl orange, and their combination)	-	Mian and Liu., 2020
HCl, KOH, and $ZnCl_2$	Glue residue	Cr (VI)	325.54 mg g^{-1}	Shi et al., 2020
$FeCl_3$ and HCl, alone or combined	Wheat straw	Ammonium	-	Wang et al., 2020
H_2O_2	Peanut shells	NH_4^+-N	123.28 mg g^{-1}	Wang et al., 2020
Dehydration (H_2SO_4), oxidation (ozone) and amination (ammonium hydroxide or Triethylenetetramine (TETA))	Watermelon peel residues	Cr (VI)	333.33 mg g^{-1}	El-Nemr et al., 2021

KMnO$_4$ and KOH	peanut shell	Fungicides (carbendazim, pyrimethanil and tebuconazole)	Carbendazim= 531.2 mg g^{-1}, pyrimethanil= 467.7 mg g^{-1}, tebuconazole= 495.1 mg g^{-1}	Lee et al., 2021
Alkali-modified (NaOH)	Straw	BPA and antibiotics (tetracycline (TC) and ofloxacin (OFL))	BPA= 71.43 mg g^{-1}, TC = 101.01 mg g^{-1}, OFL = 54.05 mg g^{-1}	Tang et al., 2022
H$_3$PO$_4$	Sour cherry stalk	Ciprofloxacin	410.06 mg g^{-1}	Sayin et al., 2021
Potassium permanganate (KMnO$_4$)	Eucalyptus	Cd (II)	31.05 mg g^{-1}	Mo et al., 2021
Fe and Si were induced for H$_3$PO$_4$ modification	Sludge	Pb (II) and methylene blue	Pb (II) = 195.75 mg g^{-1}, Methylene blue = 160.78 mg g^{-1} (NaOH modified BC)	Shahib et al., 2022
KOH and immobilization of CuO	Pinewood	Pharmaceutical compounds (CFX and CBZ)	CFX = 9.8 mg g^{-1}	Xue et al., 20–22

the adsorption capacity of MnO_2-loaded BC (BC-MnO_2) for MB was substantially higher than that of unmodified BC, with a maximum adsorption capacity of 248.96 mg/g and a removal rate of 99.73%. The modified BC also showed a decolourization rate of over 95%. The study's results suggest that MB has a strong binding affinity with MnO_2-modified BC and the adsorption kinetics and isotherm were described using the quasi-second order model and Langmuir model, respectively. Overall, they indicated that using manganese compounds in the modification of lignin-derived adsorbents holds immense promise for treating MB in industrial effluents.

Zhou et al. (2017) found that Fe/Zn-BC, a material synthesized by doping sawdust BC with iron and zinc is very effective in removing both Cu (II) and TC from water. The mechanism of removal of each substance was investigated and found to be different. Tetracycline's removal was found to be limited by liquid film diffusion, while Cu (II) was controlled by both liquid film diffusion and intra particle diffusion. The adsorption process was found to involve chemisorption for TC and physico-chemical adsorption of Cu (II). Fe/Zn-BC contains different sites with different properties (hydrophobic, hydrophilic, and iron oxide) that can be used for adsorption of Cu (II) and TC. The site competition, bridge enhancement and site recognition were also found to play important roles in the adsorption process. They showed that Fe/Zn-BC could serve as a renewable, cost-effective, and sustainable solution for removing antibiotics and heavy metals from contaminated water.

Ma et al. (2020) illustrated the effectiveness of using iron/zinc (Fe/Zn), phosphoric acid (H_3PO_4), and a combination of both (Fe/Zn + H_3PO_4) to modify SDBC in removing fluoroquinolones antibiotics (CFX, norfloxacin (NOR), and OFL) from wastewater. The results showed that Fe/Zn + H_3PO_4-SDBC had a higher surface area, total pore volume, mesoporous volume, pore diameter, and oxygen-containing functional groups. It was found to be a more effective adsorbent for CFX, NOR, and OFL, with a maximum adsorption capacity of 83.7, 39.3, and 25.4 mg/g, respectively. The analysis of the kinetic and isotherm models and characterization showed that the adsorption process was dominated by a combination of physisorption and chemisorption, including pore filling, hydrogen bonding, π–π interaction, electrostatic interaction, and functional group complexation. The rate-limiting step was found to be liquid film diffusion. The adsorption process was found to be spontaneous and endothermic. They demonstrated that Fe/Zn + H_3PO_4 modified SDBC has high adsorption capacity, making it a promising candidate for removing fluroquinolones and other antibiotics from water.

Huang et al. (2019) evaluated the effectiveness of using thiol-modified BC for removing mercury (Hg^{2+} and CH_3Hg^+) from contaminated water. The BC derived from pine-sawdust was pyrolyzed at different temperatures and then modified using 3-mercaptopropyltrimethoxysilane (3-MPTS). The structure and chemical properties of the BC were characterized and it was found that the BC pyrolyzed at low temperature (300°C) had more active sites (C=O, C-O-C, -OH, and π-bond) that could be modified by 3-MPTS. The adsorption of mercury by the thiol-modified BC was found to follow the pseudo-second-order kinetic model and the Langmuir isotherm model. The thiol-modified BC with the highest Langmuir adsorption capacity for Hg^{2+} (126.62 mg/g) and CH_3Hg^+ (60.76 mg/g) was found to be 3BS (thiol-modified BC from pyrolysis at 300°C), which was prepared using a 2% content of 3-MPTS.

The removal of mercury by the thiol-modified BC was found to be dominated by ligand exchange and complexation between the surface-active sites and mercury, with the -SH group playing a major role. The presence of natural organic matter, glucose, and humic acid were found to have little effect on the mercury removal rate by 3BS. Overall, they demonstrated the potential of thiol-modified BC for effectively eradicating mercury from contaminated water.

Luo et al. (2019) reported that BCTD, synthesized by combining ultrasonic BC and nanoscale TiO_2 (TD), can be used as an effective and inexpensive biosorbent to simultaneously remove Cd (II) and As (V) from wastewater. The sorption capacity of BCTD was found to be higher compared to other sorbents and that the adsorption capacity was above 70% at pH=5. Tt was also effective in removing both cation and anion heavy metal ions. The adsorption process was controlled by ion exchange and complexation. The results of various analyses (BET, FTIR, SEM-EDS and XPS) suggests high sorption capacities for Cd (II) and As (V) were attributed to the enhanced surface area and pore volume of the BC through ultrasonic treatment, as well as the successful support of TD on the BC surface and inner pores.

Li et al. (2021), modified BC-based pomelo peel by adding L-cysteine to enhance the hydrophilicity and surface functional groups. The modified BC (cys/ BC) resulted in high adsorption capacities of heavy metal ions such as Ag (I), Pb (II) and As (V) from an aqueous solution, with maximum sorption capacities of 618.9 mg/g, 274.5 mg/g and 34.7 mg/g, respectively. The Freundlich isotherm model explained the adsorption process for Pb (II) on cys/BC, while the Langmuir isotherm model fitted best for Ag(I) and As (V). The cys/BC composite could also be reused and the adsorption capacities of cys/BC for heavy metal ions decreased slightly after five adsorption/desorption cycles. The adsorption kinetics were found to follow the pseudo-second order equation and the adsorption process was controlled by the intraparticle diffusion for all three metals. They concluded that the modified cys/ BC composite could be a promising adsorbent for removing heavy metals from contaminated environments, using multiple adsorption mechanisms including pore adsorption, functional groups, cations-π and surface complexation.

Wang et al. (2021) used wood waste of apple branches to create apple branch BC (AB), which was physically activated using CO_2 during high-temperature pyrolysis. The BC was then modified with Mg/Al layered double hydroxide (Mg/Al-LDHs) on its surface to create the carbon containing composite material AMB. Batch adsorption experiments were conducted using AMB to remove NO_3^- from water. The study found that Mg/Al-LDHs were successfully modified on the surface of AB and the average adsorption capacity of AMB for NO_3^- was 7.43 times that of AB. The removal rate for NO_3^- improved from 13 to 83% and the pseudo-second-order and Freundlich models were used to describe the kinetics and isothermal adsorption process of AMB for NO_3^-. The theoretical maximum adsorption capacity of NO_3^- by AMB was 156.84 mg/g. In comparison to other chemical modification processes for straw or wood-based BC, AMB was found to have clear advantages for nitrate adsorption.

Jia et al. (2020) developed a lanthanum-modified platanus ball fibre BC (La-TC) as an effective and reusable adsorbent for removing and recovering phosphate from wastewater for use as an agricultural fertilizer. La-TC demonstrated exceptionally high saturated adsorption capacity for phosphate (148.11 mg/g), a wide pH adsorption stability (pH range 3–9), resistance to interference from other anions and excellent regeneration ability. An analysis using FTIR and XPS indicated that the adsorption mechanism involved electrostatic adsorption, ligand exchange and complexation. The results of a fixed bed column experiment using a La-TC for phosphate removal from actual wastewater showed promising outcomes with a high treated bed volume. The study concluded that La-TC has great potential for practical application in removing and recovering phosphate from wastewater while also promoting resource utilization of platanus ball waste.

Gao et al. (2022) uncovered a solution by treating BPA with a modified form of rice husk BC (F2BC3) impregnated with ferric chloride and pyrolyzed at 800°C, in the presence of PMS. The results showed that F2BC3 had an excellent morphology, structure, and catalytic performance. While F2BC3 was able to adsorb and remove 25.8% of BPA, the degradation rate increased to 99.5% due to the catalytic effect of PMS. The degradation of BPA was less affected by pH and more than 97% of BPA was degraded within pH 3–10. The amount of F2BC3 catalyst and PMS concentration had significant impacts on the degradation rate and high BPA concentration inhibited the reaction rate. The presence of humic acid did not affect BPA degradation, while chloride, bicarbonate, nitrate, and phosphate ions had varying effects on the degradation reaction. The degradation of BPA occurred via both non-radical and free-radical pathways, with the free-radical path being the main one. The cycling experiment showed that F2BC3 was stable and had good recycling performance. They inferred that F2BC3 has high efficiency, environmental friendliness, and economic feasibility, making it a promising material for water treatment (Table 7.2).

2.2 Physical Modification

Physical or mechanical modification methods are often preferred over chemical modification methods due to their simplicity and economic feasibility. These methods use oxidizing agents such as carbon dioxide, steam, and air to modify the physical properties of materials. Unlike chemical modification, physical modification does not involve the use of any chemicals, making it a more environmentally friendly option. However, physical modification methods are generally less effective than chemical methods in achieving certain desired properties, such as chemical stability or resistance to moisture. These modifications enhance various properties of biochar (Fig. 7.5).

Peng et al. (2018) investigated the impact of UV-modified biochar (UVBC) on hexavalent chromium (Cr (VI)) removal ability from aqueous solutions as using a pyrolysis temperature of 300°C, an irradiation time and distance of 24 hr and 40 mm, respectively. The results showed that UV irradiation increased the specific surface area of BC and added oxygen-containing functional groups to its surface, resulting in a significant enhancement of Cr (VI) removal ability. The use of UV irradiation led to changes in the properties of the BC, resulting in an increase in the number of acidic functional groups and a larger surface area. The improved removal rate of Cr (VI) was achieved through surface complexation between Cr (VI) and the surface functional groups of UVBC. The synergistic interaction of adsorption

Table 7.2 Chemical modification using Salt/ metal-doped/ metal oxide modified biochar.

Modifier	Raw biomass	Target pollutants	Adsorption capacity (Q_{max}) (mg g^{-1})	Reference
Tetradecyltrimethyl ammonium bromide (TTAB) as a cationic surfactant	Switchgrass	Red 195 A dye (RR-195A)	379.6 mg g^{-1}	Mahmoud et al., 2016
Iron and zinc doped (Fe/Zn)	Sawdust	Cu (II) and tetracycline (TC)	Cu (II) = 95.7%, TET = 94.1%	Zhou et al., 2017
Thiol-modified using 3-mercaptopropyltrimethoxysilane (3-MPTS)	Pine sawdust	Hg^{2+} and CH_3Hg^+	Hg^{2+} = 126.62 mg g^{-1}, CH_3Hg^+ = 60.76 mg g^{-1}	Huang et al., 2019
Ultrasonic BC and nanoscale TiO_2	Raw corncobs	Cd (II) and As(V)	Cd (II) = 72.62 mg g^{-1}, As(V) = 118.06 mg g^{-1}	Luo et al., 2019
Iron/zinc (Fe/Zn), phosphoric acid (H_3PO_4) or in combination (Fe/Zn + H_3PO_4)	Sludge	Fluoroquinolones antibiotics including CFX, norfloxacin (NOR) and ofloxacin (OFL)	CFX = 83.7 mg g^{-1}, NOR = 39.3 mg g^{-1}, OFL = 25.48 mg g^{-1}	Ma et al., 2020
L-cysteine	Pomelo peel (PP)	Ag (I), Pb (II), and As (V)	Ag(I) = 618.9 mg g^{-1}, Pb (II) = 274.5 mg g^{-1}, As (VI) = 34.7 mg g^{-1}	Luo et al., 2019
Mg/Al layered double hydroxide (Mg/Al-LDHs)	Woody waste of apple branch	NO_3^-	156.84 mg g^{-1}	Wang et al., 2021
Manganese compounds (KMnO4, MnSO4, and MnO2).	Lignin	Methylene blue	248.96 mg g^{-1}	Liu et al., 2021
Lanthanum-modified	Ball fibre	Phosphate	148.11 mg g^{-1}	Jia et al., 2020
Ferric chloride-modified	Rice husk	Bisphenol A	-	Gao et al., 2022

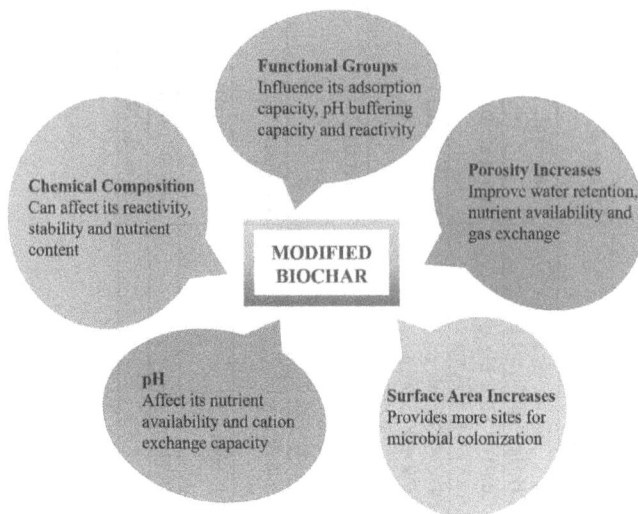

Figure 7.5 Properties of modified biochar.

and reduction processes makes UV modification an efficient method for removing anionic pollutants from wastewater.

Abedi and Mojiri (2019) examined the effectiveness of a modified BC/zeolite constructed wetland in removing contaminants from synthetic wastewater. The researchers used common reed (*Phragmites australis*) plants in two parallel cylinders (vertical subsurface flow lysimeters) as constructed wetlands (CW1 and CW2), with CW2 incorporating BC, zeolite, and gravel substrate layers, while CW1 had only substrate layer of gravels. The results of the study showed that CW2 outperformed CW1 in treating wastewater. Specifically, CW2 was able to remove all of COD, phenols, ammonia, Mn, and Pb at an optimum pH (6.3) and a retention time (57.4 hr). Additionally, the study found that N_2O emission was lower in CW2 compared to CW1. N_2O emission flux was measured in $\mu g \, m^{-2} \, h^{-1}$ and was found to be 23, 74 and 101 in 24 h, 48 hr and 72 hr retention time in CW2, respectively. Finally, the researchers measured the Bioaccumulation Factor (BF) and Translocation Factor (TF) in both CW1 and CW2. BF and TF were found to be higher than 1 in most runs in both CW1 and CW2. Overall, they provided evidence that a BC/zeolite CW can effectively treat wastewater and remove pollutants, which could be useful in addressing water resource shortages.

A novel salt-based BC was produced by incorporating silicon (Si) and manganese (Mn) onto cornstalk biomass using a one-step sintering method (Liu et al., 2020). The SiBC exhibited a smooth, non-porous and compact surface, with a relatively small surface area of 4.55 m^2/g, while the surface morphology of MnBC was rough and porous, with a greatly increased surface area of 68.13 m^2/g. Adsorption isotherms, kinetics, FTIR and XPS were used to explain the adsorption mechanism. SiBC reached adsorption equilibrium within 10 min , with an adsorption capacity of 152.61 mg/g with 97% removal rate, while MnBC took 500 min to reach adsorption equilibrium, with an adsorption capacity of 187.76 mg/g. The adsorption rate constant K2 of SiBC was about 200 times greater than that of MnBC. The adsorption process of both SiBC and MnBC confirmed to Langmuir and pseudo-second order models, indicating monolayer and chemical adsorption. The adsorption mechanisms of SiBC were dominant in precipitation and surface complexation, while those of MnBC were specialized adsorption, ion exchange and intraparticle diffusion. SiBC showed promise as an adsorbent due to its higher efficiency, larger adsorption capacity and lack of secondary pollution resulting from ion exchange in the adsorption process. (Table 7.3)

Table 7.3 Physical medication of biochar.

Modifier	Raw biomass	Target pollutants	Adsorption capacity (Q_{max}) (mg g^{-1})	Reference
Ultraviolet (UV) irradiation	Corn straw	Hexavalent chromium (Cr (VI))	20.04 mg g^{-1}	Peng et al., 2018
BC/zeolite CW	Common reed (*Phragmites australis*)	Phenols, ammonia, Pb, Mn	-	Abedi et al., 2019
SiO$_2$/MnO$_2$ by One-step sintering technique	Cornstalk	Cu (II)	187.76 mg g^{-1}	Liu et al., 2020

2.3 Physical and Chemical Combined Modification of Biochar

Physical and chemical modifications have been shown to greatly enhance the adsorption properties of BC, making it a promising candidate for removing heavy metals from water. Combining these two approaches can create modified BCs that have even greater adsorption capacities and selectivity for different heavy metals. With their ability to remove heavy metals from water, modified BCs have the potential to become a sustainable and cost-effective solution for water treatment and remediation.

Zhou et al. (2018) described the preparation and characterization of a novel adsorbent, MBCI (iminodiacetic acid magnetic BC), for removing Cd (II) from untreated water. MBCI was prepared by grafting Fe$_3$O$_4$ nanoparticles and iminodiacetic acid on the surface of BC through a coupling reaction and functional modification. The study found that the optimal pH for Cd (II) adsorption by MBCI was 6.0 and the adsorption isotherm model closely followed the surface complexation model, which was like the Langmuir model. The Langmuir model calculated the maximum adsorption capacities of Cd (II) on MBCI to be 175.29, 187.12, 191.41 and 197.96 mg/g at temperatures of 293, 303, 313 and 323 K, respectively. The adsorption kinetic process was found to occur in three steps: boundary layer diffusion, intra-particle diffusion and an equilibrium stage. The thermodynamic study suggested that the adsorption reaction was spontaneous, endothermic, and increased randomness. The magnetization saturation value of MBCI was 16.88 emu/g, indicating that it could be easily and effectively separated from the solution by magnetic forces. Therefore, MBCI has been identified as low-cost, effective, and novel adsorbent for Cd (II) removal from aqueous solution due to its unique features of outstanding adsorption performance, reusability, and separation efficiency.

Dai et al. (2020) aimed to enhance the performance of rice straw BC as an adsorbent for TC, N and P removal. The researchers used an alkali-acid combined and the magnetization method to modify the rice straw BC, which improved its performance for TC removal. The adsorption capacity of modified BC was found to be up to 98.33 mg/g which was higher than the original rice straw BC. The improved performance was attributed to the increase in specific surface area and pore volume, which led to hydrogen bonding and pore-filling effect. The pH of the solution was found to be insignificant over a pH range from 3 to 10, which means that the modified BC can be used in a wide range of pH conditions. However, the strong competition between ionic substances, such as calcium and phosphate, and TC was identified, which could suppress the sorption of TC. Despite this, the enhanced TC adsorption, strong removal of nitrogen and phosphorus, easy magnetic recovery, and good reusability in water samples make the modified rice straw BC a promising solution for wastewater treatment and the disposal of rice straw resources.

Peter et al. (2021) showed that alkali activation of softwood BC, which was pre-treated with ultrasound, enhances its physical and metal adsorption properties in water. The use of higher frequency ultrasound was found to improve the activation process by modifying the BC's surface and increasing its metal adsorption capacity. The ultrasound pre-treatment improved the accessibility of the BC's surface, while the alkali treatment enhanced the availability of surface anchoring sites, resulting in chemiosorption through ion exchange reactions. Compared to other reported wood BCs, the synthesized BC showed higher equilibrium-adsorption capacities, making them efficient candidates for heavy metal removal from wastewater. The 170 kHz pre-treated sample demonstrated an adsorption capacity of 19.99 mg/g (almost 22 times higher than non-activated sample). Isotherm and thermodynamic studies also suggested improved physicochemical properties of the BC. The study demonstrated that the combination of frequency, power, temperature, and exposure time influences the base modification on the material, with higher frequency pre-treated samples exhibiting the best adsorption properties. The two-step activation process also showed promising results in mixed metal systems, with higher selectivity towards copper ions. The study highlights the potential of using power ultrasound enhanced modifications on BC to improve its physiochemical properties and aid in the conversion of feedstock residues towards process control and design.

A new magnetic bio-adsorbent has been developed (Zahedifar et al., 2021), to remove Cd (II) and Pb (II) ions from aqueous media. Nitrogen and sulphur-containing functional groups were added to the bio-adsorbent surface, which improved its ability to absorb heavy metal ions by forming complexes with amide and thioamide functional groups. Additionally, Fe_3O_4 NPs were immobilized to enhance the adsorption capacity and facilitate the separation of heavy metals from water samples. The Langmuir model was used to determine the maximum adsorption capacity of Pb (II) and Cd (II) ions, which were found to be 103 mg/g and 106 mg/g, respectively. The kinetic data also demonstrated good agreement with the pseudo-second-order equation. Thermodynamic data, such as K^0, ΔG^0, ΔH^0 and ΔS^0, confirmed that the adsorption process was endothermic and spontaneous. The results indicated that BC-Thioamide/MNPs is an effective sorbent for removing heavy metals from contaminated wastewater. The adsorption capacity remained acceptable even after five runs of the adsorption-desorption process.

Wang et al. (2022) conducted a study to improve the effectiveness of city tail water using corn straw BC as the substrate for CW. The study found that strong oxidation methods such as $KMnO_4$ and Freeze-Thaw Cycles (FTCs), reduce the BC's carbon content and destruction of the structure, while other composite modifications had little effect on the BC's crystal structure and functional groups. However, FTCs can reduce the alkaline groups on the BC surface and when BC is modified by 8 FTCs and 0.1 $molL^{-1}$ $KMnO_4$, manganese oxide is formed and attached to the surface of BC. FTCs modification was found to increase the Specific Surface Area (SSA), pore volume and CO_2 adsorption capacity of straw BC by 28.9, 22.4 and 20.4%, respectively. BC modified by FTCs followed by $KMnO_4$ and NaOH respectively showed almost unchanged SSA and pore volume. The application of $FTCs+H_2SO_4$ modified straw BC in CW resulted in a higher removal rate of pollutants in city tail water than single FTCs modification and it improved the removal rate of Total Phosphorous (TP) in CW with plants by 20–30%. These findings have significant implications for removing BC's properties and CW's purification effectiveness in treating city tail water. (Table 7.4)

Table 7.4 Combined (Physical and chemical) modification of biochar.

Modifier	Raw biomass	Target pollutants	Adsorption capacity (Q_{max}) (mg g⁻¹)	Refs
Iminodiacetic acid magnetic BC (MBCI)	Palm fibre	Cd (II)	197.96 mg g⁻¹	Zhou et al., 2018
Alkali-acid combined and magnetization method	Rice straw	Tetracycline (TC), N, P	TC= 98.33 mg g⁻¹	Dai et al., 2020
Ultrasound pre-treatment as well as alkali (NaOH) activations	Softwood woodchips	Copper ions	19.99 mg g⁻¹	Peter et al., 2021
Functionalization with carboxylic acid, amine, amide and thioamide groups and immobilization with Fe_3O_4 NPs	Raw date leaves and stalks	Pb (II) and Cd (II)	Pb (II) = 61.25 mg g⁻¹, Cd (II) = 53.75 mg g⁻¹	Zahedifar et al., 2021
Freeze-thaw cycles (FTCs) modification and chemical modification ($KMnO_4$, NaOH and H_2SO_4)	Corn straw	CO_2 (to treat city tail water)	-	Wang et al., 2022

2.4 Coating with Bio-based Polymers

The use of chitosan as a surface modification agent for adsorbents has gained popularity due to its ability to bind to heavy metal ions through amine and hydroxyl functional groups. This feature makes it suitable for the removal of heavy metals from wastewater and industrial effluents. Chitosan, being a bio-based polymer, is also an environmentally friendly alternative to synthetic surface modification agents. Coating adsorbents with chitosan enhances their adsorption capacity, stability, and selectivity, making them highly effective in heavy metal removal applications. Additionally, chitosan-based adsorbents can be easily regenerated, making them a cost-effective solution for wastewater treatment.

Deng et al. (2017) described a study in which a new type of BC, modified with chitosan and PDMA (pyromellitic dianhydride), was developed for the purpose of removing heavy metals from water. The researchers found that the optimal pH for adsorbing heavy metals was 5.0 and that the modified BC, called CPMB, was more effective at removing lead and cadmium than unmodified BC. CPMB was also significantly better at removing copper, with a capacity 2.5 times greater than unmodified BC. The study showed that CPMB had a strong selective adsorption ability for copper in multi-metal systems. The research found that different functional groups on CPMB were responsible for removing different heavy metals, with N–C=O influencing the adsorption of Pb (II) and N-containing functional groups and C=C groups affecting the adsorption of Cd (II). CPMB shows higher adsorption capacity for Cu (II) due to its extensive N-containing and carbonyl functional groups.

Chaiyaraksa et al. (2019) synthesized chitosan-magnetic BC using water hyacinth, which is of low cost and readily available from an ecologically harmful plant and tested its effectiveness in adsorbing copper and nickel from contaminated water. The results of adsorption isotherms, kinetics and thermodynamics showed that the adsorption of Cu and Ni followed the Langmuir isotherm. The q_{max} values for Cu and Ni were 38.4615 mg/g and 0.4858 mg/g, respectively, and the E values were 0.316 kJ/mol and 1.8962 kJ/mol, respectively. The best-fitting kinetic models were intra-particle diffusion for Ni and pseudo-second-order model for Cu. The adsorption process was found to be endothermic, spontaneous at high temperatures, and non-spontaneous at low temperatures. Overall, they suggested that chitosan-magnetic BC synthesized from water hyacinth could be an effective solution for addressing heavy metal contamination in water (Table 7.5).

Table 7.5 Modification of biochar using coating with bio-based polymers.

Modifier	Raw biomass	Target pollutants	Adsorption capacity (Q_{max}) (mg g^{-1})	Reference
Chitosan and pyromellitic dianhydride (PMDA)	Rice straw	Cd, Cu and Pb	Cu (II) = 70.28 mg g^{-1}	Deng et al., 2017
Chitosan-magnetic BC	Water hyacinth (*Eichhornia crassipes*)	Copper and nickel	Cu (II) = 38.4615 mg g^{-1}, Ni (II) = 0.4858 mg g^{-1}	Chaiyaraksa et al., 2019

3. Conclusion and Future Research Directions

Emerging contaminants pose a major threat to the environment and can have harmful effects on both the aquatic and terrestrial life. While research has been conducted on the harmful effects of these contaminants, the full extent of their impact is not yet known. This is due to a number of factors, including the complex interactions that can occur between different contaminants, as well as the fact that new contaminants are constantly being introduced into the environment. There are several methods mentioned in literature such as advanced water treatment (AOP and NF/RO membrane technology) that can effectively remove ECs. However, these methods are limited because of secondary chemical pollutants and expensive nature. Therefore, cost-effectiveness and eco-friendly methods are required. It is important to continue researching and monitoring the effects of ECs on the environment and biota in order to better understand their potential risks and develop effective strategies for managing and mitigating their impact.

In this chapter recent advances in modification of BC including chemical and physical modifications to separate the ECs from water were discussed. Physical modifications of BC for separating ECs from water can provide a more environmentally friendly approach by changing the structure or properties of the BC itself, enhancing its ability to capture and remove contaminants from water. Chemical modifications of BC could involve the use of chemicals that may contribute to the overall concentration of ECs in the environment. However, it is important to note that physical modifications may not be as effective at removing certain types of contaminants as chemical modifications, and may require further research and development to optimize their performance. These modifications of BC may not be suitable for large-scale purification of water bodies, as the volume of water that needs to be treated could be too large for this approach to be practical. For example, to detect and remove microplastics in the ocean, it may not be feasible to clean the entire sea using these methods.

Banning and controlling the use of these contaminants is one such approach, and has been implemented in various countries to reduce their release into the environment. Other approaches to address the issue of ECs in the environment include the use of advanced technologies for water treatment, as well as increased public awareness and education on the importance of reducing the release of harmful chemicals into the environment. Current technologies used at wastewater treatment plants may not be sufficient to effectively remove ECs from wastewater. Therefore, the use of BC at wastewater treatment plants can be a promising approach to prevent the entry of ECs into water bodies.

In conclusion, the modification of BC represents a promising method for removing ECs from wastewater. However, the efficient and cost-effectiveness of this approach will depend on several factors that must be carefully considered. Further research is needed to develop effective detection and elimination technologies for ECs at minimal concentrations and methods to achieve maximum adsorption capacity. Even minute quantities of ECs can pose significant dangers, so it is essential to investigate new technologies for detecting and eliminating these pollutants. However, the cost of producing modified BCs should also be considered, and their profitability compared to other materials, such as ACs, needs be analyzed for each specific case.

Acknowledgments

Aman Bhalla and Savita Chaudhary gratefully acknowledges support from the Department of Science and Technology (DST), New Delhi. We would like to apologize to those scientists whose work may not have appeared in this chapter either due to the limited scope of the chapter or oversight. Jaswinder Kaur acknowledges the financial support from University Grant Commission (UGC), New Delhi vides Award No- F.No. 16-9(June2019)/2019(NET/CSIR).

Abbreviations

AB	:	Apple branch biochar
AC	:	Activated carbon
AMB	:	Apple branch derived biochar modified with Mg/Al-LDHs
AMBC	:	Amino-grafted modified biochar
AOP	:	Advanced oxidation process
BC	:	Biochar
BC-COOH	:	Carboxylic groups on pre-oxidized BC
BC-MnO$_2$	:	MnO$_2$-loaded BC
BCTD (TD)	:	Synthesized by combining ultrasonic BC and nanoscale TiO$_2$
BET	:	Brunauer–Emmett–Teller Theory
BF	:	Bioaccumulation factor
BJH	:	Method of Barrett, Joyner, and Halenda
BPA	:	Bisphenol A
BS	:	Na$_2$S modified biochar
3BS	:	*Thiol-modified BC from pyrolysis at 300°C*
CA	:	Citric acid
CAWB	:	Citric acid modified BC
CBZ	:	Carbamazepine
CFB	:	Coconut fibre BC
CFX	:	Ciprofloxacin
COD	:	Chemical oxygen demand
CPMB	:	Biochar modified with chitosan and pyromellitic dianhydride
CSBC	:	Corn stalk BC
CW	:	Constructed wetland
cys/BC	:	BC derived from pomelo peel modified by adding L-cysteine
DSC	:	Differential scanning calorimeter
ECs	:	Emerging contaminants
EDAX	:	Energy Dispersive X-Ray Analysis
F2BC3	:	BPA with a modified form of rice husk BC impregnated with ferric chloride and pyrolyzed at 800°C
Fe-H-WB	:	Wheat straw modified by FeCl$_3$ and HCl
FTCs	:	Freeze-thaw cycles
FTIR	:	Fourier transform infrared
GAC	:	Granular Activated Carbon
HTC	:	Hydrothermal carbonization
KBC	:	BC derived from waste eucalyptus wood
K-BC	:	BC modified with KOH
KW	:	Kitchen wastewater
LA	:	*Lepironia articulata*

LABC	:	*Lepironia articulata* BC
La-TC	:	Lanthanum-modified platanus ball fibre BC
MB	:	Methylene blue
MBCC	:	Modified corncob-BC
MBCI	:	Iminodiacetic acid magnetic BC
MBR	:	Membrane bioreactor
MCFB	:	Modified BC derived from coconut fiber
Melon-BO -NH$_2$ and Melon- BO-TETA	:	Made from watermelon peel via dehydration with 50% sulphuric acid to give Melon-B followed by oxidation with ozone and amination using ammonium hydroxide or Triethylenetetramine (TETA)
Mg/Al- LDHs	:	Mg/Al layered double hydroxide
MnBC	:	BC incorporated by manganese
3-MPTS	:	*3-mercaptopropyltrimethoxysilane*
MS	:	Municipal Sewage Sludge
NBC	:	BC treated with sodium hydroxide
NF	:	Nano-filtration
NOR	:	*Norfloxacin*
nZVI	:	Nano zero-valent iron
OFL	:	Ofloxacin
PBC	:	Peanut shell BC
PDMA	:	Pyromellitic dianhydride
PMS	:	Peroxymonosulphate
PP	:	Pomelo peel
PPCPs	:	Pharmaceuticals and personal-care products
PSB	:	BC derived from peanut shell
PSB$_{OX-A}$	:	Chemically-modified PSB
RO	:	Reverse osmosis
RSM	:	Resonse surface methodology
SB-TTAB bromide	:	Switchgrass BC treated with tetradecyltrimethyl ammonium
SDBC	:	Sludge derived BC
SEM	:	Surface imaging method
SiBC	:	Biochar incorporated by silicon
SSA	:	Specific surface area
TC	:	Tetracycline
TF	:	Translocation factor
TGA	:	Thermogravimetric Analysis
TOC	:	Total organic carbon
TP	:	Total phosphorous
UV	:	Ultraviolet
UVBC	:	UV-modified BC
XPS	:	X-ray photoelectron spectroscopy
Zn2GT700	:	Glue residue modified BC (2:1 ratio of ZnCl$_2$ to glue residue and pyrolyzed at 700°C)

References

Abedi, T. and Mojiri, A. 2019. Constructed wetland modified by biochar/zeolite addition for enhanced wastewater treatment. Environ. Technol. Innov. 16: 100472.

Ahmaruzzaman, Md. 2021. Biochar based nanocomposites for photocatalytic degradation of emerging organic pollutants from water and wastewater. Mater. Res. Bull. 140: 111262.

Asadullah, L. Kaewsichan and Tohdee, K. 2019. Adsorption of hexavalent chromium onto alkali-modified biochar derived from *Lepironia articulata*: A kinetic, equilibrium, and thermodynamic study. Water Environ. Res. 91: 1433.

Bilal, M., Adeel, M., Rasheed, T., Zhao, Y. and Iqbal, H.M.N. 2019. Emerging contaminants of high concern and their enzyme-assisted biodegradation—A review. Environ. Int. 124: 336.

Chaiyaraksa, C., Boonyakiat, W., Bukkontod, W. and Ngakom, W. 2019. Adsorption of copper (II) and nickel (II) by chemical modified magnetic biochar derived from *eichhornia crassipes*. EnvironmentAsia. 12: 14.

Chen, R., Zhao, X., Jiao, J., Li, Y. and Wei, M. 2019. Surface-Modified Biochar with Polydentate Binding Sites for the Removal of Cadmium. Int. J. Mol. Sci. 20: 1775.

Cheng, N., Wang, B., Wu, P., Lee, X., Xing, Y., Chen, M. et al., 2021. Adsorption of emerging contaminants from water and wastewater by modified biochar: A review Environ. Pollut. 273: 116448.

Dai. J., Meng, X., Zhang, Y. and Huang, Y. 2020. Effects of modification and magnetization of rice straw derived biochar on adsorption of tetracycline from water. Bioresour. Technol. 311: 123455.

Deng, J., Liu, Y., Liu, S., Zeng, G., Tan, X., Huang, B. et al. 2017. Competitive adsorption of Pb(II), Cd(II) and Cu(II) onto chitosan-pyromellitic dianhydride modified biochar. J. Colloid. Interface Sci. 506: 355.

Dong, H., Deng, J., Xie, Y., Zhang, C., Jiang, Z., Cheng, Y. et al. 2017. Stabilization of nanoscale zero-valent iron (nZVI) with modified biochar for Cr(VI) removal from aqueous solution. J. Hazard. Mater. 332: 79.

El-Nemr, M.A., Ismail, I.M.A., Abdelmonem, N.M., el Nemr, A. and Ragab, S. 2021. Amination of biochar surface from watermelon peel for toxic chromium removal enhancement. Chin. J. Chem. Eng. 36: 199.

Faheem, F., Bao, J., Zheng, H., Tufail, H., Irshad, S. and Du, J. 2018. Adsorption-assisted decontamination of Hg(II) from aqueous solution by multi-functionalized corncob-derived biochar RSC Adv. 8: 38425.

Gao, Y., Chen, Y., Song, T., Su, R. and Luo, J. 2022. Activated peroxymonosulfate with ferric chloride-modified biochar to degrade bisphenol A: Characteristics, influencing factors, reaction mechanism and reuse performance. Sep. Purif. Technol. 300: 121857.

Huang, Y., Xia, S., Lyu, J. and Tang, J. 2019. Highly efficient removal of aqueous Hg^{2+} and CH_3Hg^+ by selective modification of biochar with 3-mercaptopropyltrimethoxysilane. Chem. Eng. J. 360: 1646.

Jia, Z., Zeng, W., Xu, H., Li, S. and Peng, Y. 2020. Adsorption removal and reuse of phosphate from wastewater using a novel adsorbent of lanthanum-modified platanus biochar. Process Saf. Environ. Prot. 140: 221.

Lee, Y.-G., Shin, J., Kwak, J., Kim, S., Son, C., Kim, G.-Y. et al. 2021. Enhanced Adsorption Capacities of Fungicides Using Peanut Shell Biochar via Successive Chemical Modification with $KMnO_4$ and KOH. Separations 8: 52.

Li, B., Gong, J., Fang, J., Zheng, Z. and Fan, W. 2021. Cysteine chemical modification for surface regulation of biochar and its application for polymetallic adsorption from aqueous solutions. Environ. Sci. Pollut. Res. 28: 1061.

Liu, J., Cheng, W., Yang, X. and Bao, Y. 2020. Modification of biochar with silicon by one-step sintering and understanding of adsorption mechanism on copper ions. Sci. Total Environ. 704: 135252.

Liu, L., Li, Y. and Fan, S. 2019. Preparation of KOH and H_3PO_4 Modified Biochar and Its Application in Methylene Blue Removal from Aqueous Solution. Processes 7: 891.

Liu, X.-J., Li, M.-F. and Singh, S.K. 2021. Manganese-modified lignin biochar as adsorbent for removal of methylene blue. J. Mater. Res. Technol. 12: 1434.

Luo, M., Lin, H., He, Y., Li, B., Dong, Y. and Wang, L. 2019. Efficient simultaneous removal of cadmium and arsenic in aqueous solution by titanium-modified ultrasonic biochar. Bioresour. Technol. 284: 333.

Ma, Y., Li, P., Yang, L., Wu, L., He, L., Gao, F. et al. 2020. Iron/zinc and phosphoric acid modified sludge biochar as an efficient adsorbent for fluoroquinolones antibiotics removal. Ecotoxicol. Environ. Saf. 196: 110550.

Mahmoud, M.E., Nabil, G.M., El-Mallah, N.M., Bassiouny, H.I., Kumar, S. and Abdel-Fattah, T.M. 2016. Kinetics, isotherm, and thermodynamic studies of the adsorption of reactive red 195 A dye from water by modified Switchgrass Biochar adsorbent. J. Ind. Eng. Chem. 37: 156.

Mian, M.M. and Liu, G. 2020. Activation of peroxymonosulfate by chemically modified sludge biochar for the removal of organic pollutants: Understanding the role of active sites and mechanism. Chem. Eng. J. 392: 123681.

Mo, Z., Shi, Q., Zeng, H., Lu, Z., Bi, J., Zhang, H. et al. 2021. Efficient removal of Cd(II) from aqueous environment by potassium permanganate-modified eucalyptus biochar. Biomass Convers. Biorefin. DOI 10.1007/s13399-021-02079-4.

Peng, Z., Zhao, H., Lyu, H., Wang, L., Huang, H., Nan, Q. et al. 2018. UV modification of biochar for enhanced hexavalent chromium removal from aqueous solution. Environ. Sci. Pollut. Res. 25: 10808.

Peter, A., Chabot, B. and Loranger, E. 2021. Enhanced activation of ultrasonic pre-treated softwood biochar for efficient heavy metal removal from water. J Environ Manage 290: 112569.

Qiu, B., Tao, X., Wang, H., Li, W., Ding, X. and Chu, H. 2021. Biochar as a low-cost adsorbent for aqueous heavy metal removal: A review. J. Anal. Appl. Pyrolysis. 155: 105081.

Qiu, B., Shao, Q., Shi, J., Yang, C. and Chu, H. 2022. Application of biochar for the adsorption of organic pollutants from wastewater: Modification strategies, mechanisms and challenges. Sep. Purif. Technol. 300: 121925.

Rajapaksha, A.U., Chen, S.S., Tsang, D.C.W., Zhang, M., Vithanage, M., Mandal, S. et al. 2016. Engineered/designer biochar for contaminant removal/immobilization from soil and water: Potential and implication of biochar modification. Chemosphere. 148: 276.

Rathi, B.S. and Kumar, P.S. 2021. Application of adsorption process for effective removal of emerging contaminants from water and wastewater. Environ. Pollut. 280: 116995.

Sayin, F., Akar, S.T. and Akar, T. 2021. From green biowaste to water treatment applications: Utilization of modified new biochar for the efficient removal of ciprofloxacin. Sustain. Chem. Pharm. 24: 100522.

Shahib, I.I., Ifthikar, J., Oyekunle, D.T., Elkhlifi, Z., Jawad, A., Wang, J. et al. 2022. Influences of chemical treatment on sludge derived biochar; Physicochemical properties and potential sorption mechanisms of lead (II) and methylene blue. J. Environ. Chem. Eng. 10: 107725.

Shen, Y., Jiang, B. and Xing, Y. 2021. Recent advances in the application of magnetic Fe_3O_4 nanomaterials for the removal of emerging contaminants. Environ. Sci. Pollut. Res. 28: 7599.

Shi, Y., Shan, R., Lu, L., Yuan, H., Jiang, H., Zhang, Y. et al. 2020. High-efficiency removal of Cr(VI) by modified biochar derived from glue residue. J. Clean Prod. 254: 119935.

Takaya, C.A., Fletcher, L.A., Singh, S., Okwuosa, U.C. and Ross, A.B. 2016. Recovery of phosphate with chemically modified biochars. J. Environ. Chem. Eng. 4: 1156.

Tan, G., Sun, W., Xu, Y., Wang, H. and Xu, N. 2016. Sorption of mercury (II) and atrazine by biochar, modified biochars and biochar based activated carbon in aqueous solution. Bioresour. Technol. 211: 727.

Tang, L., Yu, J., Pang, Y., Zeng, G., Deng, Y., Wang, J. et al. 2018. Sustainable efficient adsorbent: Alkali-acid modified magnetic biochar derived from sewage sludge for aqueous organic contaminant removal. Chem. Eng. J. 336: 160.

Tang, Y., Li, Y., Zhan, L., Wu, D., Zhang, S., Pang, R. et al. 2022. Removal of emerging contaminants (bisphenol A and antibiotics) from kitchen wastewater by alkali-modified biochar. Sci. Total Environ. 805: 150158.

Vu, T.M., Trinh, V.T., Doan, D.P., Van, H.T., Nguyen, T.V., Vigneswaran, S. et al. 2017. Removing ammonium from water using modified corncob-biochar. Sci. Total Environ. 579: 612.

Wang, H., Teng, H., Wang, X., Xu, J. and Sheng, L. 2022. Physicochemical modification of corn straw biochar to improve performance and its application of constructed wetland substrate to treat city tail water. J. Environ. Manage. 310: 114758.

Wang, S., Ai, S., Nzediegwu, C., Kwak, J.-H., Islam, M.S., Li, Y. et al. 2020. Carboxyl and hydroxyl groups enhance ammonium adsorption capacity of iron (III) chloride and hydrochloric acid modified biochars. Bioresour. Technol. 309: 123390.

Wang, T., Zhang, D., Fang, K., Zhu, W., Peng, Q. and Xie, Z. 2021. Enhanced nitrate removal by physical activation and Mg/Al layered double hydroxide modified biochar derived from wood waste: Adsorption characteristics and mechanisms. J. Environ. Chem. Eng. 9: 105184.

Wang, X., Guo, Z., Hu, Z. and Zhang, J. 2020. Recent advances in biochar application for water and wastewater treatment: A review. Peer. J. 8: 9164.

Wang, Z., Li, J., Zhang, G., Zhi, Y., Yang, D., Lai, X. et al. 2020. Characterization of Acid-Aged. Biochar. and Its Ammonium Adsorption in an Aqueous Solution Materials 13: 2270.

Wongrod, S., Simon, S., Guibaud, G., Lens, P.N.L., Pechaud, Y., Huguenot, D. et al. 2018. Lead sorption by biochar produced from digestates: Consequences of chemical modification and washing. J. Environ. Manage. 219: 277.

Wu, W., Li, J., Niazi, N.K., Müller, K., Chu, Y., Zhang, L. et al. 2016. Influence of pyrolysis temperature on lead immobilization by chemically modified coconut fiber-derived biochars in aqueous environments. Environ. Sci. Pollut. Res. 23: 22890.

Wu, W., Li, J., Lan, T., Müller, K., Niazi, N.K., Chen, X. et al. 2017. Unraveling sorption of lead in aqueous solutions by chemically modified biochar derived from coconut fiber: A microscopic and spectroscopic investigation. Sci. Total Environ. 576: 766.

Xu, Y., Liu, Y., Liu, S., Tan, X., Zeng, G., Zeng, W. et al. 2016. Enhanced adsorption of methylene blue by citric acid modification of biochar derived from water hyacinth (Eichornia crassipes). Environ. Sci. Pollut. Res. 23: 23606.

Xue, Y., Guo, Y., Zhang, X., Kamali, M., Aminabhavi, T.M., Appels, L. et al. 2022. Efficient adsorptive removal of ciprofloxacin and carbamazepine using modified pinewood biochar—A kinetic, mechanistic study. Chem. Eng. J. 450: 137896.

Yin, Q., Ren, H., Wang, R. and Zhao, Z. 2018. Evaluation of nitrate and phosphate adsorption on Al-modified biochar: Influence of Al content. Sci. Total Environ. 631–632: 895.

Zahedifar, M., N. Seyedi, S. Shafiei and M. Basij, 2021. Surface-modified magnetic biochar: Highly efficient adsorbents for removal of Pb (II) and Cd (II). Mater. Chem. Phys. 271: 124860.

Zhang, P., Liu, S., Tan, X., Liu, Y., Zeng, G., Yin, Z. et al. 2019. Microwave-assisted chemical modification method for surface regulation of biochar and its application for estrogen removal. Process Saf. Environ. Prot. 128: 329.

Zhou, X., Zhou, J., Liu, Y., Guo, J., Ren, J. and Zhou, F. 2018. Preparation of iminodiacetic acid-modified magnetic biochar by carbonization, magnetization and functional modification for Cd(II) removal in water. Fuel 233: 469.

Zhou, Y., Liu, X., Xiang, Y., Wang, P., Zhang, J., Zhang, F. et al. 2017. Modification of biochar derived from sawdust and its application in removal of tetracycline and copper from aqueous solution: Adsorption mechanism and modelling. Bioresour. Technol. 245: 266.

CHAPTER 8

Life Cycle Risk Assessment and Fate of the Nanomaterials:
An Environmental Safety Perspective

Megha Bagariya,[1,2] *Arup Ghosh*[1,2] and *Sanjay Pratihar*[1,3,*]

1. Introduction

Nanotechnology is one of the fastest-evolving sciences in the recent past. Nowadays, nanomaterials are involved in different fields of science and industries because of their novel applications, which are increasing widely (He et al., 2015). According to reported research, future population growth-led changes in dietary habits will cause an expected 70% increase in the food demand by 2050. As one knows, the need for food is growing along with the world's population. Hence the agricultural sector needs to be developed more as both humans and animals rely heavily on it (Adisa et al., 2019). Improving food production and quality agriculture requires advanced and sustainable strategies. For this, nanotechnology-enabled products are becoming more beneficial due to their different characteristics such as agrochemical efficiency, enhancing nutrient bioavailability, and more secure packaging materials (Amenta et al., 2015). Rapid industry expansion could raise the need for materials

[1] Academy of Scientific and Innovative Research (AcSIR), Ghaziabad–201002, India.
[2] Applied Phycology and Biotechnology Division, CSIR-Central Salt and Marine Chemical Research Institute (CSMCRI) Bhavnagar, G. B. Marg, Gujarat-364002.
[3] Inorganic Materials and Catalysis Division, CSIR–Central Salt & Marine Chemicals Research Institute, G.B. Marg, Bhavnagar 364002, Gujarat, India.
* Corresponding author: spratihar@csmcri.res.in or spratihar29@gmail.com

for infrastructure, the manufacture of essential materials in various industries (such as pesticides, herbicides, insecticides, and chemical fertilizers), and R & D (Liu et al., 2012). To overcome future demands, nanomaterials have also been of great importance in a successful economy because of their rapid use in production volume for industrial and domestic reasons. Due to their fast adoption in production volumes for both residential and industrial purposes, nanomaterials have recently assumed a significant role in a thriving economy. The advantage of nanotechnology is mainly dependent on size and other unique characteristics. This will be further discussed in this chapter. The nanomaterial size ranges from 1nm to 100nm (Pacheco and Buzea, 2016; He et al., 2018; Khan et al., 2019; Wigger et al., 2020; Baig et al., 2021). Nanomaterials have incredible characteristics, such as surface area, size, chemical stability, density, melting point, and slow-release mechanism. In aquatic and terrestrial environments, nanomaterials are found naturally in the form of the finer fraction of colloidal clays and mineral particles, which could contain precipitates of aluminum, hydroxides, iron oxides, and manganese. These nanomaterials also contain different dissolved organic matter, such as humic and fulvic acid (Batley et al., 2013). As mentioned earlier, nanomaterials are used for various industrial and domestic purposes, as evidenced by their rising output volume consistently. This economic achievement carries the potential risk of negative impacts on ecosystems due to their presence in the environment (Bundschuh et al., 2018). There is some negative effect of nanomaterials on different ecosystems that has been reported by researchers (Bundschuh et al., 2018; Rajput et al., 2020). The argument over the appropriate test procedures and modelling paradigms for the behaviour of engineered nanomaterials (ENMs), a possible new class of pollutants, is gaining stream (Westerhoff and Nowack, 2013).

Numerous toxic material releases harm the environment in the short and long term. Over time, these impacts also threaten human health (Liu et al., 2012). In this regard, rigid safety testing for the extensive use of nanomaterial is a result of increased exposure to humans for food (Batley et al., 2013). Environmental Risk Assessment (ERA) is a procedure for determining the likelihood and effects of unfavourable ecological effects caused by human activity and other stressors (Kaikkonen et al., 2021). An environmental risk assessment study needs to consider the fate and behaviour of ENMs (engineered nanomaterials). The ecological concentration of these compounds and the possible dose to the ecosystem's resident organisms (Wigger et al., 2020). One of the detrimental effects of some toxic nanomaterials is the generation of Reactive Oxidative Species (ROS), which, because of their toxicological mechanism, cause cellular damage and death. Additionally, direct exposure to nanomaterials used as functional components, food additives, or nutritional components could cause risks to the health of people (He et al., 2015; He et al., 2018; Rajput et al., 2020). To determine the toxicity effects of nanomaterials, every life cycle stage of nanomaterials needs to be studied from production to release. For this purpose, the life cycle assessment is an effective method and tool for assessing the environmental impact of any product and process. However, there needs to be more comprehensive material regarding the LCA study in the nanotechnology industry. The life cycle assessment can avoid shifting forces between distinct life cycle phases and consider proposed trade across the system under the evaluation. It can fill in some gaps linked to the entire lifespan of any material or product (Guinée et al., 2001). The LCA has been used to design decisions better and avoid negative impacts in the future. The potential for LCA to support the development of safer nanotechnologies is described in literature. For instance, identifying manufacturing inputs or processes with the most significant potential for improvement. (Gilbertson et al., 2015). Additionally, the fate of nanomaterials in the environment and their risk assessment is becoming a significant concern due to the ecotoxicological effects of nanomaterials. Further toxicological analyses using different mediums have studied the environmental fate of nanomaterials. There are several processes responsible for the future of nanomaterial in the environment, such as transformation, dissolution, absorption, agglomeration, and aggregation, etc. The change of NMs is analogous to the issues of metal fate, behaviour, bioavailability, and impacts. Transformation occurs through various processes, which can be physical, chemical, or biological. ROS generation is caused by changes in nanomaterials and their surroundings via various mechanisms (He et al., 2015).

Basic information on nano-ecological risk assessment for long-term impacts and exposures could be more impressive in the nanotechnology discipline. On the other hand, societal concerns about the environmental risk posed by ENMs have grown (Semenzin et al., 2015). Researchers have identified specific technologies and models to achieve an understanding of this field. Semenzin and colleagues also mentioned nanoscale titanium dioxide (n-TiO2) in their review; It was chosen because it is widely used in consumer goods (like cosmetics) and that a wealth of information from ecotoxicological research (including endpoints for vertebrates, invertebrates, bacteria, and algae in marine, freshwater, and terrestrial compartments) is readily available in literature. In recent times, there is a need to reduce the adverse effect of nanomaterial in the environment as future environmental perspectives. Several studies have been started to make a model for nanomaterials uptake and its mechanism with organisms and ecosystems. Implementing a simulation of Nanomaterial Models (NMs') environmental fate and adoption by ecological microorganisms and cells is critical for underpinning experimental investigation, developing and encompassing concepts, improving our fundamental understanding of NM exposure and risk, and facilitating NM risk assessment. Various fate models, from more sophisticated mechanistic models to substance flow analysis models lacking nano-specificity, have been developed to study the fate of nanomaterials. In this case, NM uptake by organisms is driven by a dynamic process rather than equilibrium partitioning. Compared to the simple bioaccumulation factors used for organic compounds, biokinetic designs are better suited to model NM uptake (Baalousha et al., 2016). Several institutions proposed a case-by-case approach to improve safety as a risk assessment of NMs. They created registration or authorization procedures regulations, and specific requirements for products containing NMs (Oomen et al., 2014). For example, the EU Scientific Committee on Emerging and Newly Identified Health Risks (SCENIHR) has suggested a case-by-case approach for assessing the risk of NMs. Under the Federal Food, Drug, and Cosmetic Act (FFDCA), the Food and Drug Administration (FDA) of the United States is accountable for checking the safety of feed additives, food contact materials, and other additions before they are allowed on the market (Amenta et al., 2015; He et al., 2018; Sampathkumar et al., 2020).

The Organization for Economic Cooperation and Development (OECD) and other institutions have established standardized testing methods and reproducibility of ENMs and MNs for scientific risk assessment (Binh et al., 2015).

This chapter will provide helpful information on various aspects of nanomaterial use and a life cycle assessment of nanomaterials to understand each step and its effect on the environment. This will give an idea for risk assessment studies of nanomaterials concerning their reduction as a pollutant or toxic substance for the environment and human health. It will also provide brief information on steps taken by several research institutions and well-known laboratories towards a safer-by-design strategy for nanomaterials, in addition to different modelling techniques and future environmental safety perspectives for nanomaterials against their fate and behaviour.

Properties and Demand of Nanomaterial

The world's population will surpass eight billion in 2024. This rapid population expansion will raise food demand and strain the agri-food business. Growing agri-food demand is driving a 70% increase in global calorie consumption and a 100% increase in crop demand (Adisa et al., 2019). Under these circumstances, nanotechnology could significantly impact the agri-food sector, improving crop yield while enhancing food security, safety, and sustainability. Some distinctive properties of nanotechnology could improve animal feed, agriculture output, and environmental monitoring (Sampathkumar et al., 2020). However, recent research has revealed that some nanomaterials are hazardous to living things and may severely impact ecosystems (Gajewicz et al., 2012; Amenta et al., 2015). Nanotechnology is increasingly used in agriculture to produce better nutritional value, quality, and safe food. They are used as a nanocarrier with distinctive properties like controlled release mechanisms. As a poly nanocapsule, nanotechnology provides better ways to improve the efficient use of fertilizers, pesticides, herbicides, and plant growth regulators (He et al., 2018). Nanomaterials are classified into naturally occurring nanomaterials and manufactured (MNs) or engineered nanomaterials (ENMs). Naturally found nanomaterials result from volcanic activities, fire, and other combustion processes (Griffin et al., 2017). ENMs are specifically designed chemical substances or materials having at least one dimensionally distributed particle size in the range of 1 to 100 nm. In contrast, ENMs refer to ENPs and nanostructured materials collectively (Sharifianjazi et al., 2021). Nanomaterials have different unique characteristics (Batley et al., 2013). Due to their small size, nanomaterials exhibit distinctive phenomena allowing novel uses (Rajput et al., 2020). "Size" refers to the outside dimensions of a material's component particles, which can be unbound or in the shape of aggregates and agglomerates. The EC definition applies to all particulate NMs, whether natural, accidental, or produced, regardless of their source (Amenta et al., 2015; Xavier et al., 2021). Nanomaterials can be generated with excellent magnetic, electrical, optical, mechanical, and catalytic properties that vary considerably from their bulk materials. The properties of nanomaterials can be fine-tuned by precisely controlling their size, shape, synthesis conditions, and functionalization (Asha and Narain, 2020; Khoso et al., 2021). These characteristics have proved helpful in different fields of science, including industrial areas, agricultural and food industries, electronic and biomedical modern medical sciences, etc. (Gajewicz et al., 2012; Singh et al., 2020).

Traditional food and agricultural sectors have seen significant change due to the quick advancement of nanotechnology, including the invention of intelligent and active packaging, nanosensors, nano pesticides, and nano fertilizers, to name a few examples. Innovative nanomaterials have been developed to improve environmental monitoring, food quality and safety, and agricultural growth (He et al., 2018; Amenta et al., 2015). The creation of various nanomaterials with innovative physical and chemical properties and enhanced functioning has dramatically improved thanks to nanotechnology (Lee et al., 2016). For example, silver nanomaterials are the most important commercially among metal nanomaterials with some antimicrobial properties, followed by gold nanomaterials, which are widely used as sensors or detectors, and titanium dioxide nanomaterials, which have a wide range of uses as disinfectants, food additives, and flavour enhancers (He et al., 2018). To improve production efficiency, food quality, flavour, and safety, the food industry has recently used a number of tested nanomaterials, including inorganic (metal and metal oxide NPs) and organic (Natural Product NPs), to create appealing and novel combination products for food and safety purpose. Numerous products have reportedly been promoted over the last 10 yr (He et al., 2018). For example, Titanium dioxide (TiO_2), iron oxide (Fe_3O_4), Zinc Oxide (ZnO), Silicon Oxide (SiO_2), copper (Cu-NPs), and selenium (Se-NPs) are merely a few of the nanomaterials (NPs) that have gained much attention lately due to their non-dangerous use in the agriculture industry (Alabdallah and Hasan, 2021; Hashem et al., 2021). Different techniques have been created for nanomaterial preparation to alter their size and improve their functionality to achieve novel and sustainable use of nanomaterials. These synthesis processes are based on the top-down and bottom-up approaches. The bottom-up approach process includes the synthesis of nanomaterials from atoms and molecules (for example, nanomaterials of SnS and Ag) (Abid et al., 2022). Chemical and biological methods are examples of bottom-up approaches. The bottom-up chemical approaches use techniques like sol-gel, Chemical Vapour Deposition (CVD), chemical co-precipitation, micro-emulsions, hydrothermal method, sonochemical method, and microwave methods (Paramsivam et al., 2021), colloidal precipitation, template-assisted sol-gel, and electro-deposition (Cucurachi and Rocha, 2019). The top-down strategy involves creating nanomaterials from components at a larger scale. (Synthesis of CuO, MgO, and ZnO). The top-down approach involves physical participation methods like lithography, Physical Vapour Deposition (PVD), mechanical machining, and thermal evaporation pyrolysis (Cucurachi and Rocha, 2019), synthesis of micron-sized particles, involves breaking bulky materials into NPs (Abid et al., 2022). High-energy ball milling, atomic force manipulation, gas condensation, and aerosol droplets are a few examples of this method. Most significantly, using nanoencapsulation can reduce herbicide dosage without sacrificing effectiveness, which is good for the ecosystem (He et al., 2018). Nanomaterials with the distinct qualities were mentioned earlier. Therefore, these distinctive properties include a high specific surface area and many reactive sites on the surface because many of the elements in NPs are on the outside rather than the inside. However, because of their mobility, they also have the potential to impact the health of the earth negatively (Kumar et al., 2012) due to increased industrial use of nanomaterials, technologies for soil and water remediation, and possible applications in agriculture. Several studies have been conducted over the last several decades to demonstrate the release of nanomaterials during their various life cycle studies due to the expansion of their production and increased risk to the environment; risk assessment of nanomaterials has two key elements: hazard characterization and exposure characterization (Baalousha et al., 2016).

Life Cycle Assessment of Nanomaterials

For evolving nanotechnologies, the application of Life Cycle Assessment (LCA), a comprehensive modelling framework, to identify the escalating effects of products, processes, and technologies on the environment and human health. LCA is a comprehensive framework for determining the effects of a system or product on the environment and human health throughout its entire life cycle. Any decision that considers environmental impacts can be made using this framework for any product. These days, a wide range of bodies, from governmental institutions to businesses, apply this framework, with or without assistance from specialized research and consulting organizations (Hischier and Walser, 2012). As discussed earlier, nanomaterials are classified mainly into two types: naturally occurring nanomaterials, engineered nanomaterials, or manufactured nanomaterials. Due to their small size and unique physicochemical properties, nanomaterials found naturally have some advantages, as described by various researchers. Engineered nanomaterials (ENM) are manufactured compounds that have become a hot topic among scientists, researchers, and the public in the last 15 yr (Wigger et al., 2020). To obtain desired properties of nanomaterials, they are manipulated during manufacturing in the field of engineered nanomaterials (ENM). These changes could also be helpful for the efficient use of that material and improve the performance of many consumers and industrial products (Cucurachi and Rocha, 2019). Synthesis of nanomaterials at different levels of their life cycle release energy has material requirements, and produces pollutants in the air, water, and soil, as well as material losses and emissions. Other stages of the ENM life cycle, such as transport, use, and end-of-life, also determine the environmental effects of various kinds (such as effects related to the emissions of greenhouse gases and toxic effects), which are either directly or indirectly associated with the release of ENMs (Cucurachi and Rocha, 2019). To understand the life cycle stages at different levels, LCA has been acknowledged as a valuable tool for systematically evaluating ENMs' potential environmental effects (Linkov and Seager, 2011; Cucurachi and Rocha, 2019). Using a multifaceted approach and system-level assessments is necessary for evaluating every aspect of environmental effects associated with the life cycle of ENMs. The LCA is carried out in three phases: The initial phase entails defining the scope of the study (i.e., strategic planning and policymaking), and the second phase, referred to as the inventory stage, entails data collection, relationship identification and quantification of inputs and outputs for the product system under evaluation over its entire life cycle. The third phase includes the impact assessment phase (LCIA), where inventory data are categorized according to environmental impact categories and aggregated according to the potential ecological impacts of a product system's possible size and significance (ISO, 2006). The analyst generates conclusions from the results in the final stage of the LCA, identifies assumptions and limitations, and makes recommendations. At this stage, essential problems are highlighted, including significant life cycle areas of concern and areas of opportunity for critical environmental improvements. Additionally, the analyst considers potential sources of uncertainty, such as those brought on by a lack of data, an unrepresentative data collection process, or a difference in the geographic or temporal scopes of the data collected (Gavankar et al., 2015).

Similarly, The International Organization for Standardization (ISO) 14040 series is the basis for the life cycle assessment (LCA) methodology, which consists of four phases (Fig. 8.1): (i) goal and scope, (ii) life cycle modelling, (iii) life cycle impact evaluation, and (iv) interpretation of the life cycle. This method emerged as a tool for assessing how products and the processes connected to them affect the environment (Chong et al., 2018; Ismail et al., 2019; Nizam et al., 2021). Elementary flows and economic flows are differentiated in the LCA standard. The former provides resources with the natural environment and emissions between the technosphere (i.e., all products and innovations created by humans) and the environment. (Cucurachi and Rocha, 2019). Figure 8.2. Toxicology flow of engineered nanomaterials (ENMs) and their different pathway.

LCIA starts by categorizing each emission into the impact categories to which they lead; then the characterization factors are used to determine the corresponding impact. (CFs). CFs are calculated per unit mass emitted to a specific environmental section and provide a quantitative assessment of the impact potential related to each emission. The consensus model USEtox is a suggested practice for the aquatic environment and human health (Gilbertson et al., 2015). LCIA fate models rely

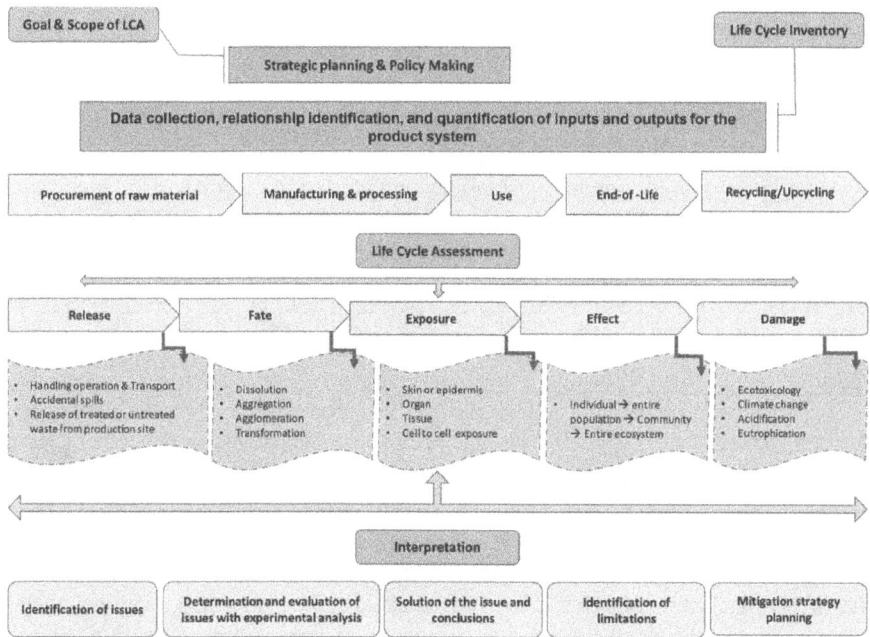

Figure 8.1 Different phases of life cycle assessment (LCA) of engineered nanomaterials (ENMs).

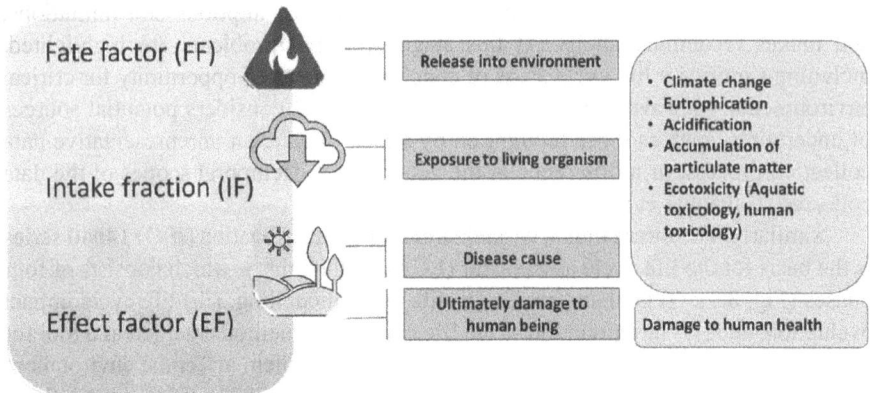

Figure 8.2 Toxicology flow of engineered nanomaterials (ENMs) and their different pathway.

on multi-media mass balance models using a fugacity methodology, such as the European-developed USES-LCA used in ReCiPe and the consensus model USEtox now accepted by TRACI (Gilbertson et al., 2015).

Nanomaterial and Plant

Nanomaterials have beneficial effects on plant growth and development in the agricultural industry. The impact mainly depends on the amount, source, and timing of nanomaterials applied to the crop (Rubilar et al., 2013; Rasheed, 2022). These days, nanomaterials have been developed to increase plant resistance to biotic and abiotic stress and avoid oxidative damage by boosting antioxidant activity (Ahmed et al., 2021).

Depending on the type of plant, the form and size of nanomaterials, their stability, and how they interact with the roots, soil, and soil microbes, manufactured nanomaterials could penetrate plants from the ground in a number of ways (Ali et al., 2021). It is additionally established that the characteristics of both the NPs and the surrounding environment substantially impact the accumulation of metal and metal oxide bases NPs by roots (Mittal et al., 2020; Ali et al., 2021). Exodermis and endodermis both occur in angiosperms. MNMs are directly transferred to the cell through the root cortex and Casparian band (hydrophobic layers of suberin and lignin), though sporadically immature root tips obstruct the apoplastic movement of the ions in the root stele. Due to this defense mechanism, MNMs penetrate the root stele by crossing a cell's, whether an endodermal or ectodermal cell or by synthetically spanning a Cell root wall at a cell outside of the endodermis or exodermis (Judy and Berstch, 2014).

The endocytosis mechanism has a significant impact on plant physiology. It allows exterior materials like lipids, membrane proteins, and other substances to enter the cell. It functions as a key process in cell-to-cell contact through various other processes, including nutrient uptake, signalling transduction, and interactions between plants and microbes (Tripathi et al., 2017). Endocytosis during MNM entry provides the correct surface chemically. Bright yellow (BY-2) Nicotiana tobacco L. cells were exposed to single-walled carbon nanotubes (SWCNTs) that were fluorescently labelled at two different temperatures and found to absorb MNMs. Confocal fluorescence imaging showed that while the MNMs were easily absorbed in 26°C, they remained inactive at 4°C even after a protracted incubation period. This finding indicates that endocytosis could be a key mechanism for plant uptake of MNMs because it has been demonstrated that endocytosis in plants is drastically reduced at low temperatures (Judy and Berstch, 2014). According to some researchers, the cell wall's thickness plays a significant role in both the entrance of nanomaterials in the plant and their absorption and translocation (Raliya et al., 2016). In comparison to controls, Hg and Fe nanomaterials can increase the uptake of N (51%), P (61%), K (27%), calcium (Ca) (53%), and magnesium (Mg) (62%) (Ahmed et al., 2021). The use of Si-NPs, on the other hand, was found to greatly improve RWC, antioxidant activity, nitrogen absorption, and photosynthesis in maize plants cultivated under drought stress (Hafez et al., 2021).

It has been noted that atrazine-capped nanocapsules have more potent herbicidal effects on mustard plants than commercially available atrazine, (Brassica juncea). It was shown that the plants under evaluation experienced a sharp decline in the net photosynthetic rate and stomatal conductance, as well as a rise in oxidative stress and weight loss. Ultimately, this caused the studied plants to lose weight and grow more slowly (He et al., 2018). Primary nutrient availability (NPK—nitrogen, phosphorus, and potassium) for plants in soil is presently a significant problem in agriculture due to insufficient transfer to the target location. To control pathogens and improve food safety, it is necessary for enhanced techniques for the controlled release and target distribution of nano-fertilizers or agrochemicals. Due to these beneficial characteristics, traditional or chemical fertilizers could be used less frequently, and the soil's health can be preserved (Adisa et al., 2019).

Nanomaterials and Microbes

Approximately 11 million tons of metal and metal oxide NPs are produced worldwide yearly, with soil resources eventually acquiring them (Keller et al., 2013; Sun et al., 2014). NPs can penetrate cells or enter bacteria through endocytosis, and many of them, particularly silver, copper, and zinc, have antimicrobial characteristics (Kumar et al., 2018). Numerous scholars have investigated the way NPs affect soil microorganisms.

Silver nanomaterials have been widely studied due to their antimicrobial activity. Several species of soil microbes are significantly influenced by factors: such as functionalization, concentration, exposure time, and texture of the soil. It has been noted that some nanomaterials showed a change in the cell wall of Bacillus cereus and Pseudomonas stutzeri, such as Ag-nanomaterials (Grun et al., 2019; Rajput et al., 2020). Commonly developed nanomaterials could hinder the usual functions of microbes like microbial biomass, fatty acid composition, and respiratory processes (Pawlet et al., 2013). Likewise, silver nanomaterials could negatively affect the microbial community and related biological reactions (Yin et al., 2020; Rajput et al., 2020).

When metallic ions in nanomaterials interact with cellular components through a number of pathways, including the generation of reactive oxygen species (ROS), the development of pores in cell membranes, cellular wall damage, destruction of DNA, and cell cycle termination, they could inhibit a variety of bacteria and fungi (Singh et al., 2019).

In microarray research, the exposure of E. coli bacteria to cerium oxide nanomaterials caused an increase in the number of oxidoreductases, which altered cellular respiration and led to oxidative stress and iron deficiency (Pelletier et al., 2010). Similar research was done on Pseudomonas aeruginosa subjected to quantum dots, and it was found that the genes for metal efflux transporters and oxidative stress were up-regulated (Yang et al. 2012).

The chemical reduction was used to generate copper nanomaterials, and the effectiveness of the anti-bacterial properties was evaluated. The E. coli strain was inhibited by copper nanomaterials, rendering them potential antibacterial agents. This copper nanomaterial was enclosed in a biopolymer, alginate, and foamed polyurethane foam antibacterial water filtration. It was discovered that polyurethane foams could function as effective antibacterial water filters with straightforward treatment methods (Hari Kumar and Arvind, 2016).

Environmental fate and behaviour

Engineered nanomaterial discharges into the environment can occur in different ways, including atmospheric emission, industrial solid or liquid waste disposal, household discharge, fuel combustion, etc. Various pathways lead to the release of ENMs depending on the intended use (indoor, outdoor, cosmetic, industrial, building material, electrical, catalyst) and the type of ENM (e.g., TiO_2, SiO_2, Ag, CeO_2, Cuco, ZnO). A tiny percentage of ENMs will be released into the environment directly because of mass flow analyses; the vast majority first pass through various managed waste facilities, such as wastewater systems, waste incinerators, and landfills (Wigger et al., 2020).

ENM changes during and after use when released into different environmental compartments like soils, surface waterways, and the atmosphere (Wigger et al., 2020). When ENM interacts with biomolecules and is introduced into the environment, it has a number of impacts. A few researchers in nanomedicine observed this binding, which provided details about the formation of a protein corona by the interaction of an ENM and a protein. This phenomenon became known as the eco-corona (Wigger et al., 2020).

Nanomaterials released into the environment have the potential to contaminate the soil, move to surface and groundwaters, infiltrate biological systems, and eventually can contaminate the food chain. Wind or stormwater drainage can carry particles from solid refuse, wastewater effluents, direct discharges, or unintentional spills into aquatic systems (Bundschuh et al., 2018).

Environmental risk is a current concern that needs to be carefully examined in the early stages of developing any new technology. Technology's ability to help society in various ways is primarily validated by nanotechnology. However, environmental risk is increasingly the foremost worry (Rajput et al., 2020). The increased use of manufactured nanomaterials has raised questions about their behaviour in the ecosystem and the need for additional research. Accidental spills during production, shipping, and product disposal result in these manufactured nanomaterials being released. Some factors can be used to determine the intrinsic properties and the environmental conditions, such as wastewater treatment effluents and sludges, landfill leachates, and waste incineration residuals (Baun et al., 2017). Depending on the features of the nanomaterial and the receiving medium, nanomaterials can change the environment in several different ways. These transformations can also involve the biodegradation of surface coatings that stabilize various formulations of nanomaterials, even though chemical and physical mechanisms are usually involved. Nanomaterials' toxic effects on algae contain adsorption on cell surfaces and interference with membrane transfer. Two processes are critical for the transfer and transformation of nanomaterials: dissolution and agglomeration. These two processes have the possibility of affecting the fate as well as the toxicity of nanomaterials. Several other processes, such as chemical transformation, aggregation, and disaggregation, determine the nanomaterial's future and ecotoxicological potential in the environment. Changes in chemical speciation, dissolution, and adsorption/desorption are essential chemical transformation processes of NPs studied in aquatic and soil ecosystems (Bundschuh et al., 2018). Agglomeration, agglomeration, sedimentation, and deposition are examples of physical processes. Dissolution and following speciation changes, redox reactions (oxidation and sulphidation), photochemical processes, and corona formation are examples of chemical processes. Agglomeration promotes the discharge of products like bio- or eco-corona (Wigger et al., 2020). Biodegradation and biomodification are two biologically mediated processes almost certainly controlled by microbes. A mental process model contains them all. Although the nature of the NMs and the environmental conditions will influence transformations, understanding and predicting these variables due to their complexity and variability is exceedingly tricky (Bundschuh et al., 2018; Wigger et al., 2020). Numerous studies have demonstrated that specific intrinsic NM properties, such as size (Tsiola et al., 2017), coating (Toncelli et al., 2017), and doping, influence dissolution, and solubility (Adeleye et al., 2018). Studies that examined the effects of outside variables like NOM on NM behaviour concluded that dissolution, particle ripening, and the precipitation of new NMs could all be enhanced (Merrifield et al., 2017). Organic materials can significantly impact NM dissolution (Luoma et al. 2016).

Due to biogeochemical changes and interactions with organic and inorganic ligands, nanomaterials released from consumer products will likely be altered after entering wastewater streams (Chae et al., 2013; Judy and Bertsch, 2014). The MNM changes in the environment, and several processes cause its dissolution, aggregation, and interaction with biomolecules. These transforming factors depend on various nanomaterial properties, including solubility, pH, ionic strength, the presence of inorganic and organic compounds, hydrophobicity, surface chemistry, and surface charges (Judy and Bertsch, 2014) (Fig. 8.3). The use of nanomaterials has significantly grown over the last few decades in various economic sectors, including industrial applications, agrochemicals, consumer goods, and medical products. However, this widespread use of nanomaterials is endangering the environment. Some research studies are being done to find the threshold concentration for the ecotoxicological impact on the terrestrial and aquatic biota (Batley et al., 2013).

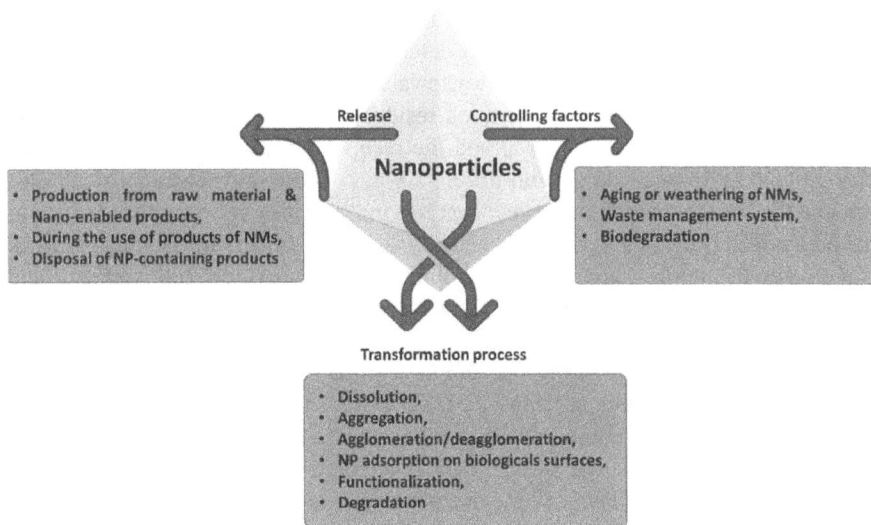

Figure 8.3 Release, controlling factors, and transformation process of engineered nanomaterials (ENMs).

The transformation of nanomaterials in the marine ecosystem is greatly influenced by physicochemical factors, including Dissolved Organic Carbon (DOC), pH, ionic strength, redox, particle-specific features (i.e., size and surface coating), and concentration (Metrevelli et al., 2016; Furtado et al., 2016). The formation of soluble AgCl(x) complexes, for example, did not affect AgNP toxicity to bacteria, whereas the building of insoluble AgCl(s) and Ag2S(s) had a significant effect. For determining environmental impacts, it is essential to comprehend how much the AgNP dissolves and the consequent Ag concentrations in aquatic systems (Furtado et al., 2016). Nanomaterials are exposed to the environment in various ways during their life cycle stages, production, manufacturing, disposal, or recycling. The release amount estimates this nanomaterial's exposure during use, the end of the similar cycle stages, and its environmental fate. The experiment was carried out at various locations of TiO_2-NM release from photocatalytic cement using a static leaching test (L/S 100) that simulated the worst-case water weathering scenario. Cement pellets released less than 0.04 w.% of the preliminary TiO_2-NMs. Photocatalytic cement is expected to be a minor source of TiO_2-NMs in the environment (Bossa et al., 2017).

Surface area exposure is a distinguishing feature of nanomaterials. The agglomeration and sedimentation of nanomaterials in aquatic environments are primarily determined by the surface area and particle size. Smaller agglomerates could dissolve or partially dissolve in the marine environment, whereas larger and colloidal agglomerates may not dissolve and affect aquatic benthic communities, potentially causing an ecotoxicological effect (Seitz et al., 2014). The different concentrations of $nTiO_2$ organisms like *Daphnia* and *Gammarus* are not affected up to 5.00 mg/L, but with more than this concentration, they were affected in their different life cycle stages. This suggests that basic variations in nTiO2's toxicity throughout its aquatic life cycle are caused by changes in these features over time (Seitz et al., 2014).

The behaviour of the nanomaterial at the cellular level could be unpredictable due to the dynamic microenvironment. For example, simultaneous interactions between a nanomaterial and components of the abiotic environment (such as DOM, solar irradiation, and a part of the target biota) can significantly alter the environmental fate of the nanomaterial and the nanotoxicological response of the cellular system due to an alteration of the physicochemical properties of the biological system and the nanomaterial at the boundaries of the suspension medium (He et al., 2015).

The particle composition primarily determines how easily nanomaterials dissolve. Silver nanomaterials needed aerobic circumstances to dissolve. According to Mitrano et al. (2015), the presence of oxide layers in some environments causes particles to form around them and release Ag+. This process is also influenced by many other factors, such as surface coating, particle size, shape, aggregation state, and physicochemical variables like pH, temperature, and dissolved organic matter (Metreveli et al., 2016; Bundschuh et al., 2018).

In any medium, nanomaterials can be combined through the processes of homo and hetero agglomeration. They produce homo aggregates of nanomaterials that exhibit a positive correlation between their concentration and the substance in question. Due to the low chances of collisions makes homo-aggregation less probable. There needs to be more knowledge e regarding the ecotoxicological impact of homo-aggregate nanomaterial collisions on the environment. When a high concentration of nanomaterials was used in a laboratory-based study, the critical criterion of a low chance of collision was disregarded (Metreveli et al., 2016). Natural colloids with nanomaterial relevance have greater relevance to the environment than hetero aggregation. A crucial step in the sedimentation process to remove cerium oxide nanomaterials from the aqueous phase is hetero aggregation (Quik et al., 2014).

AgNP dissolution and agglomeration have two effects on the transformation in marine ecosystems: the creation of the soluble AgCl (x) complex demonstrates the toxicity of AgNP to bacteria, and the insoluble AgCl (S) and Ag2S (s) complex lessens the toxicity of bacteria. The degree of these changes will be influenced by physicochemical factors like dissolved organic carbon (DOC), pH, ionic strength, redox, and the particles' size, surface coating, and concentration (Furtado et al., 2016).

Higher organisms can consume nanomaterials directly, and aquatic and terrestrial organisms can assemble nanomaterials within the food web (Batley et al., 2013). At least a few gaps need to be filled, including those related to the toxicity of nanomaterials to mammal cells, tissues, and organs and their long-term effects on human health, the migration of nanomaterials into food, their degradation or eventual environmental fate, and their bioaccumulation and impacts on ecosystems (He et al., 2018). In people, excessive production and generation of ROS can cause autophagy, neuronal damage, severe DNA damage, and possibly mutagenesis, carcinogenesis, and age related illnesses (He et al., 2015; He et al., 2018; Rajput et al., 2020). Changes in particle size, surface charge, and chemical form will be essential regulators of bioavailability because of nanomaterials' fate as they accomplish in the environment (Batley et al., 2013).

When nanomaterials engage with organic material, they bind to biomolecules, which create a corona that regulates how cells interact with them, how they spread throughout the body, and how tissue accumulates in organs. The biological distribution of nanomaterials through the air-blood barrier in the pulmonary system, lung retention, and toxicity are all influenced by particle size (Pacheco and Buzea, 2016).

The word "nano-eco interaction" was established to describe how abiotic elements and the environmental nanomaterial interface in aquatic and terrestrial ecosystems interact. (Physicochemical parameters – Surfactants, Dissolved Organic matter, etc.). This interaction oversees changing or altering various characteristics of released or discharged nanomaterials (He et al., 2015). Different naturally occurring nanomaterials can be found in aquatic and terrestrial habitats, such as colloidal clay particles, iron, manganese hydroxide, dissolved organic matter (fulvic acid and humic acid), fibrillar colloids, and other microorganisms. Greater knowledge of colloidal behaviour and environmental risk assessment is required to understand the fate and behaviour of manufactured nanomaterials and naturally occurring nanomaterials (Batley et al., 2013; He et al., 2015).

In comparison to nano-eco interaction, nano-bio interaction is more complicated. These nano-bio interactions involve the interplay of nanomaterials with biomolecules and relationships at the cellular level. This interface greatly influences the environmental destiny of the nanomaterial. The protein corona is a phenomenon of denaturation of proteins, particle enveloping, and the biocatalytic processes. According to reports, it could be responsible for alterations in phase, free energy transfers, particle aggregation, surface reconstruction, and dissolution of nanomaterials. Highly abundant proteins include immunoglobulin G (IgG), fibrinogen, apolipoproteins, serum albumin, serotransferrin, prothrombin, alpha-fetoprotein, and kininogen-1 have been discovered to be present in NMs (He et al., 2015).

Many nanomaterials in natural environments exhibit colloidal properties, and when they are exposed to aggregation and sedimentation, they could generate nanomaterials with sizes ranging from 100 to 1000 nm. The variables which regulate accumulation include surface charge, particle size, ionic strength, pH, cation composition of the solution, and particle shape, which could influence steric interactions. Rapid particle aggregation with a nearly neutral charge is seen (Batley et al., 2013; He et al., 2015).

Residues and deposits of nanomaterial in more amounts significantly affect in different ways: i) it increases the negative association between living things and NPs, ii) it affects the macroscopic properties of soil, for example., fulvic acid and humic acid content (Ben-Moshe et al., 2013) iii) it influences microbial activity, diversity as well as plant growth (Kumar et al., 2012; Rajput et al., 2020).

Besides air or water, soil is a crucial nanomaterial-receiving site. The accumulation and release of nanomaterials in the agricultural sector is an increasing concern. This is mainly due to the significant relationship between the biological effects on crops and symbiotic organisms and the actual exposure concentration (Gogos et al., 2016). Additionally, various point and non-point sources (industrial spills, landfills, and when sewage sludge is applied as a fertilizer) allow the entry of these nanomaterials into the soil, where they can then find their way into the aquatic ecosystem (Rajput et al.,2020).

The interplay of nanomaterials with abiotic components of the environment can be determined by the knowledge of sources, pathways, transformation, transportation, and sinks for NM. This knowledge also enabled the identification of threats and the potential for bioaccumulation, particularly within the aquatic food web and food chain (Luoma et al., 2014).

Several distinctive properties of nanomaterials, such as metal shedding and disintegration, can cause environmental toxicity. Such a novel and environmentally friendly strategy would need to be used to reduce the environmental/ecological toxicity of nanomaterials. They modify a nanomaterial's properties through safer-by-design nanomaterial synthesis (Lin et al., 2018). The overlapping of the conduction band energy level and redox potential in the atmosphere has led to oxidative stress and redox equilibrium. This interfering process caused ecotoxicity at the cellular level in the ecosystem. The balance between the conduction band energy level and redox potential should be preserved to minimize environmental ecotoxicity (Lin et al., 2018).

It has been noted that, with or without the addition of biological amenities, nanomaterials of ZnO and CuO can interact with several biological systems components, such as Na, Ca, P, and Cl, and create the iron corona (Xu et al., 2012).

Every year, nanomaterials have increased, leading to unsafe disposal practices that release nanomaterials into the atmosphere and advance the development of certain metal oxides, including Ag, Al, Ce, Cu, Fe, Si, Ti, and Zn. Natural resources have become contaminated because of this form of accumulation, and there are a lot of NPs in it (Keller et al., 2013).

Risk Assessment of Nanomaterial

Risk assessment studies of ENMs or MNs have been affected by some consequences related to the assessment methods or techniques as they differ from purely conventional chemicals in considerably changing their physicochemical properties (Fadeel et al., 2018). The optimal risk assessment strategy for nanomaterials and how to deal with the unique characteristics of ENM, for instance, has been explained in several published works (Hristozov et al., 2016). Many models could be used to evaluate quantitative and qualitative environmental dangers (Sorensen et al., 2019). The optimal method for determining the risks posed by nanomaterials, how to incorporate their unique qualities, and, most critically, which changes to make to test systems at present to obtain repeatable and acceptable results for regulators, are all hotly debated in literature (Petersen et al., 2015). Identifying the Environmental, Health, and Safety (EHS) risks associated with nanotechnologies is a regulatory prerequisite for their implementation. These risks are difficult to evaluate today due to the unknown surroundings and how manufactured nanomaterials (MNs) will interact with people and the environment. Implementing appropriate frameworks and tools is essential to produce reliable scientific data on risk and exposure to mitigate this unpredictability. A dynamic area of study is the development of such methods to simplify the Risk Assessment (RA) of MNs (Fig. 8.4) (Hristozov et al., 2016).

To establish risk assessment, the Predicted Environmental Concentrations (PECs) were compared which indicated no-effect concentrations (PNECs)

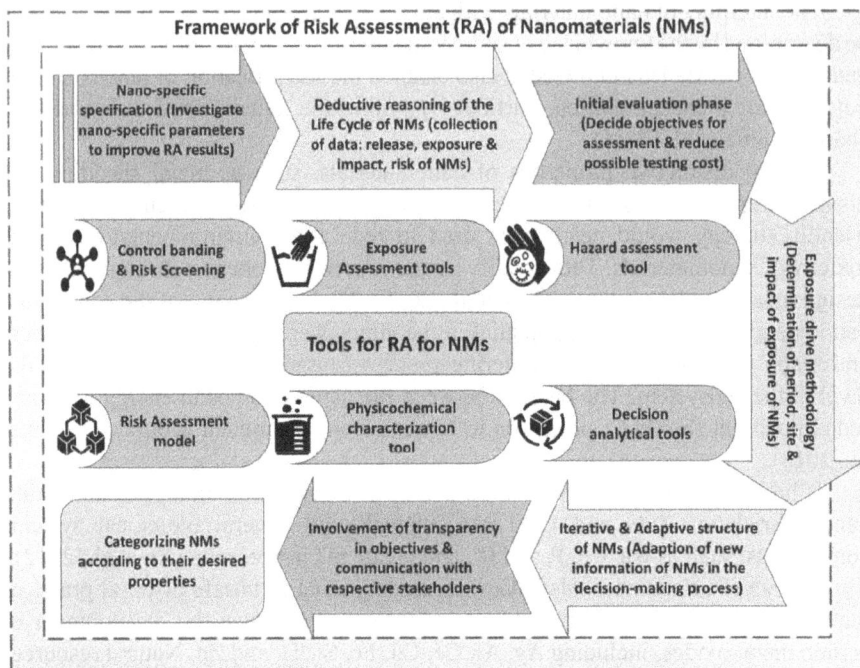

Figure 8.4 Framework and tools for risk assessment of engineered nanomaterials (ENMs).

because PEC necessitates measuring or forecasting the concentration and size of nanomaterials depending on where and how they will travel after being discharged into the environment. According to the initial efforts at life cycle assessment, the direct entry of nanomaterials into aquatic systems by spillage, discharge, atmospheric deposition, or soil runoff was negligible compared to the proportion that goes into the sewage treatment and could be in STP discharge waters. Estimates of total product use and release rates should be acquired to calculate mass flows, which can be related to the exposure period (Batley et al., 2013; Arvidsson, 2018). The formulation of trustworthy guidelines for using nanomaterials in the environment and the computation of no-effect concentrations are made possible as only a small amount of toxicity data has been released. A chronic PNEC value of 7.9 g/L for n-C60 and 40 g/L for TiO_2 disseminated by sonication was found based on most data for n-C60 and nano TiO_2 in aquatic systems (Batley et al., 2013).

The distinct elemental makeup of nanomaterials plays a significant role in their distribution in environments and living entities, including the human body, via the skin, lungs, and gastrointestinal tracts. They can also penetrate and communicate with tissues and fine capillaries and travel among organs, cells, and sub-cellular structures in a similar manner. According to pharmacokinetic research, numerous cell types contain different kinds of nanomaterials, including mitochondria, lipid vesicles, fibroblasts, nuclei, and macrophages (Gajewicz et al., 2012).

One should be able to separate the pertinent information while considering the statistical correlation between the variables in the evaluation to identify the hazard-relevant molecular features (signatures). In addition to helping in identifying and categorizing ENM concerns, omics-based techniques can be used to create biomarkers for exposure or effect (Fadeel et al., 2018).

Ideally, exposure assessment specifies the evaluation's sources, pathways, routes, and uncertainties. Data on the production volume, potential industrial applications, anticipated use in everyday goods, suggested ways for disposal and recycling at the end of their cycle lives, behaviour, environmental transfer, fate, and distribution are all required for exposure assessment (Gajewicz et al., 2012).

Analytical methods have been developed to study materials at the nanoscale, but the requirements for sample properties often need to be met. Dynamic light scattering calculates particle size distributions in complex suspensions, but its limitations have been explored (Wigger et al., 2020). ENMs have wide use in various sectors due to their chemical properties, but their variations make identifying specific features causing toxicity challenging. Recent advances in hazard and risk assessment of ENMs have included systems biology approaches, high-throughput screening platforms, and novel risk assessment and management tools (Fadeel et al., 2018).

In silico modelling, the technique is widely used to develop strategies for using software to capture, analyze and integrate biological and medical data from various sources in the risk assessment study of nanomaterials against human health becoming more essential due to the nanomaterial toxicity at a different level. *In-silico* modelling concepts are also useful for determining the nanomaterial concentration in the environment. Two other models are also frequently used to know nanomaterial concentration in the background and assume the release of nanomaterials in the environment. The Species Sensitivity Distribution (SSD) approach suggested by Gottschalk et al. (2013) is the only tool capable of performing an uncertainty analysis; it uses a Monte Carlo probabilistic approach to produce a compared SSD with probability distributions of PECs for the determination of environmental hazards. The SSD is estimated using the Species Sensitivity Weighted Distribution (SSWD) technique. It blends species relevance and trophic level abundance criteria to assess the validity of ecotoxicological data and their applicability for ecological risk assessment (Semenzin et al., 2015). TEARR (the Tool for MNs-Application Pair Risk Ranking) is a semi-quantitative method created by the U.S. Army Corps of Engineers to rank MNs according to their risk to human health. It considers various input factors, such as structural and chemical properties, environmental fate indicators, reactivity indicators, application-specific data, and use patterns (Jacobson et al., 2015). Material Flow Analysis (MFA) and Environmental Fate Models (EFM) are widely used models for the mentioned purpose (Adam and Nowack, 2017). The life cycle of nano-enabled products, which includes their development, formulation, manufacture, use, and end-of-life, is the first stage of the MFA model. It assumes both transfers to the following process step and potential environmental release (Wigger et al., 2020). Baalousha et al. (2016) evaluated the currently available models for the aquatic and terrestrial compartments (including biota uptake). In contrast, Williams et al. (2019) reviewed the presently available models for the sunken room.

A platform for EU-funded initiatives addressing the safety of nanomaterials and nanotechnologies is the EU Nanosafety Cluster (www.nanosafetycluster.eu). The main objectives are optimizing the effect, developing policies, and fostering collaboration among these projects. ENM hazard and risk assessment developments, such as systems biology methods, high throughput testing mediums, and new risk assessment and risk management tools, have been created recently (Fadeel et al., 2018). Developing common terminologies, standards, and unified infrastructures pertinent to stakeholders' requirements in the nanosafety field is another significant challenge for the nanosafety community (Fadeel et al., 2018).

Two complementary web-based risk evaluation tools for ENM were created by the EU-funded projects FP7-SUN and FP7-GUIDEnano: the GUIDEnano Tool and the SUN Decision Support System (SUNDS). The SUNDS software system can calculate the risks to consumers, workers, and the environment from ENM in industrial products throughout a product's life. When the stakes are very significant, SUNDS suggests appropriate risk management strategies and provides details on their prices concerning the advantages of nanotechnology. Risk management can be demonstrated by lowering the risk to below threshold levels or looking into practical substitutes for the substance (Subramanium et al., 2016; Fadeel et al., 2018). The application of these technologies could represent a substantial improvement in assessing and mitigating risks associated with nanotechnology-enabled products over their entire life cycle (Dekkers et al., 2016).

Several countries around the world are looking into whether existing legal frameworks are appropriate for handling nanotechnologies and are using various techniques to guarantee the security of nano goods in food, feed, and agriculture. Although it is generally acknowledged that current risk assessment and test methodologies apply to NMs, several factors, including sample preparation, characterization, dosimetry, effect endpoints, exposure data, and models, call for creating standardized and validated methodologies (Amenta et al., 2015).

Official bodies have issued risk assessment guidance focusing on the properties of NMs. A "Guidance on the Risk Assessment of the Application of Nanoscience and Nanotechnologies in the Food and Feed Chain" was published in 2011 by the European Food Safety Authority (EFSA)," which provided a practical method for evaluating possible risks. Before completely transforming NMs into non-nanoforms, local effects and absorption should be addressed. Based on a comparison to their non-nano counterpart, the International Life Science Institute (ILSI) developed a systematic, tiered strategy for assessing the safety of NMs in food. It guides applicants to generate data on physicochemical characterization and testing approaches to identify and characterize hazards resulting from nanomaterial properties (Amenta et al., 2015).

The FDA has issued several guidance papers addressing nanotechnology issues to assist the industry, such as "Considering Whether an FDA-Regulated Product Involves the Application of Nanotechnology". The guidance states that the FDA takes into account the material's size (about 1 nm to 100 nm) or any characteristics or phenomena linked to the material's exterior dimension(s) when deciding whether a product contains NMs (Amenta et al., 2015). Asian countries actively use NM production and regulations, with standards and certification systems for nano-enabled goods. The Food Safety and Standards Act is India's primary component of food safety legislation. The government introduced the Nano Science and Technology Initiative (NSTI) in 2001 and the "Nano Mission" program in 2007. Japan and Korea actively participate in the OECD Working Party on Manufactured Nanomaterials.

The National Centre develops national standards in nanotechnology for Nanoscience and Technology (NCNST) and the Commission on Nanotechnology Standardization in China. The South African Nanotechnology Initiative (SANi) was founded in 2002, and a 10-yr nanotechnology strategy has been developed.

Thus, correct nanotoxicology evaluation research is essential to ensure the secure engineering, handling, and application of nanomaterials in food and agricultural goods. Additionally, modern toxicology techniques provide some knowledge that chemists can use to improve their sustainable design for large-scale use (He et al., 2018).

Environmental Safety Perspectives

From the perspective of environmental safety, some advanced remediation techniques have evolved for environmental pollution. In this regard, *in situ* remediation techniques (thermal treatment, pump-and-treat, chemical oxidation, and bioremediation) are expensive, less time- and effort-consuming. A new cleanup technique, nano remediation, was developed that is more affordable, effective, environmentally, socially, and economically healthy (Corsi et al., 2018).

To effectively apply nanotechnology, a thorough eco-safe predictive assessment approach should be used, with some essential crucial aspects: Estimate the behaviour of ENMs in the remediated media, determine potential toxicological targets, take into account the features of the polluted media/area and its surroundings, and provide a mechanism-based assessment of ecotoxicity in multiple species (Corsi et al., 2018).

Preparing safer design materials in the nanotechnology sector should necessitate proper criteria, principles, and methods. This research can lead to the development of safer-by-design nanomaterials, lowering their toxicity potential. The safer-by-design strategy for them includes modifying the properties of nanomaterial design and developing various strategies to minimize or prevent nanomaterial exposure (Lin et al., 2018). Nanomaterials have been connected to several diseases that can manifest immediately or years later. As it is impossible to extrapolate a nanomaterial's toxicity from its bulk properties, identifying hazardous nanomaterials requires a case-by-case strategy. Consequently, considerable caution is needed when managing and using nanomaterials in applications (Lin et al., 2018).

The term "Safer-by-Design" is relatively new in the field of nanotechnology, having first been addressed in literature around 2008; comparable approaches had earlier been implemented to promote a conscious design to avoid the potential adverse effects of chemicals and pharmaceuticals (Lin et al., 2018). Due to the growing body of experimental evidence from environmental health and safety studies, designing safer nanomaterials and nanostructures has become increasingly essential. Materials scientists can now better grasp how physicochemical properties relate to hazard/safety profiles (Lin et al., 2018).

Interaction between the nanomaterials and living organisms is affected by several important factors, such as electrostatic and van der Waals forces, surface characteristics, charge/charge density, functional groups, and surface defects. To check the toxicity profile of nanomaterials, it is essential to observe nanomaterial charge density and distribution. Due to the positively charged nanomaterials being more hazardous than negatively charged ones, there is great attraction toward negatively charged cell membranes (Ma et al., 2013; Lin et al., 2018). Plate-shaped Ag nanomaterials were shown to be highly toxic due to the high level of crystal defects on the nanomaterial surface (Lin et al., 2018).

Recently there have been several reported research works on safer-by-design approaches, including particle doping and surface passivation. The first effort to reduce the hazard potential of nanomaterials due to metal dissolution was demonstrated through Fe doping of ZnO nanomaterials (Xie et al., 2011). Fe was incorporated into the lattice structure of ZnO nanomaterials via the flame spray pyrolysis. The presence of a stable dopant, in this case, Fe, in the ZnO lattice reduced the rate of Zn ion dissolution considerably for the doped particles compared to the undoped ZnO. It should be noted that particle doping may not be a universal safer-by-design approach for every nanomaterial. Particle doping has demonstrated in some instances to increase toxicity. For example, Fe doping could shift the photoactivation spectrum of TiO_2 nanomaterials from UV to visible, which is a desirable characteristic for photocatalysts. However, the photoinduced toxicity of Fe-doped TiO_2 was more significant than undoped TiO_2 (Lin et al., 2018).

Cell-based assays are the standard method for testing compounds for cytotoxicity *in vitro* due to their ease of use, sensitivity, and low cost; however, when used with nanomaterials, their limitations can produce misleading results. A real-time cell monitoring tool suitable for mammalian cell lines has been created based on cell impedance. Traditional cytotoxicity experiments were used to validate cytotoxicity evaluations. These findings demonstrate that the cell impedance measuring method is more effective and appropriate for assessing the cytotoxicity of nanomaterials for environmental safety screening (Tripathi et al., 2017).

Carbon nanotubes and nanomaterials based on Rare Earth Oxide (REO) have recently been illustrated to cause fibrogenic effects in cells and animal lungs (Lin et al., 2018). The physicochemical characteristics of carbon nanotubes that led to toxicity generation were surface hydrophobicity and a high aspect ratio. Amphiphilic copolymers, such as Pluronic P108, have proven to be a promising safer-by-design approach for these nanomaterials, effectively coating the surface and shielding nanotubes from the membrane (Lin et al., 2018).

The Biannual ECOtoxicology Meeting 2016 (BECOME) in Livorno, Italy, included a session on "Future views and ecotoxicological assessment of nano remediation carried out to infected sediments and soils." Only some subjects were mentioned, and beginning from the modern-day country of the artwork of nano remediation, which represents a step forward in pollutants control, the subsequent guidelines were proposed. The following points should be taken into consideration when designing ENMs for environmental remediation: (i) eco-safety should be a top priority; (ii) ENMs for environmental remediation must undergo a predictive safety assessment; (iii) greener, more innovative, and environmentally friendly nanostructured materials should be encouraged; and (iii) ENMs that adhere to the strictest environmental safety standards will support industrial competitiveness, innovation, and sustainability (Corsi et al., 2018).

Developing sustainable and safe handling techniques to create and regulate nanomaterials through legislation or non-binding suggestions and guidance is essential. The regulation of NMs is one of the many laws focused on the EU or other countries (Arts et al., 2014). Many expert organizations, including the EU Scientific groups and Agencies, the Organisation for Economic Cooperation and Development (OECD), the International Standard Organization (ISO), and the US Food and Drug Administration (FDA), are active in this field and have acknowledged the need for additional guidance to assess potential risks and recommendations to ensure the safe use of NMs (Amenta et al., 2015; Kaikkonen, et al., 2021).

2. Conclusion

Understanding the fate and behaviour of nanomaterials in the environment is extremely important before their wide commercial applications in various fields due to their unique characteristics. In the past 10–15 yr, however some progress has been made to understand the environmental safety of nanomaterials (NMs), there is more research needed to fill the gap. In this regard, few testing strategies/ protocols have been provided to estimate the involved in NMs, their exposure to the organisms in the environment, and in some cases detrimental long-term effects for at least a few abundant first-generation NMs. To access the fate, transfer, and toxicological aspects of the NMs over time, in-depth mechanistic studies should be performed. Thus, toxicity modelling assessment of NMs in the environment, their fate, transfer, and long term effect must be studied intensively to identify the primary mechanism involved with the fate and toxicity of NMs. Besides that, the increased use of nanomaterials in various fields could have toxicological consequences. Several governmental and non-governmental research organizations are developing different modelling and analytical techniques to measure the effect and behaviour of nanomaterials. To mitigate these toxicological effects, one should create safer-by-design materials for environmental safety prospects for various ecosystems and humans.

References

Abid, N., Khan, A.M., Shujait, S., Chaudhary, K., Ikram, M., Imran, M. and Maqbool, M. 2022. Synthesis of nanomaterials using various top-down and bottom-up approaches, influencing factors, advantages, and disadvantages: A review. Advances in Colloid and Interface Science 300: 102597.

Adam, V. and Nowack, B. 2017. European country-specific probabilistic assessment of nanomaterial flows towards landfilling, incineration, and recycling. Environmental Science: Nano 4(10): 1961–1973.

Adeleye, A.S., Pokhrel, S., Mädler, L. and Keller, A.A. 2018. Influence of nanomaterial doping on the colloidal stability and toxicity of copper oxide nanomaterials in synthetic and natural waters. Water Research 132: 12–22.

Adisa, I.O., Pullagurala, V.L.R., Peralta-Videa, J.R., Dimkpa, C.O., Elmer, W.H., Gardea-Torresdey, J.L. et al. 2019. Recent advances in nano-enabled fertilizers and pesticides: a critical review of mechanisms of action. Environmental Science: Nano 6(7): 2002–2030.

Adisa, I.O., Pullagurala, V.L.R., Peralta-Videa, J.R., Dimkpa, C.O., Elmer, W.H., Gardea-Torresdey, J.L. et al. 2019. Recent advances in nano-enabled fertilizers and pesticides: a critical review of mechanisms of action. Environmental Science: Nano 6(7): 2002–2030.

Adisa, I.O., Pullagurala, V.L.R., Peralta-Videa, J.R., Dimkpa, C.O., Elmer, W.H., Gardea-Torresdey, J.L. et al. 2019. Recent advances in nano-enabled fertilizers and pesticides: a critical review of mechanisms of action. Environmental Science: Nano 6(7): 2002–2030.

Ahmed, H.M., Roy, A., Wahab, M., Ahmed, M., Othman-Qadir, G., Elesawy, B.H. et al. 2021. Applications of nanomaterials in agrifood and pharmaceutical industry. Journal of Nanomaterials 2021: 1–10.

Alabdallah, N.M. and Hasan, M.M. 2021. Plant-based green synthesis of silver nanomaterials and its influential role in abiotic stress tolerance in crop plants. Saudi Journal of Biological Sciences 28(10): 5631–5639.

Ali, S., Mehmood, A. and Khan, N. 2021. Uptake, translocation, and consequences of nanomaterials on plant growth and stress adaptation. Journal of Nanomaterials 2021: 1–17.

Amenta, V., Aschberger, K., Arena, M., Bouwmeester, H., Moniz, F.B., Brandhoff, P. et al. 2015. Regulatory aspects of nanotechnology in the agri/feed/food sector in EU and non-EU countries. Regulatory Toxicology and Pharmacology 73(1): 463–476.

Arts, J.H., Hadi, M., Keene, A.M., Kreiling, R., Lyon, D., Maier, M. et al. 2014. A critical appraisal of existing concepts for the grouping of nanomaterials. Regulatory Toxicology and Pharmacology 70(2): 492–506.

Arvidsson, R., Baun, A., Furberg, A., Hansen, S.F. and Molander, S. 2018. Proxy measures for simplified environmental assessment of manufactured nanomaterials. Environmental Science & Technology 52(23): 13670–13680.

Asha, A.B. and Narain, R. 2020. Nanomaterials properties. pp. 343–359. *In*: Polymer Science and Nanotechnology. Elsevier.

Baalousha, M., Cornelis, G., Kuhlbusch, T.A.J., Lynch, I., Nickel, C., Peijnenburg, W.J.G.M. et al. 2016. Modeling nanomaterial fate and environmental uptake: current knowledge and future trends. Environmental Science: Nano 3(2): 323–345.

Baalousha, M., Cornelis, G., Kuhlbusch, T.A.J., Lynch, I., Nickel, C., Peijnenburg, W.J.G.M. et al. 2016. Modeling nanomaterial fate and uptake in the environment: current knowledge and future trends. Environmental Science: Nano 3(2): 323–345.

Baig, M.M., Zulfiqar, S., Yousuf, M.A., Shakir, I., Aboud, M.F.A. and Warsi, M.F. 2021. DyxMnFe2-xO4 nanomaterials decorated over mesoporous silica for environmental remediation applications. Journal of Hazardous Materials 402: 123526.

Batley, G.E., Kirby, J.K. and McLaughlin, M.J. 2013. Fate and risks of nanomaterials in aquatic and terrestrial environments. Accounts of Chemical Research 46(3): 854–862.

Baun, A., Sayre, P., Steinhaeuser, K.G. and Rose, J. 2017. Regulatory relevant and reliable methods and data for determining the environmental fate of manufactured nanomaterials. NanoImpact 8: 1–10.

Ben-Moshe, T., Frenk, S., Dror, I., Minz, D. and Berkowitz, B. 2013. Effects of metal oxide nanomaterials on soil properties. Chemosphere 90(2): 640–646.

Binh, C.T.T., Peterson, C.G., Tong, T., Gray, K.A., Gaillard, J.F. and Kelly, J.J. 2015. Comparing acute effects of a nano-TiO2 pigment on cosmopolitan freshwater phototrophic microbes using high-throughput screening. PLoS One 10(4): e0125613.

Bossa, N., Chaurand, P., Levard, C., Borschneck, D., Miche, H., Vicente, J. et al. 2017. Environmental exposure to TiO2 nanomaterials incorporated in building material. Environmental Pollution 220: 1160–1170.

Bundschuh, M., Filser, J., Lüderwald, S., McKee, M.S., Metreveli, G., Schaumann, G.E. et al. 2018. Nanomaterials in the environment: where do we come from, where do we go to? Environmental Sciences Europe 30(1): 1–17.

Chae, S.R., E.M. Hotze, A.R. Badireddy, S. Lin, J.O. Kim and M.R. Wiesner. 2013. Environmental implications and applications of carbon nanomaterials in water treatment. Water Science and Technology 67(11): 2582–2586.

Chae, Y. and An, Y.J. 2016. Toxicity and transfer of polyvinylpyrrolidone-coated silver nanowires in an aquatic food chain consisting of algae, water fleas, and zebrafish. Aquatic Toxicology 173: 94–104.

Chong, W.C., Chung, Y.T., Teow, Y.H., Zain, M.M., Mahmoudi, E. and Mohammad, A.W. 2018. Environmental impact of nanomaterials in composite membranes: Life cycle assessment of algal membrane photoreactor using polyvinylidene fluoride–composite membrane. Journal of Cleaner Production 202: 591–600.

Corsi, I., Winther-Nielsen, M., Sethi, R., Punta, C., Della Torre, C., Libralato, G. et al. 2018. Ecofriendly nanotechnologies and nanomaterials for environmental applications: Key issue and consensus recommendations for sustainable and eco-safe nano remediation. Ecotoxicology and Environmental Safety 154: 237–244.

Cucurachi, S. and Rocha, C.F.B. 2019. Life-cycle assessment of engineered nanomaterials. pp. 815–846. *In*: Nanotechnology in Eco-efficient Construction. Woodhead Publishing.

Dekkers, S., Oomen, A.G., Bleeker, E.A., Vandebriel, R.J., Micheletti, C., Cabellos, J. and Wijnhoven, S. W. 2016. Towards a nonspecific approach for risk assessment. Regulatory Toxicology and Pharmacology 80: 46–59.

Fadeel, B., Farcal, L., Hardy, B., Vázquez-Campos, S., Hristozov, D., Marcomini, A. et al. 2018. Advanced tools for the safety assessment of nanomaterials. Nature Nanotechnology 13(7): 537–543.

Finkbeiner, M., Inaba, A., Tan, R., Christiansen, K. and Klüppel, H.J. 2006. The new international standards for life cycle assessment: ISO 14040 and ISO 14044. The International Journal of Life Cycle Assessment 11: 80–85.

Furtado, L.M., Bundschuh, M. and Metcalfe, C.D. 2016. Monitoring the fate and transformation of silver nanomaterials in natural waters. Bulletin of Environmental Contamination and Toxicology 97: 449–455.

Gajewicz, A., Rasulev, B., Dinadayalane, T.C., Urbaszek, P., Puzyn, T., Leszczynska, D. and Leszczynski, J. 2012. Advancing risk assessment of engineered nanomaterials: application of computational approaches. Advanced Drug Delivery Reviews 64(15): 1663–1693.

Gajewicz, A., Schaeublin, N., Rasulev, B., Hussain, S., Leszczynska, D., Puzyn, T. et al. 2015. Towards understanding mechanisms governing cytotoxicity of metal oxides nanomaterials: Hints from nano-QSAR studies. Nanotoxicology 9(3): 313–325.

Gavankar, S., Anderson, S. and Keller, A.A. 2015. Critical components of uncertainty communication in life cycle assessments of emerging technologies: nanotechnology as a case study. Journal of Industrial Ecology 19(3): 468–479.

Gilbertson, L.M., Wender, B.A., Zimmerman, J.B. and Eckelman, M.J. 2015. Coordinating modeling and experimental research of engineered nanomaterials to improve life cycle assessment studies. Environmental Science: Nano 2(6): 669–682.

Gogos, A., Moll, J., Klingenfuss, F., van der Heijden, M., Irin, F., Green, M.J. et al. 2016. Vertical transport and plant uptake of nanomaterials in a soil mesocosm experiment. Journal of Nanobiotechnology 14(1): 1–11.

Gottschalk, F., Sun, T. and Nowack, B. 2013. Environmental concentrations of engineered nanomaterials: review of modeling and analytical studies. Environmental Pollution 181: 287–300.

Griffin, S., Masood, M.I., Nasim, M.J., Sarfraz, M., Ebokaiwe, A.P., Schäfer, K.H. et al. 2017. Natural nanomaterials: a particular matter inspired by nature. Antioxidants 7(1): 3.

Grün, A.L., Manz, W., Kohl, Y.L., Meier, F., Straskraba, S., Jost, C. et al. 2019. Impact of silver nanomaterials (AgNP) on soil microbial community depending on functionalization, concentration, exposure time, and soil texture. Environmental Sciences Europe 31(1): 1–22.

Guinée, J.B., Huppes, G. and Heijungs, R. 2001. Developing an LCA guide for decision support. Environmental Management and Health 12(3): 301–311.

Hafez, E.M., Osman, H.S., Gowayed, S.M., Okasha, S.A., Omara, A.E.D., Sami, R. et al. 2021. Minimizing the adverse impacts of water deficit and soil salinity on maize growth and productivity in response to the application of plant growth-promoting rhizobacteria and silica nanomaterials. Agronomy 11(4): 676.

Harikumar, P.S. and Aravind, A. 2016. Antibacterial activity of copper nanomaterials and copper nanocomposites against Escherichia coli bacteria. Int. J. Sci. 2: 83–90.

Hashem, A.H., Abdelaziz, A.M., Askar, A.A., Fouda, H.M., Khalil, A.M., Abd-Elsalam, K.A. et al. 2021. Bacillus megaterium-mediated synthesis of selenium nanomaterials and their antifungal activity against Rhizoctonia solani in faba bean plants. Journal of Fungi 7(3): 195.

He, X., W.G. Aker, P.P. Fu and H.M. Hwang. 2015. Toxicity of engineered metal oxide nanomaterials mediated by nano–bio–eco–interactions: A review and perspective. Environmental Science Nano 2: 564–582.

He, Y., Liu, J.C., Luo, L., Wang, Y.G., Zhu, J., Du, Y. et al. 2018. Size-dependent dynamic structures of supported gold nanomaterials in CO oxidation reaction condition. Proceedings of the National Academy of Sciences 115(30): 7700–7705.

Hischier, R. and Walser, T. 2012. Life cycle assessment of engineered nanomaterials: State of the art and strategies to overcome existing gaps. Science of the Total Environment 425: 271–282.

Hristozov, D.R., Zabeo, A., Foran, C., Isigonis, P., Critto, A., Marcomini, A. et al. 2014. A weight of evidence approach for hazard screening of engineered nanomaterials. Nanotoxicology 8(1): 72–87.

Hristozov, D., Gottardo, S., Semenzin, E., Oomen, A., Bos, P., Peijnenburg, W. et al. 2016. Frameworks and tools for risk assessment of manufactured nanomaterials. Environment International 95: 36–53.

Ismail, H. and Hanafiah, M.M. 2019. An overview of LCA application in WEEE management: Current practices, progress, and challenges. Journal of Cleaner Production 232: 79–93.

Jacobson, K.H., Gunsolus, I.L., Kuech, T.R., Troiano, J.M., Melby, E.S., Lohse, S.E. et al. 2015. Lipopolysaccharide density and structure govern the extent and distance of nanomaterial interaction with actual and model bacterial outer membranes. Environmental Science & Technology 49(17): 10642–10650.

Judy, J.D. and Bertsch, P.M. 2014. Bioavailability, toxicity, and fate of manufactured nanomaterials in terrestrial ecosystems. Advances in Agronomy 123: 1–64.

Kaikkonen, L., Parviainen, T., Rahikainen, M., Uusitalo, L. and Lehikoinen, A. 2021. Bayesian networks in environmental risk assessment: A review. Integrated Environmental Assessment and Management 17(1): 62–78.

Keller, A.A., McFerran, S., Lazareva, A. and Suh, S. 2013. Global life cycle releases of engineered nanomaterials. Journal of Nanomaterial Research, 15: 1–17.

Khan, I., Saeed, K. and Khan, I. 2019. Nanomaterials: Properties, applications, and toxicities. Arabian Journal of Chemistry, 12(7): 908–931.

Khoso, W.A., Haleem, N., Baig, M.A. and Jamal, Y. 2021. Synthesis, characterization, and heavy metal removal efficiency of nickel ferrite nanomaterials (NFN's). Scientific Reports 11(1): 3790.

Kumar, P., Senthamil Selvi, S., Lakshmi Prabha, A., Prem Kumar, K., Ganeshkumar, R.S. and Govindaraju, M. 2012. Synthesis of silver nanomaterials from *Sargassum tentorium* and screening phytochemicals for its antibacterial activity. Nano Biomed. Eng. 4(1): 12–16.

Kumar, S., Shukla, A., Baul, P.P., Mitra, A. and Halder, D. 2018. Biodegradable hybrid nanocomposites of chitosan/gelatin and silver nanomaterials for active food packaging applications. Food Packaging and Shelf Life 16: 178–184.

Lead, J.R., Batley, G.E., Alvarez, P.J., Croteau, M.N., Handy, R.D., McLaughlin, M.J. et al. 2018. Nanomaterials in the environment: behavior, fate, bioavailability, and effects—an updated review. Environmental Toxicology and Chemistry 37(8): 2029–2063.

Lee, J., Yang, J., Kwon, S.G. and Hyeon, T. 2016. Nonclassical nucleation and growth of inorganic nanomaterials. Nature Reviews Materials 1(8): 1–16.

Lin, S., Yu, T., Yu, Z., Hu, X. and Yin, D. 2018. Nanomaterials safer-by-design: an environmental safety perspective. Advanced Materials 30(17): 1705691.

Linkov, I. and Seager, T.P. 2011. Coupling multi-criteria decision analysis, life-cycle assessment, and risk assessment for emerging threats.

Liu, X., Wang, D. and Li, Y. 2012. Synthesis and catalytic properties of bimetallic nanomaterials with various architectures. Nano Today 7(5): 448–466.

Luoma, S.N., Stoiber, T., Croteau, M.N., Römer, I., Merrifeld, R. and Lead, J.R. 2016. Effect of cysteine and humic acids on the bioavailability of Ag from Ag nanomaterials to a freshwater snail. NanoImpact 2: 61–69.

Ma, N., Ma, C., Li, C., Wang, T., Tang, Y., Wang, H. et al. 2013. Influence of nanomaterial shape, size, and surface functionalization on cellular uptake. Journal of Nanoscience and Nanotechnology 13(10): 6485–6498.

Merrifield, R.C., Arkill, K.P., Palmer, R.E. and Lead, J.R. 2017. A high-resolution study of dynamic changes of Ce2O3 and CeO2 nanomaterials in complex environmental media. Environmental Science & Technology 51(14): 8010–8016.

Metreveli, G., Frombold, B., Seitz, F., Grün, A., Philippe, A., Rosenfeldt, R.R. et al. 2016. Impact of the chemical composition of ecotoxicological test media on the stability and aggregation status of silver nanomaterials. Environmental Science: Nano 3(2): 418–433.

Mitrano, D.M., Motellier, S., Clavaguera, S. and Nowack, B. 2015. Review of nanomaterial aging and transformations through the life cycle of nano-enhanced products. Environment International 77: 132–147.

Mittal, D., Kaur, G., Singh, P., Yadav, K. and Ali, S.A. 2020. Nanomaterial-based sustainable agriculture and food science: Recent advances and future outlook. Frontiers in Nanotechnology 2: 579954.

Nizam, N.U.M., Hanafiah, M.M. and Woon, K.S. 2021. A content review of life cycle assessment of nanomaterials: Current practices, challenges, and future prospects. Nanomaterials 11(12): 3324.

Oomen, A.G., Bos, P.M., Fernandes, T.F., Hund-Rinke, K., Boraschi, D., Byrne, H.J. et al. 2014. Concern-driven integrated approaches to nanomaterial testing and assessment–report of the NanoSafety Cluster Working Group 10. Nanotoxicology 8(3): 334–348.

Pacheco, I. and Buzea, C. 2016. Nanomaterial Toxicity. pp. 273–324. *In*: Advanced Environmental Analysis.

Paramasivam, G., Palem, V.V., Sundaram, T., Sundaram, V., Kishore, S.C. and Bellucci, S. 2021. Nanomaterials: Synthesis and applications in theranostics. Nanomaterials 11(12): 3228.

Pawlett, M., Ritz, K., Dorey, R.A., Rocks, S., Ramsden, J. and Harris, J.A. 2013. The impact of zero-valent iron nanomaterials upon soil microbial communities is context-dependent. Environmental Science and Pollution Research 20: 1041–1049.

Pelletier, D.A., Suresh, A.K., Holton, G.A., McKeown, C.K., Wang, W., Gu, B. et al. 2010. Effects of engineered cerium oxide nanomaterials on bacterial growth and viability. Applied and Environmental Microbiology 76(24): 7981–7989.

Petersen, E.J., Diamond, S.A., Kennedy, A.J., Goss, G.G., Ho, K., Lead, J. et al. 2015. Adapting OECD aquatic toxicity tests for use with manufactured nanomaterials: key issues and consensus recommendations. Environmental Science & Technology 49(16): 9532–9547.

Quik, J.T., Velzeboer, I., Wouterse, M., Koelmans, A.A. and Van de Meent, D. 2014. Heteroaggregation and sedimentation rates for nanomaterials in natural waters. Water Research 48: 269–279.

Rajput, V., Minkina, T., Mazarji, M., Shende, S., Sushkova, S., Mandzhieva, S. et al. 2020. Accumulation of nanomaterials in the soil-plant systems and their effects on human health. Annals of Agricultural Sciences 65(2): 137–143.

Raliya, R., Franke, C., Chavalmane, S., Nair, R., Reed, N. and Biswas, P. 2016. Quantitative understanding of nanomaterial uptake in watermelon plants. Frontiers in Plant Science 7: 1288.

Rasheed, T. 2022. Magnetic nanomaterials: Greener and sustainable alternatives for the adsorption of hazardous environmental contaminants. Journal of Cleaner Production 132338.

Rubilar, O., Rai, M., Tortella, G., Diez, M.C., Seabra, A.B. and Durán, N. 2013. Biogenic nanomaterials: copper, copper oxides, copper sulfides, complex copper nanostructures and their applications. Biotechnology Letters 35: 1365–1375.

Sampathkumar, K., Tan, K.X. and Loo, S.C.J. 2020. Developing nano-delivery systems for agriculture and food applications with nature-derived polymers. Iscience 23(5): 101055.

Seitz, F., Rosenfeldt, R.R., Schneider, S., Schulz, R. and Bundschuh, M. 2014. Size-, surface-and crystalline structure composition-related effects of titanium dioxide nanomaterials during their aquatic life cycle. Science of the Total Environment 493: 891–897.

Semenzin, E., Lanzellotto, E., Hristozov, D., Critto, A., Zabeo, A., Giubilato, E. et al. 2015. Species sensitivity weighted distribution for ecological risk assessment of engineered nanomaterials: The n-TiO2 case study. Environmental Toxicology and Chemistry 34(11): 2644–2659.

Sharifianjazi, F., Irani, M., Esmaeilkhanian, A., Bazli, L., Asl, M.S., Jang, H.W. et al. 2021. Polymer incorporated magnetic nanomaterials: Applications for magnetoresponsive targeted drug delivery. Materials Science and Engineering: B 272: 115358.

Singh, A., Gautam, P. K., Verma, A., Singh, V., Shivapriya, P. M., Shivalkar, S. et al. 2020. Green synthesis of metallic nanomaterials as effective alternatives to treat antibiotics resistant bacterial infections: A review. Biotechnology Reports 25: e00427.

Singh, J., Kumar, V., Kim, K.H. and Rawat, M. 2019. Biogenic synthesis of copper oxide nanomaterials using plant extract and its prodigious potential for photocatalytic degradation of dyes. Environmental Research 177: 108569.

Sørensen, S.N., Baun, A., Burkard, M., Dal Maso, M., Hansen, S.F., Harrison, S. et al. 2019. Evaluating environmental risk assessment models for nanomaterials according to requirements along the product innovation Stage-Gate process. Environmental Science: Nano 6(2): 505–518.

Subramanian, V., Semenzin, E., Hristozov, D., Zabeo, A., Malsch, I., McAlea, E. et al. 2016. Sustainable nanotechnology decision support system: bridging risk management, sustainable innovation, and risk governance. Journal of Nanomaterial Research 18: 1–13.

Sun, T.Y., Gottschalk, F., Hungerbühler, K. and Nowack, B. 2014. Comprehensive probabilistic modeling of environmental emissions of engineered nanomaterials. Environmental Pollution 185: 69–76.

Toncelli, C., Mylona, K., Kalantzi, I., Tsiola, A., Pitta, P., Tsapakis, M. et al. 2017. Silver nanomaterials in seawater: A dynamic mass balance at part per trillion silver concentrations. Science of the Total Environment 601: 15–21.

Tripathi, D.K., Singh, S., Singh, S., Pandey, R., Singh, V.P., Sharma, N.C. et al. 2017. An overview on manufactured nanomaterials in plants: uptake, translocation, accumulation, and phytotoxicity. Plant Physiology and Biochemistry 110: 2–12.

Tripathi, D.K., Tripathi, A., Singh, S., Singh, Y., Vishwakarma, K., Yadav, G. et al. 2017. Uptake, accumulation, and toxicity of silver nanomaterial in autotrophic plants, and heterotrophic microbes: a concentric review. Frontiers in Microbiology 8: 07

Tsiola, A., Pitta, P., Callol, A.J., Kagiorgi, M., Kalantzi, I., Mylona, K. et al. 2017. The impact of silver nanomaterials on marine plankton dynamics: Dependence on coating, size, and concentration. Science of the Total Environment 601: 1838–1848.

Wagner, S., Gondikas, A., Neubauer, E., Hofmann, T. and von der Kammer, F. 2014. Spot the difference: engineered and natural nanomaterials in the environment—release, behavior, and fate. Angewandte Chemie International Edition 53(46): 12398–12419.

Westerhoff, P. and Nowack, B. 2013. Searching for global descriptors of engineered nanomaterial fate and transport in the environment. Accounts of Chemical Research 46(3): 844–853.

Wigger, H., Kägi, R., Wiesner, M. and Nowack, B. 2020. Exposure and possible risks of engineered nanomaterials in the environment—Current knowledge and directions for the future. Reviews of Geophysics 58(4): e2020RG000710.

Williams, R.J., Harrison, S., Keller, V., Kuenen, J., Lofts, S., Praetorius, A. et al. 2019. Models for assessing engineered nanomaterial fate and behaviour in the aquatic environment. Current Opinion in Environmental Sustainability 36: 105–115.

Xavier, M., Parente, I.A., Rodrigues, P.M., Cerqueira, M.A., Pastrana, L. and Gonçalves, C. 2021. Safety and fate of nanomaterials in food: The role of *in vitro* tests. Trends in Food Science & Technology, 109: 593–607.

Xie, Y., He, Y., Irwin, P. L., Jin, T. and Shi, X. 2011. Antibacterial activity and mechanism of action of zinc oxide nanomaterials against Campylobacter jejuni. Applied and Environmental Microbiology 77(7): 2325–2331.

Xu, P., Zeng, G.M., Huang, D.L., Feng, C.L., Hu, S., Zhao, M.H. et al. 2012. Use of iron oxide nanomaterials in wastewater treatment: a review. Science of the Total Environment 424: 1–10.

Yang, G., Xie, J., Hong, F., Cao, Z. and Yang, X. 2012. Antimicrobial activity of silver nanomaterial impregnated bacterial cellulose membrane: effect of fermentation carbon sources of bacterial cellulose. Carbohydrate Polymers 87(1): 839–845.

Yin, I.X., Zhang, J., Zhao, I.S., Mei, M.L., Li, Q. and Chu, C.H. 2020. The antibacterial mechanism of silver nanomaterials and its application in dentistry. International Journal of Nanomedicine 2555–2562.

CHAPTER 9

Technologies for Treatment of Emerging Contaminants

Pricila Nass, Patrícia Caetano, Richard Machado,
*Leila Queiroz Zepka and Eduardo Jacob-Lopes**

1. Introduction

The shortcoming of public policies and the precariousness of sanitation services, added to the uncontrolled population growth, have been identified as the main factors responsible for the decreased quality of water resources (Xu et al., 2022).

Considering these aspects, until two decades ago concerns about water quality focused on contaminants that caused color, odor, turbidity, and microorganisms that could alter water properties. Today, the preoccupation is greater because even the treated effluent can contain other harmful components, such as emerging contaminants, which have a low concentration can cause damage to the environment (Puri et al., 2023).

Emerging Contaminants (ECs) represent a wide range of compounds of natural or anthropic origin, such as pesticides, pharmaceuticals, personal care products, products from water disinfection processes, and cyanotoxins, whose effects on the environment and human health are still little understood (Wee and Ahmad, 2019).

Therefore, the input of ECs into the aquatic environment is a matter of concern, as these compounds tend to bioaccumulate in organisms, through the biomagnification process, due to their recalcitrant properties and lipophilic characteristics (Yan et al., 2014).

Thus, several studies have evaluated the toxicity of ECs, results suggest that the presence of these compounds in the environment could be related to the origin of resistant bacteria and the disruption of the endocrine system of organisms. For this reason, the occurrence of ECs in the environment, even in low concentrations, is associated with potential harm to the ecosystem and human health (Xu et al., 2020).

Department of Food, Technology and Science, Federal University of Santa Maria (UFSM), Santa Maria, RS, Brazil.
* Corresponding author: ejacoblopes@gmail.com

The occurrence of these compounds in fresh, salty/brackish, groundwater, and even public supply waters has already been reported in several countries with different concentrations. Thus, the removal of ECs is generally not efficient by conventional processes present in water treatment plants (Chaturvedi et al., 2021; Yao et al., 2021).

Based on this understanding, microalgae have been reported for ECs removal, including *Chlamydomonas reinhardtii, Chlorella vulgaris, Chlorella pyrenoidosa, Coelastrum* sp., *Tetraesmus dimorphus* and *Scenedesmus obliquus* (Zhou et al., 2022). Although ECs inhibit their growth, improving the removal efficiency of emerging contaminants by microalgae is a pressing issue to be solved (Alvarez-Gonz et al., 2023).

Based on this, the chapter summarizes the emerging contaminants, conventional treatment processes for emerging contaminants removal, as well as the application of microalgae in bioremediation, and finally, regulatory aspects.

2. Emerging Contaminants

The occurrence of ECs in the environment is a worldwide issue. Emerging contaminants consist of a range of natural and synthetic chemical compounds. These, have physical and chemical properties such as persistent volatility, or lipophilic, and can affect both the ecosystem and the health and quality of life of living beings (Puri et al., 2023).

Present in wastewater, ECs have low concentrations ($ng.L^{-1}$ to $\mu g.L^{-1}$) and diversity of compounds, complicating detection and analysis procedures, in addition to creating challenges for water treatment processes (Rizzo et al., 2019).

Several groups of substances have been considered ECs, including pesticides, pharmaceuticals, personal care products, products from water disinfection processes, and cyanotoxins. Table 9.1 summarizes research data on ECs in the environment and their effects on humans and biota.

Pesticides are compounds intended for agriculture to prevent or reduce the effects caused by pests, diseases, or weeds. These substances are synthetic organic compounds with low molecular weight, generally with low solubility in water and high biological activity. The term includes all insecticides, fungicides, herbicides, and organic compounds used as growth regulators, defoliants, or desiccants (Degrendele et al., 2022).

Another contaminant is pharmaceuticals, biologically active chemical substances synthesized in order to produce physiological responses in humans and animals. These compounds can cause harmful effects on aquatic fauna, and cause various morphological and even metabolic damages. However, there is particular concern about antibiotics. Research has shown that these contaminants when discarded into the environment can cause biological toxicity, induction of antibiotic resistance in pathogenic bacteria, and genotoxicity, which can be defined as the ability of some chemical substances to produce genetic alterations (Nieto-Juárez et al., 2021).

Personal use products include cosmetics, fragrances, repellent insecticides, and sunscreens. Many compounds used in these products are fat-soluble and therefore have a high potential for bioaccumulation (Ebele et al., 2020).

Table 9.1 Occurrence and effects of emerging contaminants in the environment.

Class	Environmental matrice	Concentration detected (ng.L⁻¹)	Analytical technique	Effects on human and biota
Pesticides	Drinking water	0,2–2600	LC-MS/MS HPLC-UV HRGC-ECD GC-ECD	Deleterious effects on fish gills; feminization of aquatic organisms; reproductive and sexual systems of humans are severely affected
Pharmaceuticals	Drinking water	18,5	LC-MS/MS	Genotoxicity, neurotoxicity, and oxidative stress in molluscs; reduced algae community growth; disruption with hormones
Personal care products	Drinking water	18–135,5	LC-MS/MS GC-MS/MS	Induce vitellogenin production in juvenile rainbow trout, reduction on plasma testosterone by over 50% in goldfish, affecting follicular growth, fertilization, and implantation in females
Water disinfection products	Drinking water	100–41000	GC-ECD	Dysregulation of thyroid hormones and adverse kidney health in humans

Some compounds for industrial use are also classified as emerging contaminants. Among them bisphenol A, alkylphenols, polychlorinated biphenyls, phthalates, and perfluorinated compounds. Most have liposoluble characteristics and some of them are even classified as persistent organic pollutants (Jha et al., 2021).

Considered a contaminant of natural origin, cyanotoxins are secondary metabolites of cyanobacteria, whose occurrence is aggravated due to the release of wastewater into water bodies that increase a load of nutrients, considerably increasing the cellular reproduction of cyanobacteria. Known cyanotoxins have three main targets: hepatotoxins, neurotoxins and dermatotoxins (Mutoti et al., 2023).

Finally, water pollution with ECs represents a threat to the environment, however, the removal of these compounds is usually not efficient by conventional processes. Thus, there is a need for alternatives, linked to conventional technology for the removal of these substances.

3. Conventional Treatment Processes

In the last decade, many contaminants have been discovered in wastewater, surface water and even drinking water that should be treated to ensure the safety of the environment. Emerging contaminants include many synthetic or natural substances, detected in natural environments, from domestic, commercial, and industrial sources (Garcia-Rodríguez et al., 2014; Ahmed et al., 2017; Tran et al., 2018).

Given the current case , contaminants have received increasing attention for their ability to cause potential ecological effects, causing changes in marine environments, affecting environmental organisms and the sustainability of aquatic ecosystems.

In view of this, some treatments are used to decontaminate the ECs (Lima, 2018; Rout et al., 2021).

According to Ranjit et al. (2021), wastewater treatment allows the disposal of anthropic and industrial effluents without creating risks to human health or damage to the natural environment. The conventional treatment of effluents, carried out in Sewage Treatment Stations (STPs), occurs through a combination of physical, chemical, and biological processes, called preliminary, primary, secondary, tertiary and sludge treatment, which aim to eliminate these ECs. When applied in sequence, they increase the degree of treatment that is related to the type of contaminant to be removed (Crini and Lichtfouse, 2019).

The treatments applied in the STPs are based on the removal of pollutants such as suspended solids, oils, gases, organic materials, dissolved salts, heavy metals, dyes, nitrogen, and microorganisms. Thus, for each type of existing contaminant, there is a specific method, that is, the choice of method will depend, therefore, on the characteristics of the effluent, not only in terms of costs, but also in terms of practicality, reliability, environmental impact, production of sludge and its potential to form toxic waste. Based on this, Fig. 9.1 shows the treatments used in the STPs (Choi et al., 2022).

The levels of the steps applied in the treatment of effluents aim at the degree of removal of these pollutants. The primary treatment is responsible for the physical removal of effluents through grids and retention boxes, while the secondary removes fine solids in suspension, dispersed solids and organics that are removed by volatilization, biodegradation, and incorporation into the sludge. The purpose of

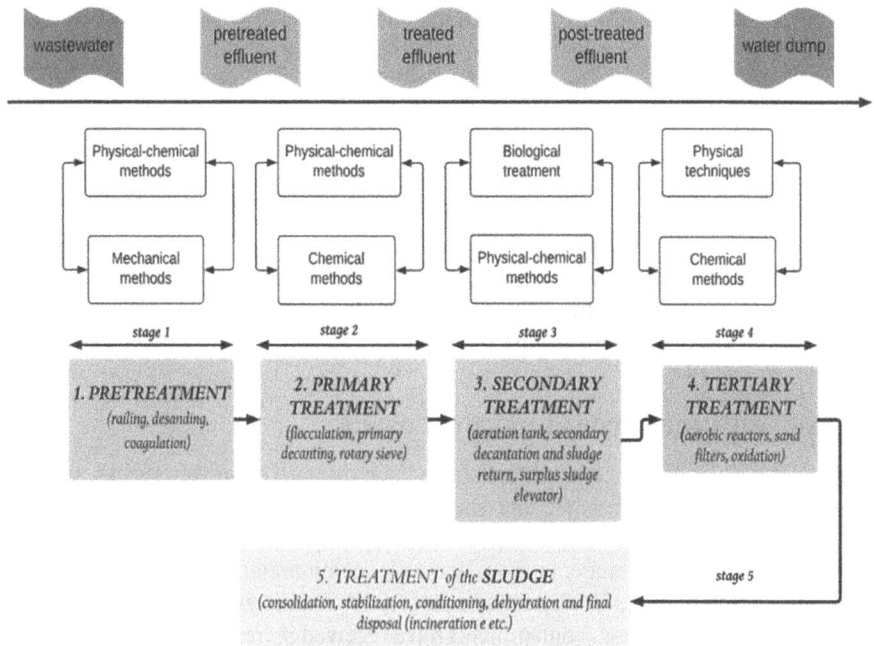

Figure 9.1 Treatment steps carried out in the STPs to control the decontamination of effluents and disposal of excess sludge.

tertiary or advanced treatment is to improve the quality of water discharged into natural waters, using a number of biological, physical, and chemical treatments (Zinicovscaia, 2016; Guillossou et al., 2019).

Sometimes, treatments applied to STPs result in a by-product with great potential for a pathogenic source, is called sludge. It is a solid or semi-solid material formed by a set of compounds such as inorganic and organic contaminants, hydrocarbons, microbial pollutants (pathogenic bacteria, etc.), which is the result of the process applied to wastewater, and can be formed by any other material dumped in the biological and chemical operation units during the treatment action (Lee et al., 2018).

Complex by nature, sludge treatment aims to reduce its volume, as it contains a large amount of water and organic materials that need stabilization. It can be classified as primary sludge formed by suspended solids, secondary sludge from waste removed from secondary decanters or settling ponds, and physical-chemical sludge generated during treatment with flocculants or coagulants (Santos and Lopes, 2022). After removing this contaminant, the sludge is reused, which can be done by different treatment methods applied in the destruction of harmful contaminants, so that it can be used as a source of energy and fertilizer (Puyol et al., 2017; Lanko et al., 2020; Chen et al., 2022).

Among the alternatives, linked to the treatment applied in the STPs for the removal of ECs are the adsorption processes with activated carbon, advanced oxidative processes and membrane filtration (Guo et al., 2021).

Among the advanced physical-chemical treatments, the adsorption processes are widely applied in the elimination of ECs. The most applied adsorbents are activated carbon, as they do not generate toxic products and have a high adsorption capacity (Guo et al., 2021).

Advanced oxidative processes are recognized as an alternative in the oxidation of ECs in the most diverse matrices, being adaptable to the most varied industrial processes. However, these oxidative processes are not selective and can lead to the formation of a wide range of by-products, which promotes the need for evaluating the toxicity of the effluent (Ajiboye et al., 2020).

Regarding membrane separation, reverse osmosis and nanofiltration processes can promote the removal of ECs. The efficiency of the application depends on the physicochemical properties of the target compounds, as well as on the operating conditions and the properties of the membranes. When comparing these two processes, it is observed that nanofiltration is less efficient than reverse osmosis, which is capable of promoting an almost complete removal of ECs, although the high energy consumption does not favor its use (Simmons et al., 2011).

Finally, while the removal effect of advanced treatment methods is satisfactory, it is difficult to apply these processes on a large scale due to their high energy consumption, uncontrollable degradation products, and low recovery rate. Thus, microalgae research on the removal of ECs from wastewater has received increasing attention, due to lower operating costs and higher bioenergy recovery (Zhou et al., 2022).

4. Microalgae in Bioremediation

There is a need for more sustainable and effective methods and microalgae have come up as a potential candidate for environment-friendly technologies. Wastewater treatment using microalgae, then assimilating high pollutant concentrations, exhibiting excellent energy-conversion efficiencies and with concomitant production of bioactive compounds (Beigbeder et al., 2021).

Photosynthetic microorganisms, but with metabolic versatility, are capable of recovering and recycling nutrients mainly inorganic Nitrogen (N) and Phosphate (P) contained in wastewater while producing biomass. Therefore, bioremediation using microalgae has been of growing interest, as it not only removes the pollutants but also purifies the water by producing oxygen (Maneechote et al., 2023). As an example, it is possible to produce about 1 and 10 kg.m^{-3} of dry biomass-based waste sewage and manure, respectively. This means that microalgae have been successfully used as a waste-based nutrient cycling technology, as illustrated in Fig. 9.2 (Sartori et al., 2022), thus making wastewater treatment a circular bioeconomy (Palafox-Sola et al., 2023).

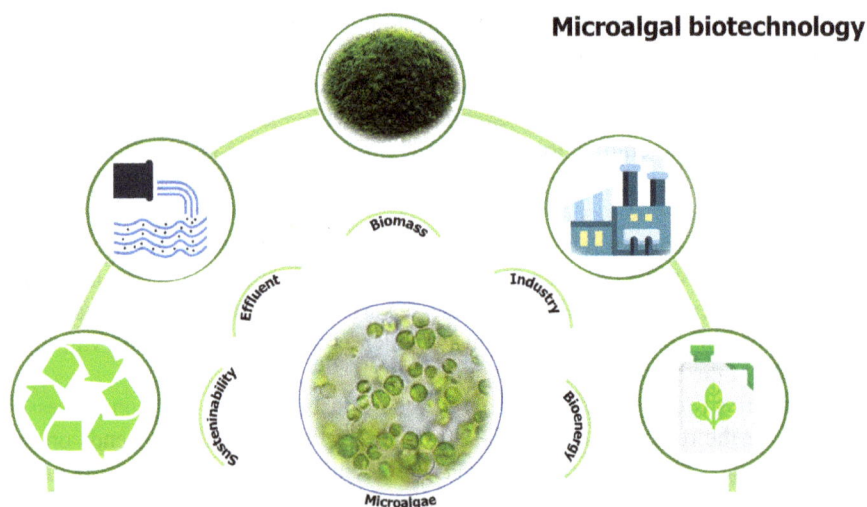

Figure 9.2 Microalgae are used as a waste-based nutrient recycling technology.

Traditional wastewater treatment methods as physical and chemical treatments are not very efficient in removing these contaminants, completely therefore a few of these pollutants are becoming persistent in the aquatic environments. Treatment processes generate several waste streams (i.e., sludge, chemical precipitates) that are difficult to dispose of and that cannot be re-circulated within the same process or revalorized. Moreover, while conventional wastewater treatment plants require costly aeration to the microbial communities, microalgae can uptake CO_2 from the atmosphere and provide aerobic microorganisms with oxygen (Nanda et al., 2021; Ajeng et al.,, 2022; Palafox-Sola et al., 2023).

Some species of microalgae more researched in the process of bioremediation, are *Chlorella* sp., *Scenedesmus* sp., *Nannochloropsis* sp., and *Desertifilum* spp., which have been confirmed to assimilate high pollutant concentrations and exhibit excellent energy conversion efficiencies. For example, it was reported earlier that 81.7% chemical oxygen demand and 96.2% were removed, with the production of 3078 mg.L^{-1} accumulating high carbohydrate content used for biobutanol production by Neochloris aquatica CL-M1 (Wang et al., 2017; Qu et al., 2019).

Microalgae are emerging as an effective bioremediation platform for heavy metals. In the study carried out by Tambat et al. (2023), between two microalgal strains, the *Chlorella sorokiniana* exhibited the highest growth rate and maximum removal of vanadium in mixotrophic cultive. The mixotrophic mode led to a higher growth rate as compared to autotrophic mode and increased removal efficiency by 292 and 66% for *Chlorella sorokiniana* and Picochlorum oklahomensis, with maximum biomass and lipid yield ranging between 2.5 and 3.0 g.L^{-1} and 26.6–29.5%, respectively. These findings support the practical feasibility of combined microalgal purification and energy production systems.

Lastly, several technologies are being developed to treat effluents. However, the conditions, types, and concentrations of ECs, in parallel with current legislative restrictions should be considered.

5. Regulatory Aspect

Emerging pollutant emissions are a global environmental problem due to limited regulations and the worldwide consensus that establishes legislation in this regard. The regulatory frameworks for emerging contaminants bring measures of continuous monitoring and regulation of maximum permissible limits in environmental matrices. To this end, several countries have implemented methodologies for ECs (Salimi et al., 2017; Wang et al., 2020).

In the United States, the Environmental Protection Agency (EPA) is the body responsible for monitoring and ensuring compliance with laws and regulations. The US EPA regulates 129 priority ECs. As for the European Union, it introduced a list of 33 priority ECs for surface waters (Barbosa et al., 2016; Bopp et al., 2019; Ramírez-Malule et al., 2020).

The Australian drinking water guidelines determine priority ECs that pose a greater risk to consumers, including perfluorinated chemicals (perfluorooctanoic acid, perfluoroctane sulfonic acid, and their salts). The guidelines are developed in accordance with EU and WHO guidelines and supplementary documents (Naidu et al., 2016).

In Brazil, there are no records of official programs focused on the problem of emerging contaminants. However, discussions in different sectors of society have increased significantly in recent years and academic research has contributed significantly, providing numerous subsidies for decision-making, which has aroused the interest of regulators, sanitation companies, and government agencies (Zini and Mariliz, 2021).

Although the legislation is frequently revised, it is still limited in terms of emerging contaminants and their concentration patterns, making it possible that water considered to be potable is contaminated by substances not yet regulated that can be harmful to human health and ecosystems.

6. Conclusion

Unquestionably alarming, is the fact that ECs are widely distributed in the environment. These include different groups of contaminants such as pesticides, pharmaceuticals, personal care products, products from water disinfection processes, and cyanotoxins. The current status of conventional methods for removing emerging contaminants is low effectiveness, requiring advanced treatment for their elimination. Among the alternatives, microalgae show good performance for the removal of ECs. Finally, the need to regulate ECs with systematic and enforceable legislation that should be implemented globally.

References

Ahmed, M.B., Zhou, J.L., Ngo, H.H., Guo, W., Thomaidis, N.S. and Xu, J. 2017. Progress in the biological and chemical treatment technologies for emerging contaminant removal from wastewater: a critical review. Journal of Hazardous Materials 323: 274–298.

Ajeng, A.A., Rosli, N.S.M., Abdullah, R., Yaacob, J.S., Qi, N.C. and Loke, S.P. 2022. Resource recovery from hydroponic wastewaters using microalgae-based biorefineries: A circular bioeconomy perspective. Journal of Biotechnology 360: 11–22.

Ajiboye, T.O., Kuvarega, A.T. and Onwudiwe, D.C. 2020. Recent strategies for environmental remediation of organochlorine pesticides. Applied Sciences 10(18): 6286.

Alvarez-Gonz, A., Uggetti, E., Serrano, L. and Gorchs, G. 2023. The potential of wastewater grown microalgae for agricultural purposes: Contaminants of emerging concern, heavy metals and pathogens assessment. Environ. Pollut. 1(324): 121399.

Barbosa, M.O., Moreira, N.F.F., Ribeiro, A.R., Pereira, M.F.R. and Silva, A.M.T. 2016. Occurrence and removal of organic micropollutants: An overview of the watch list of EU Decision 2015/495. Water Research 94: 257–79.

Beigbeder, J.B., Sanglier, M., de Medeiros Dantas, J.M. and Lavoie, J.M. 2021. CO_2 capture and inorganic carbon assimilation of gaseous fermentation effluents using *Parachlorella kessleri* microalgae. Journal of CO_2 Utilization 50: 101581.

Bopp, S.K., Kienzler, A., Richarz, A.N., Linden, S., Paini, A., Parissis, N. et al. 2019. Regulatory assessment and risk management of chemical mixtures: challenges and ways forward. Critical Reviews in Toxicology 49(2): 174–89.

Chaturvedi, P., Shukla, P., Giri, B., Chowdhary, P., Chandra, R., Gupta, P. et al. 2021. Prevalence and hazardous impact of pharmaceutical and personal care products and antibiotics in environment: A review on emerging contaminants. Environmental Research 194: 110664.

Chen, Y., Lin, M. and Zhuang, D. 2022. Wastewater treatment and emerging contaminants: Bibliometric analysis. Chemosphere 133932.

Choi, S., Yoom, H., Son, H., Seo, C., Kim, K., Lee, Y. 2022. Removal efficiency of organic micropollutants in successive wastewater treatment steps in a full-scale wastewater treatment plant: Bench-scale application of tertiary treatment processes to improve removal of organic micropollutants persisting after secondary treatment. Chemosphere 288: 132629.

Crini, G., and Lichtfouse, E. 2019. Advantages and disadvantages of techniques used for wastewater treatment. Environmental Chemistry Letters 17(1): 145–155.

Degrendele, C., Klánová, J., Prokeš, R., Přibylová, P., Šenk, P., Šudoma, M. et al. 2022. Current use pesticides in soil and air from two agricultural sites in south Africa: Implications for environmental fate and human exposure. Science of the Total Environment 807.

Ebele, A.J., Oluseyi, T., Drage, D., Harrad, S. and Abdallah, M. 2020. occurrence, seasonal variation and human exposure to pharmaceuticals and personal care products in surface water, groundwater and drinking water in lagos state, Nigeria. Emerging Contaminants 6: 124–32.

Elalami, D., Carrere, H., Monlau, F., Abdelouahdi, K., Oukarroum, A. and Barakat, A. 2019. Pretreatment and co-digestion of wastewater sludge for biogas production: Recent research advances and trends. Renewable and Sustainable Energy Reviews 114: 109287.

Garcia-Rodríguez, A., Matamoros, V., Fontàs, C. and Salvadó, V. 2014. The ability of biologically based wastewater treatment systems to remove emerging organic contaminants—a review. Environmental Science and Pollution Research 21(20): 11708–11728.

Guillossou, R., Le Roux, J., Mailler, R., Vulliet, E., Morlay, C., Nauleau, F. et al. 2019. Organic micropollutants in a large wastewater treatment plant: what are the benefits of an advanced treatment by activated carbon adsorption in comparison to conventional treatment? Chemosphere 218: 1050–1060.

Guo, Y., Wen, Y., Li, G. and An, T. 2021. Recent advances in voc elimination by catalytic oxidation technology onto various nanoparticles catalysts: A critical review. Applied Catalysis B: Environmental 281: 119447.

Jha, G., Kankarla, V., Lennon, E., Pal, S., Sihi, D., Dari, B. et al. 2021. Per-and Polyfluoroalkyl Substances (PFAS) in Integrated Crop–Livestock Systems: Environmental Exposure and Human Health Risks. International Journal of Environmental Research and Public Health 18: 23.

Lanko, I., Flores, L., Garfí, M., Todt, V., Posada, J.A., Jenicek, P. et al. 2020. Life cycle assessment of the mesophilic, thermophilic, and temperature-phased anaerobic digestion of sewage sludge. Water 12(11): 3140.

Lee, L.H., Wu, T.Y., Shak, K.P.Y., Lim, S.L., Ng, K.Y., Nguyen, M.N. et al. 2018. Sustainable approach to biotransform industrial sludge into organic fertilizer via vermicomposting: A mini-review. Journal of Chemical Technology and Biotechnology 93(4): 925–935.

Lima, E.C. 2018. Removal of emerging contaminants from the environment by adsorption. Ecotoxicology and Environmental Safety 150: 1–17.

Maneechote, W., Cheirsilp, B., Angelidaki, I., Suyotha, W. and Boonsawang, P. 2023. Chitosan-coated oleaginous microalgae-fungal pellets for improved bioremediation of non-sterile secondary effluent and application in carbon dioxide sequestration in bubble column photobioreactors. Bioresource Technology 372: 128675.

Mutoti, I. M., Edokpayi,J., Mutileni, N., Durowoju, O. and Munyai, F. 2023. Cyanotoxins in groundwater; occurrence, potential sources, health impacts and knowledge gap for public health. Toxicon 226: 107077.

Naidu, R., Jit, J., Kennedy, B. and Arias, V. 2016. Emerging contaminant uncertainties and policy: The chicken or the egg conundrum. Chemosphere 154: 385–90.

Nanda, M., Chand, B., Kharayat, S., Bisht, T., Nautiyal, N., Deshwal, S. et al. 2021. Integration of microalgal bioremediation and biofuel production: a 'clean up' strategy with potential for sustainable energy resources. Current Research in Green and Sustainable Chemistry 4: 100128.

Nieto-Juárez, J.I., Torres-Palma, R., Botero-Coy, A. and Hernández, F. 2021. Pharmaceuticals and environmental risk assessment in municipal wastewater treatment plants and rivers from Peru. Environment International 155.

Palafox-Sola, M.F., Yebra-Montes, C., Orozco-Nunnelly, D. A., Carrillo-Nieves, D., González-López, M. E. and Gradilla-Hernández, M.S. 2023. Modeling growth kinetics and community interactions in microalgal cultures for bioremediation of anaerobically digested swine wastewater. Algal Research 102981.

Puri, M., Gandhi, K. and Kumar, M. 2023. Emerging environmental contaminants: a global perspective on policies and regulations. Journal of Environmental Management 332: 117344.

Puyol, D., Batstone, D.J., Hülsen, T., Astals, S., Peces, M. and Krömer, J.O. 2017. Resource recovery from wastewater by biological technologies: opportunities, challenges, and prospects. Frontiers in Microbiology 7: 2106.

Qu, W., Zhang, C., Zhang, Y. and Ho, S.H. 2019. Optimizing real swine wastewater treatment with maximum carbohydrate production by a newly isolated indigenous microalga *Parachlorella kessleri* QWY28. Bioresource Technology 289: 121702.

Ramírez-Malule, H., Quiñones-Murillo, D. and Manotas-Duque, D. 2020. Emerging contaminants as global environmental hazards. A bibliometric analysis. Emerging Contaminants 6: 179–93.

Ranjit, P., Jhansi, V. and Reddy, K.V. 2021. Conventional Wastewater Treatment Processes. pp. 455–479. *In*: Advances in the Domain of Environmental Biotechnology. Springer, Singapore.

Rizzo, L., Malato, S., Antakyali, D., Beretsou, V., Đolić, M., Gernjak, W. et al. 2019. Consolidated vs new advanced treatment methods for the removal of contaminants of emerging concern from urban wastewater. Science of the Total Environment 655: 986–1008.

Rout, P.R., Zhang, T.C., Bhunia, P. and Surampalli, R.Y. 2021. Treatment technologies for emerging contaminants in wastewater treatment plants: A review. Science of the Total Environment 753: 141990.

Salimi, M., Esrafili, A., Gholami, M., Jafari, A., Kalantary, R., Farzadkia, M. et al. 2017. Contaminants of emerging concern: A review of new approach in AOP technologies. Environmental Monitoring and Assessment 189(8).

Santos, M.T. and Lopes, P.A. 2022. Sludge recovery from industrial wastewater treatment. Sustainable Chemistry and Pharmacy 29: 100803.

Sartori, R.B., Deprá, M.C., Dias, R.R., Lasta, P., Zepka, L.Q. and Jacob-Lopes, E. 2022. Microalgae-based systems applied to the dual purpose of waste bioremediation and bioenergy production. pp. 127–145. *In*: Algal Biotechnology. Elsevier.

Simmons, F.J., Kuo, D. and Xagoraraki, I. 2011. Removal of human enteric viruses by a full-scale membrane bioreactor during municipal wastewater processing. Water Research 45(9): 2739–50.

Tambat, V.S., Patel, A.K., Chen, C.W., Raj, T., Chang, J.S., Singhania, R.R. et al. 2023. A sustainable vanadium bioremediation strategy from aqueous media by two potential green microalgae. Environmental Pollution 323: 121247.

Tran, N.H., Reinhard, M. and Gin, K.Y.H. 2018. Occurrence and fate of emerging contaminants in municipal wastewater treatment plants from different geographical regions—a review. Water research, 133: 182–207.

Wang, Y., Ho, S.H., Cheng, C.L., Nagarajan, D., Guo, W.Q., Lin, C. et al. 2017. Nutrients and COD removal of swine wastewater with an isolated microalgal strain Neochloris aquatica CL-M1 accumulating high carbohydrate content used for biobutanol production. Bioresource Technology 242: 7–14.

Wang, Z., Walker, G., Muir, D. and Nagatani-Yoshida, K. 2020. Toward a global understanding of chemical pollution: A first comprehensive analysis of national and regional chemical inventories. Environmental Science and Technology 54(5): 2575–84.

Wee, S.Y. and Ahmad, Z.A. 2019. Occurrence and public-perceived risk of endocrine disrupting compounds in drinking water. Npj Clean Water 2: 1.

Xu, H., Berres, A., Liu, Y., Allen-Dumas, M. and Sanyal, J. 2022. An overview of visualization and visual analytics applications in water resources management. Environmental Modelling and Software 153: 105396.

Yan, Z., Yang, X., Lu, G., Liu, J., Xie, Z. and Wu, D. 2014. Potential environmental implications of emerging organic contaminants in Taihu Lake, China: Comparison of two ecotoxicological assessment approaches. Science of the Total Environment 470–471: 171–79.

Yang, W., Cai, C., Guo, Y., Wu, H., Guo, Y. and Dai, X. 2022. Diversity and fate of human pathogenic bacteria, fungi, protozoa, and viruses in full-scale sludge treatment plants. Journal of Cleaner Production 380: 134990.

Yao, C., Yang, H. and Li, Y. 2021. A review on organophosphate flame retardants in the environment: occurrence, accumulation, metabolism and toxicity. Science of the Total Environment 795: 148837.

Zhou, J.L., Yang, L., Huang, K., Chen, D. and Gao, F. 2022. Mechanisms and application of microalgae on removing emerging contaminants from wastewater: A review. Bioresource Technology 364: 128049.

Zini, L.B. and Mariliz G. 2021. Chemical contaminants in Brazilian drinking water: A systematic review. Journal of Water and Health 19(3): 351–69.

Zinicovscaia, I. 2016. Conventional methods of wastewater treatment. Cyanobacteria for Bioremediation of Wastewaters 17–25.

CHAPTER 10

Modern Applications and Current Status of ECs in the Industries

Manviri Rani,[1,] Keshu,[1,2] Shikha Sharma,[2] Rishabh[2] and Uma Shanker[2,]*

1. Introduction

Numerous pollutants that have low concentrations but cause the environment great harm, such as antibiotics and endocrine disruptors, are gradually gaining attention as human needs for the environment continue to rise are considered as Emerging Contaminants (ECs) (Cheng et al., 2021). Any natural and synthesized chemical that is not frequently found in the ecosystem (and is mostly unregulated) yet has the potential to have known or suspected detrimental ecological and human health effects is considered as an ECs (Gogoi et al., 2018; US EPA, 2012; DoD, 2011). They could have an industrial ancestry or come from effluents in municipal (home), agricultural, hospital, or laboratory settings. In 2004, there were more than eight million natural and synthetic materials accessible in the world; yet, only 126 of these compounds were listed by the US Environmental Protection Agency (US EPA) as priority contaminants, and only 3% of them were subject to regulation (Daughton, 2004; Hoenicke et al., 2007). US EPA has a high likelihood of producing unique compounds for additional development and research given that the American Chemical Society (CAS Registry) now has over 180 million chemicals indexed (American Chemical Society, 2021). The chemicals in question are predominantly sourced from three major groups: a) Pharmaceuticals b) Personal Care Products and

---

[1] Department of Chemistry, Malaviya National Institute of Technology Jaipur, Rajasthan-302017-India.
[2] Department of Chemistry, Dr B R Ambedkar National Institute of Technology, Jalandhar, Punjab, India-144011.
* Corresponding authors: manviri.chy@mnit.ac.in; Shankeru@nitj.ac.in

c) Endocrine Disrupting Compounds. They can also include nanomaterials (NMs), ECs metabolites, illegal drugs, engineered genes and many more. Pharmaceutical compounds from different categories of veterinary antibiotics and human, over-the-counter medications, even some sex and steroid hormones have been found in wastewater. The consumption of pharmaceutical products is estimated to amount to hundreds of kilotons annually, making the use of developing toxins crucial given how much better life and health are as a result (Ogunwole et al., 2021). The utilization of various ECs increases due to the growing human population without any apparent reduction. However, the detrimental effects caused by the increased consumption of ECs have been extensively documented, and their biodiversity and biological accumulation should not be disregarded (Zenker et al., 2014). Chemicals prevalent in consumer items, such as galaxolide and tonalide, are also included in Personal Care Products (PCPs). Bath soaps, sunscreens, lipsticks, detergents, dental care items, lotions and many more are different examples of PCPs they are mainly meant to improve daily life (Samal et al., 2021). PCPs are primarily divided into five classes: perfumes (e.g., musks), insect repellents (e.g., DEET), UV filters (e.g., methylbenzylidene camphor), preservatives (e.g., parabens), and disinfectants (e.g., triclosan). Due to PCPs' external use on the body rather than their internal use as pharmaceuticals, which are prone to metabolic changes, these do not. As a result, large amounts of PCPs enter the ecosystem via daily household activities (Ternes et al., 2004). Due to their androgenic or estrogenic properties, Endocrine Disrupting Substances (EDCs) can cause harm to the endocrine system even at low concentrations. Possible issues from the existence of these emerging toxins in the environment include aberrant physiological functions and reproductive dysfunction, an increase in cancer cases, the emergence of microorganisms resistant to antibiotics, and which could lead to enhanced chemical combination toxicity. Drinking, ground, and surface water discharge all include ECs (Yang et al., 2014; Samaras et al., 2013). One of the primary discharge pathways for contaminants from non-point and point sources, industry, storm as well as domestic wastewater, and water treatment facilities is municipal wastewater (Ternes et al., 2004). ECs are frequently found in pharmaceuticals, cosmetics, antibiotics, endocrine-disrupting substances, disinfection by-products, persistent organic pollutants, and other industrial chemicals (Bo et al., 2015). The details of the toxicological impact of ECs are mentioned in Table 10.1. These ECs are long-lasting and persistent in the environment. More than 30 different types of ECs have been discovered in urban rainfall, freshwater, treated wastewater, agricultural rainwater, and untreated wastewater in earlier research. Among these, personal care items, medications, and artificial sweeteners were found in many water samples (Tran et al., 2019).

Despite the relatively smaller amount of ECs in water, after being accumulated by organisms, they could have negative effects on the environment and human health through the food chain (Gomes et al., 2017). Bioactive compounds known as Pharmaceutical Contaminants (PC) are used to prevent, treat, or cure illnesses. They are among the most relevant EC types for the pharmaceutical sector (Mahapatra et al., 2022; Samal and Trivedi, 2020). Pharmaceutical-related compounds in the environment are assumed to come from both personal care products and medications.

Table 10.1 Detailed toxicological impacts of emerging contaminants on various organisms.

S. no.	Organism	Emerging contaminant	Type of compound	Class of EC's	Toxicity effects	Reference
1.	Human	Triclosan	Antimicrobial agent	Personal care products (PCPs)	Causes breast cancer	Gee et al., 2008
2.	*Carassius auratus* (Gold fish)	Galaxolide (HHCB)	Synthetic musk	Personal care Product	Causes Oxidation stress	Chen et al., 2012
3.	*Pseudokirchneriella subcapitata* (Micro algae)	Triclosan and triclocarban	Antimicrobial reagents	Personal care products	Growth inhibitors	Yang et al., 2009
4.	*Vibrio fischeri* (Gram negative bacteria)	paracetamol	Analgesic	Pharmaceuticals	Bioluminence inhibition	Nunes et al., 2014
		Sulphamonome-thoxine	Antibiotics	Pharmaceuticals		Huang et al., 2014
		Chloramphenicol Diclofenac	NSAIDs	Pharmaceuticals		Czech et al., 2014
5.	*Cyprinus carpio* (Fish)	Paracetamol	Analgesics	Pharmaceuticals	Cytoplasmic and nuclear deformities	Sharma et al., 2019
6.	Zebra fish	Cyclophosphamide and Ifosfamide	Anti-cancer drugs	Pharmaceuticals	Genomic instability and DNA strands breaks	Novak et al., 2017
7.	*Oncorhynchus myskiss* (Rainbow trout fish)	Diclofenac	NSAIDs	Pharmaceuticals	Renal lesions and gill alterations	Schwaiger et al., 2004
		Carbamazepine	Anti-epileptic drug	Pharmaceuticals	Oxidation stress	Li et al, 2010
		Metoprolol	β-blocker	Pharmaceuticals	Liver damage	Brausch et al., 2012
8.	*Oreochromis niloticus* (Fish)	Oxytetracycline Florfenicol	Antibiotics	Pharmaceuticals	Extensive DNA damage in erythrocytes on acute and chronic exposure	Botelho et al., 2015

Table 10.1 contd. ...

... *Table 10.1 contd.*

S. no.	Organism	Emerging contaminant	Type of compound	Class of EC's	Toxicity effects	Reference
9.	*Daphnia magna* (Planktonic Crustacean)	Propranolol	β-blockers	Pharmaceuticals	Changes in reproductive system, in growth and heart malfunctioning	Oliveira et al., 2016
		Metoprolol	β-blockers	Pharmaceuticals	Decrease in growth, heart rate abnormalities, number of offspring's, changes in neonates	Dietrich et al., 2010
		Sulphamonomethoxine	Antibiotics	Pharmaceuticals	Reproductive changes	Huang et al., 2014
		Chloramphenicol	Antibiotics	Pharmaceuticals	Oxidative tissue damage	Czech et al., 2014
		Ibuprofen Diclofenac	NSAIDs	Pharmaceuticals	Mortality and reproductive changes	Du et al., 2016
		Paracetamol	Analgesic	Pharmaceuticals	Mortality Immobalization Fecundity and Growth impairments Oxidative stress	Nunes et al., 2014

2. Classification of Emerging Contaminants

Pesticides, brominated flame retardants, personal care and pharmaceuticals products are widely used resulting in the increased number of emerging contaminants. Ineffective conventional wastewater treatment, recycling, toxic wastewater from agricultural activity and the use of sludge all significantly contribute to this (Shahid et al., 2021). Organic pollutants that have been found in a wide range of personal care items are known to cause reproductive problems and endocrine disruption. Among other substances, these developing pollutants include methoxycinnamate, nano-ZnO, hydroxytoluene and butyl paraben (Rathi et al., 2021). Regular use of perfumes, preservatives, and disinfectants releases various types of organic pollutants. The identified developing pollutants are released into drinking water and wastewater by everyday use of disinfectants, perfumes, and preservations. These new pollutant bioaccumulations have detrimental consequences on the ecosystem and the well-being of people (Shahid et al., 2021). The physical and chemical characteristics that allow pesticides to be used as bactericides, insecticides, fungicides, and herbicides are reflected in their structural makeup. These pesticides contain substances such phosphamidone, chlorpyriphos, alachlor, dichlorodiphenyltrichloroethane and simazine that are likely to cause significant environmental harm and have a negative impact on abiotic as well as biotic methods (Sehrawat et al., 2021). Pesticide pollutants that bioaccumulate in human tissues alter the cytogenetic processes, neurotoxic and genotoxic. Additionally, they could cause an hormonal imbalance that results in thyroid dysfunction, ovary tumours, and reproductive issues (Venkidasamy et al., 2021). ECs from pesticides are likely released in environment and discharged from their application sites into surface water. Pentabromodiphenyl ether, decabromodiphenyl ether and tetrabromodiphenyl are examples of developing pollutants that are typically applied to polymeric materials including electronics, furniture, and textiles to stop or delay the spread of a fire. According to Charbonnet et al. (2020), their detrimental consequences on humans include diminished cognitive function, endocrine disturbance, cancer development and reproductive harm. The classification of ECs is shown in Fig. 10.1.

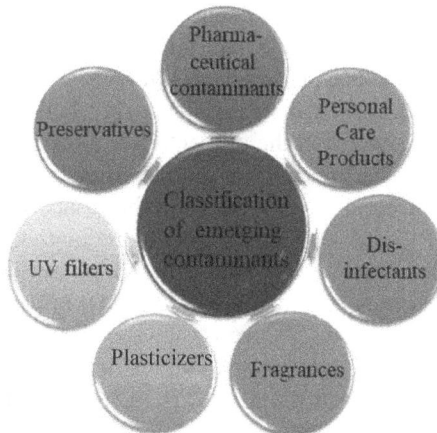

Figure 10.1 Classification of ECs.

2.1 Pharmaceutical Contaminants (PCs)

The pharmaceutical industry is a broad category of biological chemicals used to treat illnesses and infections. Detection of oestrogen, painkillers, birth control medicines and other medications in aquatic living beings is quite alarming (Bhushan et al., 2020). Pharmacologically active pollutants produced by PPCPs are resistant to degradation and persistent in aqueous media. PCs are manufactured chemicals with a specific aim that are intended to be absorbed and distributed throughout the human body (Chen et al., 2002). The use of PPCPs depends on a number of variables, including the nation's socioeconomic status, the location and region, the availability of healthcare services, and seasonal variations. Anti-inflammatory medications, beta-blockers, antibiotics, lipid regulators, non-steroidal, antipyretics, psychostimulants, antiretrovirals and steroid hormones are commonly used in both human and animal medicine considered as PCs (Shahid et al., 2021; Rathi et al., 2021). Due to excretion by people and animals, insufficient WWT, effluent from drug manufacturing facilities and pit latrines, this category of developing pollutants predominates in the aquatic environment. Additionally, incorrect disposal of outdated stock, particularly leachate from pharmaceutical landfilling, poses a serious threat to groundwater contamination and surface water contamination through runoff (Peng et al., 2019; Wood et al., 2017; Shahid et al., 2021). The ineffectiveness of WWTP and unauthorized disposal of waste is the primary source and the most practical method for allowing the release of these chemicals into various natural water compartments, especially rivers (Peng et al., 2014). Anti-retroviruses (ARVs) are typical pharmaceuticals that are widely used to treat the acquired immune deficiency syndrome and human immunodeficiency virus (HIV). In the use of ARVs, South Africa is the top country with over seven million people who have been diagnosed positive and are receiving ARV medication. Given the widespread use of ARV medication and the identification of these chemicals in European surface water and water treatment plants influent and effluent, it naturally follows that South African water supplies are likely to be severely affected due to larger population consumption (Wood et al., 2016). In a study published by Wood et al. (2017), the ARV compounds were found in the environmental water in South Africa using UHPLC-QTOF and UHPLC-MS/MS. In research to determine the presence of ARVs in surface waters, Adeola and Forbes, (2022) found that South African wastewater treatment facilities had influent amount ranging from 670 to 34 000 ngL^{-1} and effluent concentration ranges of 540 to 34 000 ngL^{-1}. As per the environmental amount set by the nation's HIV load, it is believed that efavirenz and nevirapine, which were detected in the effluents at the maximum concentration of 140 gL^{-1}, were the most prevalent and permanent in surface water. Several bacterial infections like tuberculosis, salmonellosis, gonorrhoea, and pneumonia have been treated and prevented with the help of a number of antibiotics. Antibiotics are used to treat bacterial illnesses in humans and animals on a global scale on a yearly basis of 100,000 to 200,000 tonnes (Chaturvedi et al., 2021). In particular, South Africa faces unusual difficulties such ineffective wastewater treatment, a high prevalence of tuberculosis, drug-resistant tuberculosis, and restricted entry to drinkable water, especially in rural areas (Wood et al., 2017). Non-steroidal anti-inflammatory medicines (NSAIDs) are a wide category of drugs used for treating inflammation, pain, and fever having analgesic and antipyretic properties. These metabolites have long been used in veterinary and human medicine, but their presence in many water streams has also been acknowledged as a small but growing source of pollutants (Waleng and Nomngongo, 2022; Izadi et al., 2020). The NSAIDs indicate endocrine disruptor potency in human health, chronic ecotoxic effects on aquatic ecosystem living creatures, and poor effects on water quality (Tyumina et al., 2020). Pharmaceutical NSAIDs that have been discovered in groundwater, soil, drinking water, sediments, surface water and wastewater, are more likely to degrade in aerobic environmental conditions due to photolysis and biotransformation processes (Meja-Garca et al., 2020). According to research studies, some of the NSAID developing pollutants were detected using HPLC-MS/MS and LC-ESI-MS (Matongo et al., 2015; Agunbiade and Moodley, 2016), For physiological functions such as, synthetic steroid hormones, sexual maturity, control for animal breeding, stress responses, using natural osmoregulation, contraception used daily and veterinary pharmaceuticals (Ojoghoro et al., 2021; Chen et al., 2019). These medications are usually not fully metabolized by the human body, thus they are excreted through urination and end up in wastewater or terrestrial runoff. Similar to other emerging pharmaceutical pollutants, steroid hormones enter the environment mostly through wastewater treatment plant effluents that are dumped into rivers untreated (Ojoghoro et al., 2021; Iancu et al., 2019). Environmental water bodies have been shown to contain several naturally occurring steroid hormones, such as 17-estradiol and 17-ethinylestradiol, 17-estriol, and 17-estrone, as well as synthetic steroids that support menopause supplements (Bhandari et al., 2015; Ojoghoro et al., 2021). Lipid regulators and beta-blockers have been widely used for the treatment of cardiovascular diseases such as coronary artery disease, heart attack and excessive blood pressure stroke (Wood et al., 2016; Maszkowska et al., 2014; Zhang et al., 2020). They have demonstrated significant degradability into a range of dangerous metabolites, and environmental factors and hydrolysis activities, redox reactions and photolysis in an aquatic setting which can speed up this process (Zhang et al., 2020). In South African dams and surface water, bezafibrate and atenolol are common detected by using HPLC-DAD and HPLC-MS/MS (Matongo et al., 2015; Agunbiade and Moodley, 2014).

2.2 Personal Care Products (PCPs)

UV screens and scents (musk chemicals) are examples of common PCPs. Detergents, personal insecticides, lotions, cleaning supplies and deodorants that are usually distributed through the skin include: (sunscreens), Wastewater from households (Roosens et al., 2007; Peck, 2006). Cleaning agents used in residential and commercial applications contain alkylphenol ethoxylates (AEOs), the majority of which are nonylphenol ethoxylates (NPEOs) (La Farre et al., 2008). The most popular example of NPEOs is 40-nonylphenol, which has attracted the attention of the research community. It was recently discovered in the Vaal River in South Africa at quantities as high as 80 ngL^{-1} (Chokwe et al., 2015). Synthetic Musk Compounds (SMCs), which have a wide range of chemical structures and are frequently found in nature, are used in many personal care items (Roosens et al., 2007). Since their initial development at the turn of the 20th century, these substances have been routinely produced for use as detergents, fragrance in lotions, soaps, and cosmetic products (Lee et al., 2010). Musk Xylene (MX), Galaxolide (HHCB), Musk Ketone (MK) and Tonalide (AHTN) are some of the SMCs that are most frequently found; they have been found in PCPs like perfumes, deodorants, and lotions at levels as high as, 22,000 mg/g, 26 mg/g, and 8000 mg/g, respectively (Roosens et al., 2007). According to Lee et al. (2010), SMCs typically occur at levels of influent (3690–7330 ngL^{-1}), effluent (960–2960 ngL^{-1}) and surface water of Sewage Treatment Works (STW) 150–16700 ngL^{-1}. Only lately have ultraviolet (UV) filters such 2-phenylbenzimidazole-5-sulphonic acid (PBSA) and benzophenone-4 (BP-4) have been identified in drinking water, STW influent and effluent (Rodil et al., 2008). According to Rodil et al. (2012), bathing and other water-related recreational activities are thought to be the primary seasonal sources of UV filter residues. In all sampled STW influents in northwest Spain, BP-4 and PBSA were at moderate concentrations of 2100 ng/L and 200 ng/L, respectively (Rodil et al., 2012). At the same time, treated STW effluent included up to 600 ng/L BP-4 and 20 ng/L PBSA, with median values of 1200 ng/L BP-4 and 240 ng/L PBSA (Rodil et al., 2012). With the exception, of BP-4, which has been found up to 60 ng/L in northwest Spain, detection of UV filters in drinking water normally has concentrations below 10 ng/L (Rodil et al., 2012). When compared to the relatively substantial amount of study done explaining the aquatic toxicity of human pharmaceuticals, the impacts of PCPs on water ecosystems continue to be difficult (Brausch and Rand, 2011). It is hotly contested whether trace-level PCPs are toxic to aquatic life and is covered in more length elsewhere. Further investigation is necessary, especially to evaluate potential endocrine disruption brought on by UV-filter exposure and PCP bioaccumulation in the aquatic ecosystem (Brausch and Rand, 2011). The toxicity impact of pharmaceutical and PCPs are mentioned in Table 10.2.

2.2.1 Disinfectants

According to McAvoy et al. (2002), triclocarban (TCC) and the biphenyl ethers triclosan (TCS) are used as antimicrobials in deodorants, soaps, toothpastes, plastics and skin creams. According to frequency and concentration statistics, TCC and TCS are considered as the top 10 organic pollutants detected in different water matrixes (Halden and Paull, 2005; Kolpin et al., 2002). According to Lopez and Ra (1980), TCS were detected in WWTP effluent at levels higher than 10 l gL^{-1}. TCS was one of the most frequently discovered chemicals, with surface water concentrations as high as 2.3 lg L^{-1}, according to a USGS research that examined 95 compounds in surface water across the country (Kolpin et al., 2002). TCS and its methyl derivative methyl triclosan (M-TCS) have been discovered in surface water, water from WWTP and fish tissue throughout the world (Dougherty et al., 2010; Loraine and Pettigrove, 2006; Benotti et al., 2009; United States, Boyd et al., 2004). According to this investigation, TCS was identified in surface water at levels as high as 74 ng L^{-1} (Lindstrom et al., 2002), which is greater than M-TCS, although both chemicals were found at levels as high as 11 ng L^{-1} and 650 ng L^{-1} in WWTP effluent. Due to its structural resemblance to the non-steriodal oestrogen diethylstilbestrol, evidence suggests TCS is slightly estrogenic (Ishibashi et al., 2004; Foran et al.,

Table 10.2 A summary of toxicity imposed by pharmaceutical and personal care products in different countries, their concentration in environmental matrices, and respective class.

S. no	PPCPs	Class	Chemical formula	Presence in environment	Country	Concentration	References
1.	Metformin	Anti-diabetic (antihyperglycemic agent)	$C_4H_{11}N_5$	Sea water	Saudi Arabia	, 7– >3000 ngL^{-1}	Ali et al., 2017
2.	Acetaminophen/ Paracetamol	Analgesic and antipyretic	$C_8H_9NO_2/C_8H_9NO_2$	Sea water, surface water, sediment/ Municipal and hospital WWTPs	Saudi Arabia, China, Greece	< LOQ-2379 ngL^{-1}, 71.95 ngL^{-1}, 23.98 ngL^{-1}/20.6 ngL^{-1}, 9.3 μgL^{-1}	Ali et al., 2017; Kosma et al., 2010
3.	Carbamazepine	Anticonvulsant	$C_{15}H_{12}N_2O$	Surface water, sediment	China, Germany	271.02 μgL^{-1}, 54.20 μgL^{-1}, 6.3 μgL^{-1}	Zhang et al. 2018a,b
4.	Caffeine/Nicotine	Stimulant	$C_8H_{10}N_4O_2/C_{10}H_{14}N_2$	Sea water, River water/treated and untreated WWTPs	Saudi Arabia, China, England, UK	62–> 3000 ngL^{-1}, 865 ngL^{-1}/ 85.7–3919.3 ngL^{-1}	Ali et al. 2017; Baker and Kasprzyk-Hordern, 2013
5.	Morphine/ Dihydrocodeine	Opioid analgesics	$C_{17}H_{19}NO_3/ C_{18}H_{23}NO_3$	River water, treated and untreated WWTPs	England, UK	59.1–371.2 $ngL^{-1}/$ 118.2–226.6 ngL^{-1}	Baker and Kasprzyk-Hordern, 2013
6.	Methylparaben/ Ethylparaben	Preservative (endocrine disrupter)	$C_8H_8O_3$	River water, reservoirs	Brazil, China	1192.39 mg L^{-1}, 8 mg L-1, 27.50 mg L^{-1}/12.5 ngL^{-1}	Galinaro et al., 2015; Peng et al., 2014
7.	Clofibric acid/ Benzafibrate/ Gemfibrozil	Lipid regulator	$C_{10}H_{11}ClO_3/ C_{19}H_{20}ClNO_4/ C_{15}H_{22}O_3$	Ground water	Spain	73.9 ngL-1/ND-7.57 ngL^{-1}, ND- 25.8 ngL^{-1}	López-Serna et al., 2013
8.	Ibuprofen/ Diclofenac	NSAID	$C_{13}H_{18}O_2/ C_{14}H_{11}O_2Cl_2NO_2$	Ground water/ ground water and sea water	Spain/ Spain, France, Saudi Arabia	0.16–200 $\mu gL^{-1}/$ 0.184–225 μgL^{-1}, 9.7 μgL^{-1}, > 3 μgL^{-1}	Ali et al., 2017; López-Serna et al., 2013; Vulliet and Cren-Olivé, 2011

2000). The medaka (O. latipes) fin length and sex ratios have been linked to TCS exposure (Foran et al., 2000). Additionally, after 35 d of exposure, TCS demonstrated a in decrease sperm counts and VTG synthesis in Gambusia and encouraged O. latipes males to produce VTG (Raut and Angus, 2010; Ishibashi et al., 2004). Rana catesbeiana and Xenopus laevis have both been tested for TCS' endocrine effects because to the substance's structural resemblance to hormone of thyroid (Veldhoen et al., 2006; Fort et al., 2010). TCS had no impact on the metamorphosis of X. laevis (Fort et al., 2010) and only generated marginal effects in R. catesbeiana (Veldhoen et al., 2006), indicating that TCS has a negligible impact on Thyroid Hormone (TH) and amphibian development. No studies have examined the possible endocrine disruption caused by other disinfectants with comparable structures.

2.2.2 Fragrances

Fragrances are thought to be one of the most pervasive environmental pollutants and are possibly the PCP class that has received the most research (Daughton and Ternes, 1999). The most popular scents are imitation musks. deodorants, soaps, and detergents, among other goods, using synthetic musks as perfumes. Examples of synthetic musks are nitro musks synthesized in 1800s and then in 1950s polycyclic musks were introduced (Daughton and Ternes, 1999). Musk Xylene (MX) and Musk Ketone (MK) are the two kinds of nitro musks that are used the most frequently, although Musk Ambrette (MA), Musk Tibetene (MT) and Musk Moskene (MM) were used very little (Daughton and Ternes, 1999). Nitro musks, however, are rapidly being phased out due to their environmental persistence and probable harm to aquatic creatures (Daughton and Ternes, 1999). According to Daughton and Ternes (1999), polycyclic musks are being used more frequently compared to nitro musks, with toxalide, celestolide (ABDI) and galaxolide (HHCB) used mainly, while cashmeran (DPMI), phantolide (AHMI), and traseloide (ATII) have been explored very little. Indole, acetophenone, camphor, isoborneol, dlimonene, ethyl citrate, skatol, and isoquinolone, are up to eight other fragrances that have been detected in water matrices but all of them have only been found in a small number of samples, of ethyl citrate (Glassmeyer et al., 2005; Kolpin et al., 2002). In surface waters across the US, ethyl citrate, a tobacco ingredient, has regularly been found (Kolpin et al., 2004; Glassmeyer et al., 2005). With each of these substances tested separately, acute, and long-term toxicity is not anticipated; however, more research is required.

2.2.3 Insect Repellents

The most widely used active component in insect repellents is N,N-diethyl-m-toluamide (DEET), which has been repeatedly found in various waters bodies in the United States (Glassmeyer et al., 2005). The 1940s saw the development of DEET, which blocks an insect's capacity to recognize lactic acid on its host (Davis, 1985). According to the USEPA (1998), presently DEET is approved for use in more than 200 different products and is used on a yearly basis in excess of two million kg (Kolpin et al., 2002; Glassmeyer et al., 2005; Sandstrom et al., 2005; Quednow and Püttmann, 2009). DEET has lower BCF and is probably not deposited into aquatic creatures, unlike many other PCPs (such as perfumes and UV filters), despite its persistentence in the aquatic environment (Costanzo et al., 2007). DEET has consistently been found in wastewater and surface water with moderate values of about 0.2 lg L^{-1} and 55 ng L^{-1}, respectively. 1,4-Dichlorobenzene is the only additional insect repellent found in effluents of WWTP or surface water. In the US, surface water receiving considerable inputs of WWTP effluent has been found to contain 1,4-dichlorobenzene at values up to 0.28 lg L^{-1} (Glassmeyer et al., 2005). Like DEET, 1,4-dichlorobenzene has also been the subject of a preliminary risk evaluation. While invertebrates, particularly D. magna, tend to be most susceptible to short-term exposures, fish are more responsive for long term doses (Boutonnet et al., 2004). According to ambient concentrations, it is doubtful if freshwater and marine creatures experience any acute or long-term consequences (Boutonnet et al., 2004). Additionally, 1,4-dichlorobenzene has a low potential for bioaccumulation and there is currently no evidence that it can affect the endocrine system (Boutonnet et al., 2004).

2.2.4 Preservatives

Alkyl-p-hydroxybenzoates, sometimes known as parabens, are antimicrobial preservatives used in personal care products, medicines, cosmetics, and food (Daughton and Ternes, 1999). Seven kinds of parabens that are now in use are butyl, methyl, propyl, ethyl, benzyl, isopropyl and isobutyl. According to Soni et al. (2005), approximately seven thousand kg of parabens were used in toiletries and cosmetics in 1987 alone, and during the past 20 yr, it was anticipated that this quantity would rise. The most widely used form of methylpropylparaben is usually combined with other preservatives to boost their preservative benefits. Only a small number of studies have looked at paraben levels in surface water and WWTP to date. As seen in paraben, surface water had the highest amounts of parabens (15 to 400 ng L^{-1}), the effluent had lesser quantities, between 50–85 ng L^{-1} (Benijts et al., 2004). According to toxicity data and environmental trace propylparaben, benzyl- and butyl- appear to have the potential to be harmful to aquatic life. Parabens, particularly benzyl-, butyl-, and propylparabens, have been shown to cause low-level estrogenic reactions in humans, according to Dobbins et al. (2009), who also found that they only provide a minimal risk to aquatic creatures. Parabens can elicit estrogenic responses at a lesser amount, according to *in vitro* investigations using yeast and fish MCF-7 cell lines (Routledge et al., 1998). Additionally, parabens have been shown to cause VTG synthesis in fish when comes in contact to lesser concentrations (Inui et al., 2003; Bjerregaard et al., 2008). Therefore, exposure to parabens at low concentrations has the potential to have estrogenic effects at levels that are crucial for the environment. Additional research has been done on the effects of parabens on serum testosterone and spermatogenesis in male rats, among other sexual endpoints (Oishi, 2002). However, neither butyl- nor propylparaben had any effect on serum testosterone. Instead, they both significantly inhibited spermatogenesis. Butyl-, isobutyl-, and benzylparaben all exhibit estrogenic action, albeit to a much lesser extent than oestrogen itself, according to a review of paraben endocrine activity in rats by Golden et al. in 2005. These findings suggest that paraben exposure may have negative effects on aquatic life. However, given that the impact of the amount are typically more than thousands of times higher than those found in surface water, initial data on the environmental amount suggest that there is only a very small risk to aquatic organisms.

2.2.5 UV Filters

The use of UV filters has increased as concerns over the effects of ultraviolet (UV) radiation on people has grown. UV filters can either be organic or inorganic micropigments are only organic type compounds which will be covered in this study. To protect from UV radiation, UV filters are used in cosmetics and sunscreen goods. Sunscreens and cosmetics usually contain three to eight different UV filters, which collectively can make up more than 10% of all products (Schreurs et al., 2002). In the US, there are currently 16 substances approved for use as agents in sunscreen and more than 25 UV filters approved for use in plastics, cosmetics, and other products (Reisch, 2005; Fent et al., 2008). There are two ways that UV filters enter the environment: indirectly through the WWTP effluent and directly through sloughing off while swimming and engaging in other leisure activities. The bioaccumulation of UV filters is widely known, and recent research has also suggested that estrogenic action could be present. Five UV-B and UV-A sunscreens have the efficiency to exert estrogenic effects, according to *in vitro* tests using fish MCF-7 cell lines (Kunz and Fent, 2006; Schlumpf et al., 2001). Ten UV filters which have estrogenic effects was determined using a recombinant yeast experiment with rainbow trout ERa (Kunz et al., 2006). The only other chemical that indicated estrogenicity below 1 mg L^{-1} exposure was 40 hydroxybenzophenone, making benzophenone-1 the most effective UV filter (Kunz et al., 2006c). Multiple UV filters could have estrogenic effects and negatively impact fecundity and reproduction, according to aquatic studies conducted on fish (P. promelas and O. mykiss). A detailed risk profile cannot be created because of the minimal number of species used to determine hazardous effects.

2.3 Plasticizers

Plasticizers are substances that make a material more malleable. They are frequently used in manufacturing of packaging, plastic products, thermal printing paper, drinking bottles, water pipe lining, epoxy resins, food packaging, dental sealants, implanted medical devices, plastic food containers, mobile phones, CDs, DVDs and eyeglass lenses (Barraza, 2013;Rani et al., 2022; Keshu et al., 2021; Keshu et al., 2022). In Pearl River, China three plasticizers Bisphenol-F (BPF), bisphenol-S (BPS) and bisphenol-A (BPA) are found in surface waters and identified up to 1110 ngL^{-1}, 135 ngL^{-1} and 98 ngL^{-1} (Yamazaki et al., 2015). Additionally, BPA has been found in up to 50 ngL^{-1} (average: 9 ng/L) of the 291 tap waters in France (Colin et al., 2014). Annual production of BPA is approximately two million metric tonnes which is one of the greatest production rates of any PPCP in the world (Maia et al., 2009; Lang et al., 2008). The market for BPA has proved to be harmful for ongoing expansion despite legislative limits on its use (Merchant Research and Consulting, 2015). Due to serious safety concerns in the scientific and public sectors, in 2011 the European Commission limited the application of toxic BPA as a monomer for the synthesis of polycarbonates (Grignard et al., 2012). According to Merchant Research and Consulting (2015), however, since 2011, the overall amount of BPA produced globally climbed by 5.25% annually, from 4.4 to 4.6 million tonnes in one year. According to Merchant Research and Consulting (2015), the major countries producing BPA were China, South Korea, Japan, and Taiwan. In response to concerns surrounding the environmental toxicity of BPA and the substance's widespread usage in thermal printer paper, the USEPA produced a report detailing 19 potential BPA alternatives (mostly other plasticizers) and their associated toxicity (USEPA, 2014). 4,4'-(1-Phenylethylidene)bisphenol, 2,20-bis(4-hydroxy-3-methylphenylphenyl) propane, 2,40-bis(hydroxyphenyl) sulfone, and other substances are potential substitutes (USEPA, 2014). But it is important to keep in mind that the USEPA study discusses alternative substitutes with potential toxicity that is either possibly greater than that recorded for BPA than BPF, or potentially similar to that described for BPA, as in the case of BPS, BPC, and BPAP. Despite the fact that BPS is now a common plasticizer, permanence in surface, its environmental presence in the ground, and drinking waters have not been fully evaluated (Grignard et al., 2012). BPA (75 ng/L; river wate, 43 ng/g; sediment) detected in various environmental matrices (Yang et al., 2014). In greater London, the UK, plasticizers like BPS and BPa have also been found in runoff water in rainy periods with amounts 50 and 2410 ng/L, respectively. As alternatives to plasticizers controlled by agencies entering the market, research is crucial to determining their dispersal in the environment as well as their destiny in suspended material, biota, and sediment.

2.4 Additional Compounds

Researchers from the United States Geologic Survey have discovered three more PCPs in surface water there. The flavouring CH_3OH was discovered at the greatest amounts (1.31 gL^{-1}), whereas most frequently the benzophenone (67.5% of samples) detected in samples (Glassmeyer et al., 2005; Kolpin et al., 2002). Both studies (Marchini et al., 1992; Ura et al., 2002) examined the chronic and acute toxicity of benzophenone for Caenorhabditis elegans and P. promelas found that benzophenone is safer than aquatic organisms. Although it has only been found in small amounts and in a limited number of environmental samples, the other substance found in water is methyl salicylate, which is used to flavour and liniment wintergreen (Glassmeyer et al., 2005).

3. Hazard Assessment

A preliminary risk assessment was performed using hazard quotients for PCPs for which data on ambient concentration and chronic toxicity were available. The ratio that can be used to calculate the likelihood of unfavourable outcomes is called the hazard quotient. It is calculated by subtracting the exposure amount from the toxicological standard value (Suter, 2007). Hazard quotients above one signify possible impacts. Literature has revealed that triclocarban and triclosan are mainly responsible for producing chronic effects based on hazard quotients larger than one. In Kansas, the USA study, 0.12 lg L^{-1} TCS exposed for 96 hr towards the growth of algal assemblage was markedly inhibited and showed a hazard quotient of nearly 19 (Wilson et al., 2003). In addition, TCC exhibited a hazard quotient of 10.9 or higher for Americamysis bahia following a 28-d contact to 0.13 lg L^{-1} (TCC Consortium, 2002). As a result, the actual risk is probably significantly lesser than expected, which is consistent with early risk evaluations made elsewhere (Lyndall et al., 2010;

Costanzo et al., 2007). Hazard quotients could not be determined for UV filters since there are not enough long-term *in vivo* data available. Due to the lesser availability of *in vivo* studies of marine organisms, hazard quotients for putative endocrine impacts were also not estimated; nonetheless, research on mammals proved endocrine effects of UV filters. Non-Extractable Residues (NERs) are a major cause of PPCP dissipation to other natural sources (Boxall et al., 2004). Increased biomagnification bioavailability and bioaccumulation of PPCPs in soils, plants, aquatic mammals, animals, and humans are mostly caused by NERs. In fact, it has been demonstrated that PCPPs have the capacity to amplify and bioaccumulate in top species of the food chain, including large mammals, top predators, birds of prey and carnivorous fish. As per the report, arctic charr, polar foxes, arctic birds (glaucous gall), polar bears, orcas and whales are among the most vulnerable species to PCPPs (Garric et al., 2013). The harmful impact of ECs on human as well as on environment is shown in Fig. 10.2.

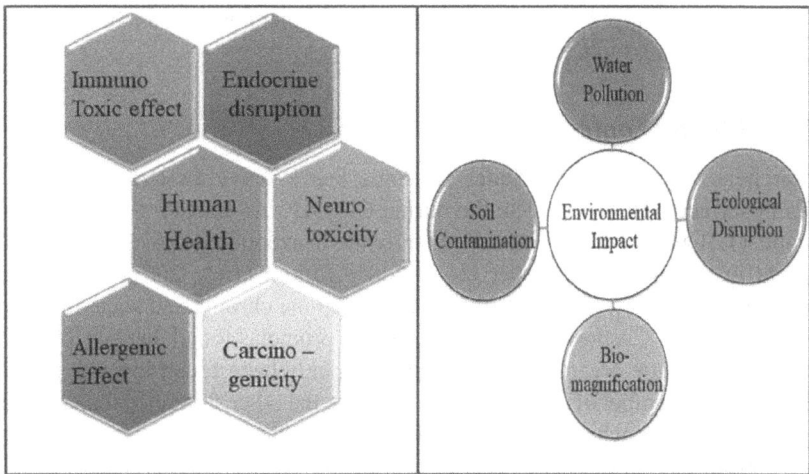

Figure 10.2 Harmful effect of ECs on human health and environment.

4. Current Dilemma and Discussion

One of the most fortunate discoveries of the modern era was the application of antibiotic and pharmaceutical therapy in the health care industry. This discovery gave rise to a vital network for complex and refined clinical intervention that allowed one to effectively extend people's lives all over the world. Therefore, the unintentional presence in the environment has increased as a result of the use, disposal, and excretion of ECs in marine systems. Bacteria have therefore developed similarly to other organisms in order to survive, developing complex defences against the onslaught of antimicrobial chemicals. According to studies, the creation of diverse ARB and ARGs has been significantly influenced by sources including the household, pharmaceutical industry, veterinary, municipal wastes, aquaculture, agriculture-related, veterinary, and anti-fouling paints effluents. These substances can induce maximum biological activity at low concentrations for an extended length of time.

This chapter emphasizes the accumulating negative effects of personal care items, pharmaceuticals and endocrine disruptor in aquatic and soil systems and people in entire cycle of life. Metabolites products of these dangerous substances are still present in various matrices of the ecosystem, despite the use of numerous modern degradation procedures. In our research, we discovered that PPCPs significantly altered bacterial communities and other living species in the terrestrial and aquatic environments. Prevalence of antibiotics including amoxicillin, tetracycline, beta-lactams, cefepime, erythromycin, ceftazidime, aztreonam, cefotaxime, sulphonamides, cefpodoxime, and many more in waste as well as natural water bodies has generated an obstruction to a number of microbial populations. Similarly, personal care products have a number of extremely harmful consequences. However, it is clear from the chapter that much more research is necessary to evaluate the impact of PPCPs on living beings. As the less studied organic class contaminants have displayed hazardous reactions. Most toxicological researches have shown that hormones, antidepressants, anti-inflammatory drugs, and analgesics can be found in water resources. A frequent component of food and personal care products, synthetic husks/fragrances, has not been the subject of very many investigations. In a similar vein, research into antioxidants needs to be speeded up. Publication reveals a knowledge gap regarding comprehensive knowledge of PPCPs' bioaccumulation and antibacterial activities. This analysis also highlighted the requirement for additional studies on PPCPs in aquatic resources of India. China is the first most populous nation followed by India and its aquatic bodies are known to be a major source of organic pollutants, ARGs, and ARB.

5. Conclusions

Pharmaceutical, personal care, and other new toxins will inevitably contaminate water as long as they are used as essential elements of a contemporary, healthy society. According to several studies, such pollutants are almost always present in the aquatic ecosystem but are infrequently found in some drinking waters. By using personal care products as intended, people discharge unaltered materials into the environment. ECs discharged in higher amounts than many other substances, including pharmaceuticals, but there has not been a lot of research to determine their environmental concentrations and potential toxicity. Regarding the factors and circumstances affecting the ecosystem destiny and weakening of ECs, a great deal remains to be understood. Some pollutants appear to bioaccumulate, more strongly attach to suspended particles or water sediments, volatilize, react with other natural components, and break down into unknown or insufficiently recognized metabolites, each of which may have different behaviours and fates. This idea has special difficulties for identifying environmental contaminants because some of them could only be present in specific aquatic matrices, which creates a great deal of confusion about where to look for them. PCPs are persistent substances that call for both acute and chronic investigations because they are persistent compounds that are continuously supplied by everyday use in the environment. Understanding the potential consequences and jeopardy of the discharge of ECs into various water matrices requires further research in both chronic and acute toxicity.

Acknowledgments

One of the authors Dr Manviri Rani is grateful for the funding from DST-SERB, New Delhi (Sanction order no. SRG/2019/000114) and TEQIP-III MNIT Jaipur, India. Dr Uma Shanker wishes to thank TEQIP-III, NIT Jalandhar for partial funding. Keshu, Rishabh, and Shikha are thankful to the Ministry of Education, New Delhi, India for providing funding.

References

Adeola, A.O. and Forbes, P.B. 2022. Antiretroviral drugs in African surface waters: prevalence, analysis, and potential remediation. Environ. Toxicol. Chem. 41(2): 247–262.

Agunbiade, F.O. and Moodley, B. 2014. Pharmaceuticals as emerging organic contaminants in Umgeni River water system, KwaZulu-Natal, South Africa. Environ. Monit. Assess. 186: 7273–7291.

Agunbiade, F.O. and Moodley, B. 2016. Occurrence and distribution pattern of acidic pharmaceuticals in surface water, wastewater, and sediment of the Msunduzi River, Kwazulu-Natal, South Africa. Environ. Toxicol. Chem. 35(1): 36–46.

Ali, A.M., Rønning, H.T., Arif, W.M. Al, Kallenborn, R. and Lihaibi, S. Al, 2017. Occurrence of pharmaceuticals and personal care products in effluent-dominated Saudi Arabian coastal waters of the Red Sea. ECSN. https://doi.org/10.1016/j. chemosphere.2017.02.095.

Baker, D.R., Kasprzyk-Hordern, B., 2013. Spatial and temporal occurrence of pharmaceuticals and illicit drugs in the aqueous environment and during wastewater treatment: new developments. Sci. Total Environ. 454–455, 442–456.

Barraza, L. 2013. A new approach for regulating bisphenol A for the protection of the public's health. The Journal of Law, Medicine & Ethics, 41: 9–12.

Benijts, T., Lambert, W. and De Leenheer, A. 2004. Analysis of multiple endocrine disruptors in environmental waters via wide-spectrum solid-phase extraction and dual-polarity ionization LC-ion trap-MS/MS. Anal. Chem. 76(3): 704–711.

Benotti, M.J., Trenholm, R.A., Vanderford, B.J., Holady, J.C., Stanford, B.D. and Snyder, S.A. 2009. Pharmaceuticals and endocrine disrupting compounds in US drinking water. Environ. Sci. Technol., 43(3): 597–603.

Bhandari, R.K., Deem, S.L., Holliday, D.K., Jandegian, C.M., Kassotis, C.D., Nagel, S.C. et al. 2015. Effects of the environmental estrogenic contaminants bisphenol A and 17α-ethinyl estradiol on sexual development and adult behaviors in aquatic wildlife species. Gen. Comp. Endocrinol. 214: 195–219.

Bhushan, S., Rana, M.S., Raychaudhuri, S., Simsek, H. and Prajapati, S.K. 2020. Algae-and bacteria-driven technologies for pharmaceutical remediation in wastewater. pp. 373–408. *In*: Removal of Toxic Pollutants Through Microbiological and Tertiary Treatment. Elsevier.

Bjerregaard, P., Hansen, P.R., Larsen, K.J., Erratico, C., Korsgaard, B. and Holbech, H. 2008. Vitellogenin as a biomarker for estrogenic effects in brown trout, Salmo trutta: laboratory and field investigations. Environ. Toxicol. Chem.: Int J. 27(11): 2387–2396.

Bo, L., Shengen, Z. and Chang, C.C. 2016. Emerging pollutants-part II: Treatment. Water Environment Research 88(10): 1876–1904.

Botelho, R.G., Christofoletti, C.A., Correia, J.E., Ansoar, Y., Olinda, R.A.D. and Tornisielo, V.L. 2015. Genotoxic responses of juvenile tilapia (Oreochromis niloticus) exposed to florfenicol and oxytetracycline. Chemosphere 132: 206–212.

Boutonnet, J.C., Thompson, R.S., De Rooij, C., Garny, V., Lecloux, A. and Van Wijk, D. 2004. 1, 4-Dichlorobenzene marine risk assessment with special reference to the OSPARCOM region: North Sea. Environ. Monit. Assess. 97: 103–117.

Boxall, A.B. 2004. The environmental side effects of medication: How are human and veterinary medicines in soils and water bodies affecting human and environmental health?. EMBO reports, 5(12): 1110–1116.

Boyd, G.R., Palmeri, J.M., Zhang, S. and Grimm, D.A. 2004. Pharmaceuticals and personal care products (PPCPs) and endocrine disrupting chemicals (EDCs) in stormwater canals and Bayou St. John in New Orleans, Louisiana, USA. Sci. Total Environ. 333(1–3): 137–148.

Brausch, J.M. and Rand, G.M. 2011. A review of personal care products in the aquatic environment: environmental concentrations and toxicity. Chemosphere 82(11): 1518–1532.

Brausch, J.M., Connors, K.A., Brooks, B.W. and Rand, G.M. 2012. Human pharmaceuticals in the aquatic environment: a review of recent toxicological studies and considerations for toxicity testing. Rev. Environ. Contam. Toxicol. 218: 1–99.

Charbonnet, J.A., Weber, R. and Blum, A. 2020. Flammability standards for furniture, building insulation and electronics: Benefit and risk. Emerg. Contam. 6: 432–441.

Chaturvedi, P., Shukla, P., Giri, B.S., Chowdhary, P., Chandra, R., Gupta, P. and Pandey, A. 2021. Prevalence and hazardous impact of pharmaceutical and personal care products and antibiotics in environment: A review on emerging contaminants. Environ. Res. 194: 110664.

Chen, F., Gao, J. and Zhou, Q. 2012. Toxicity assessment of simulated urban runoff containing polycyclic musks and cadmium in Carassius auratus using oxidative stress biomarkers. Environ. Pollut. 162: 91–97.

Chen, J., Liu, Y.S., Deng, W.J. and Ying, G.G. 2019. Removal of steroid hormones and biocides from rural wastewater by an integrated constructed wetland. Sci. Total Environ. 660: 358–365.

Chen, X.Q., Cho, S.J., Li, Y. and Venkatesh, S. 2002. Prediction of aqueous solubility of organic compounds using a quantitative structure–property relationship. J. Pharm. Sci. 91(8): 1838–1852.

Cheng, N., Wang, B., Wu, P., Lee, X., Xing, Y., Chen, M. and Gao, B. 2021. Adsorption of emerging contaminants from water and wastewater by modified biochar: A review. Environ. Pollut. 273: 116448.

Chokwe, T.B., Okonkwo, J.O., Sibali, L.L. and Ncube, E.J. 2015. Alkylphenol ethoxylates and brominated flame retardants in water, fish (carp) and sediment samples from the Vaal River, South Africa. Environ Sci. Pollut. Res. 22: 11922–11929.

Colin, A., Bach, C., Rosin, C., Munoz, J. F. and Dauchy, X. 2014. Is drinking water a major route of human exposure to alkylphenol and bisphenol contaminants in France?. Arch. Environ. Contam. Toxicol., 66, 86-99.

Costanzo, S.D., Watkinson, A.J., Murby, E.J., Kolpin, D.W. and Sandstrom, M.W. 2007. Is there a risk associated with the insect repellent DEET (N, N-diethyl-m-toluamide) commonly found in aquatic environments?. Sci. Total Environ. 384(1–3): 214–220.

Czech, B., Jośko, I. and Oleszczuk, P. 2014. Ecotoxicological evaluation of selected pharmaceuticals to Vibrio fischeri and Daphnia magna before and after photooxidation process. Ecotoxicol. Environ. Saf. 104: 247–253.

Daughton, C.G. and Ternes, T.A. 1999. Pharmaceuticals and personal care products in the environment: agents of subtle change? Environmental Health Perspectives 107(suppl 6): 907–938.

Daughton, C.G. 2004. Non-regulated water contaminants: emerging research. Environ. Impact Assess. Rev. 24(7–8): 711–732.

Davis, E.E. 1985. Insect repellents: concepts of their mode of action relative to potential sensory mechanisms in mosquitoes (Diptera: Culicidae). Journal of Medical Entomology 22(3): 237–243.

de Oliveira, L.L.D., Antunes, S.C., Gonçalves, F., Rocha, O. and Nunes, B. 2016. Acute and chronic ecotoxicological effects of four pharmaceuticals drugs on cladoceran Daphnia magna. Drug and Chemical Toxicology 39(1): 13–21.

Dietrich, S., Ploessl, F., Bracher, F. and Laforsch, C. 2010. Single and combined toxicity of pharmaceuticals at environmentally relevant concentrations in Daphnia magna–A multigenerational study. Chemosphere 79(1): 60–66.

Dinh, Q., Moreau-Guigon, E., Labadie, P., Alliot, F., Teil, M.J., Blanchard, M. et al. 2017. Fate of antibiotics from hospital and domestic sources in a sewage network. Sci. Total Environ. 575: 758–766.

Dobbins, L.L., Usenko, S., Brain, R.A. and Brooks, B.W. 2009. Probabilistic ecological hazard assessment of parabens using Daphnia magna and Pimephales promelas. Environ. Toxicol. Chem. 28(12): 2744–2753.

Dougherty, J.A., Swarzenski, P.W., Dinicola, R.S. and Reinhard, M. 2010. Occurrence of herbicides and pharmaceutical and personal care products in surface water and groundwater around Liberty Bay, Puget Sound, Washington. J. Environ. Qual. 39(4): 1173–1180.

Du, J., Mei, C.F., Ying, G.G. and Xu, M.Y. 2016. Toxicity thresholds for diclofenac, acetaminophen and ibuprofen in the water flea Daphnia magna. Bull. Environ. Contam. Toxicol. 97: 84–90.

Fent, K., Kunz, P.Y. and Gomez, E. 2008. UV filters in the aquatic environment induce hormonal effects and affect fertility and reproduction in fish. Chimia 62(5): 368–368.

Foran, C.M., Bennett, E.R. and Benson, W.H. 2000. Developmental evaluation of a potential non-steroidal estrogen: triclosan. Marine Environ. Res. 50(1–5): 153–156.

Fort, D.J., Rogers, R.L., Gorsuch, J.W., Navarro, L.T., Peter, R. and Plautz, J.R. 2010. Triclosan and anuran metamorphosis: no effect on thyroid-mediated metamorphosis in Xenopus laevis. Toxicol. Sci. 113(2): 392–400.

Galinaro, C.A., Pereira, F.M. and Vieira, E.M. 2015. Determination of parabens in surface water from Mogi Guaçu River (São Paulo, Brazil) using dispersive liquid-liquid microextraction based on low density solvent and LC-DAD. J. Braz. Chem. Soc. 26: 2205–2213.

Garric, J., Férard, J.F. and Blaise, C. 2013. Emerging issues in ecotoxicology: pharmaceuticals and personal care products (PPCPs).

Gee, R.H., Charles, A., Taylor, N. and Darbre, P.D. 2008. Oestrogenic and androgenic activity of triclosan in breast cancer cells. J. Appl. Toxicol. Int. J. 28(1): 78–91.

Glassmeyer, S.T., Furlong, E.T., Kolpin, D.W., Cahill, J.D., Zaugg, S.D., Werner, S.L. et al. 2005. Transport of chemical and microbial compounds from known wastewater discharges: potential for use as indicators of human fecal contamination. Environ. Sci. Technol. 39(14): 5157–5169.

Gogoi, A., Mazumder, P., Tyagi, V.K., Chaminda, G.T., An, A.K. and Kumar, M. 2018. Occurrence and fate of emerging contaminants in water environment: a review. Groundw. Sustain. Dev. 6: 169–180.

Golden, R., Gandy, J. and Vollmer, G. 2005. A review of the endocrine activity of parabens and implications for potential risks to human health. Crit. Rev. Toxicol. 35(5): 435–458.

Gomes, A.R., Justino, C., Rocha-Santos, T., Freitas, A.C., Duarte, A.C. and Pereira, R. 2017. Review of the ecotoxicological effects of emerging contaminants to soil biota. J. Environ. Sci. Health A J ENVIRON SCI HEAL A, Part A 52(10): 992–1007.

Grignard, E., Lapenna, S. and Bremer, S. 2012. Weak estrogenic transcriptional activities of Bisphenol A and Bisphenol S. Toxicol *In Vitro*. 26(5): 727–731.

Halden, R.U. and Paull, D.H. 2005. Co-occurrence of triclocarban and triclosan in US water resources. Environ. Sci. Technol. 39(6): 1420–1426.

Hoenicke, R., Oros, D.R., Oram, J.J. and Taberski, K.M. 2007. Adapting an ambient monitoring program to the challenge of managing emerging pollutants in the San Francisco Estuary. Environ. Res. 105(1): 132–144.

Huang, D.J., Hou, J.H., Kuo, T.F. and Lai, H.T. 2014. Toxicity of the veterinary sulfonamide antibiotic sulfamonomethoxine to five aquatic organisms. Environ. Toxicol. Pharmacol. 38(3): 874–880.

Iancu, V.I., Radu, G.L. and Scutariu, R. 2019. A new analytical method for the determination of beta-blockers and one metabolite in the influents and effluents of three urban wastewater treatment plants. Anal. Methods. 11(36): 4668–4680.

Inui, M., Adachi, T., Takenaka, S., Inui, H., Nakazawa, M., Ueda, M. et al. 2003. Effect of UV screens and preservatives on vitellogenin and choriogenin production in male medaka (Oryzias latipes). J. Toxicol. 194(1–2): 43–50.

Ishibashi, H., Matsumura, N., Hirano, M., Matsuoka, M., Shiratsuchi, H., Ishibashi, Y. et al. 2004. Effects of triclosan on the early life stages and reproduction of medaka Oryzias latipes and induction of hepatic vitellogenin. Aquat. Toxicol. 67(2): 167–179.

Izadi, P., Izadi, P., Salem, R., Papry, S. A., Magdouli, S., Pulicharla, R. et al. 2020. Non-steroidal anti-inflammatory drugs in the environment: Where were we and how far we have come?. Environ. Pollut. 267: 115370.

Keshu, Rani, M. and Shanker, U. 2022. Efficient removal of plastic additives by sunlight active titanium dioxide decorated Cd–Mg ferrite nanocomposite: Green synthesis, kinetics and photoactivity. Chemosphere 290: 133307.

Keshu, Rani, M., Yadav, J., Chaudhary, S. and Shanker, U. 2021. An updated review on synthetic approaches of green nanomaterials and their application for removal of water pollutants: Current challenges, assessment and future perspectives. J. Environ. Chem. Eng. 9(6): 106763.

Kolpin, D.W., Furlong, E.T., Meyer, M.T., Thurman, E.M., Zaugg, S.D., Barber, L.B. et al. 2002. Pharmaceuticals, hormones, and other organic wastewater contaminants in US streams, 1999–2000: A national reconnaissance. Environ. Sci. Technol. 36(6): 1202–1211.

Kolpin, D.W., Skopec, M., Meyer, M.T., Furlong, E.T. and Zaugg, S.D. 2004. Urban contribution of pharmaceuticals and other organic wastewater contaminants to streams during differing flow conditions. Sci. Total Environ. 328(1–3): 119–130.

Kosma, C.I., Lambropoulou, D.A. and Albanis, T.A. 2010. Occurrence and removal of PPCPs in municipal and hospital wastewaters in Greece. J. Hazard Mater. 179: 804–817.

Kunz, P.Y., Galicia, H.F. and Fent, K. 2006. Comparison of *in vitro* and *in vivo* estrogenic activity of UV filters in fish Toxicol. Sci. 90(2): 349–361.

Kunz, P.Y., Galicia, H.F. and Fent, K. 2006c. Comparison of *in vivo* estrogenic activity of UV filters in fish. Toxicol. Sci. 90: 349–361.

López-Serna, R., Jurado, A., Vázquez-Súñe, E., Carrera, J., Petrović, M. and Barceló, D. 2013.Occurrence of 95 pharmaceuticals and transformation products in urban groundwaters underlying the metropolis of Barcelona, Spain. Environ. Pollut. 174: 305–315.

La Farre, M., Pérez, S., Kantiani, L. and Barceló, D. 2008. Fate and toxicity of emerging pollutants, their metabolites and transformation products in the aquatic environment. TrAC, Trends Anal. Chem. 27(11): 991–1007.

Lang, I.A., Galloway, T.S., Scarlett, A., Henley, W.E., Depledge, M., Wallace, R.B. et al. 2008. Association of urinary bisphenol A concentration with medical disorders and laboratory abnormalities in adults. Jama 300(11): 1303–1310.

Lee, I.S., Lee, S.H. and Oh, J.E. 2010. Occurrence and fate of synthetic musk compounds in water environment. Water Res. 44(1): 214–222.

Li, H., Helm, P.A. and Metcalfe, C.D. 2010. Sampling in the Great Lakes for pharmaceuticals, personal care products, and endocrine-disrupting substances using the passive polar organic chemical integrative sampler. Environ. Toxicol. Chem. Intern. J. 29(4): 751–762.

Lindström, A., Buerge, I.J., Poiger, T., Bergqvist, P.A., Müller, M.D. and Buser, H.R. 2002. Occurrence and environmental behavior of the bactericide triclosan and its methyl derivative in surface waters and in wastewater. Environ. Sci. Technol. 36(11): 2322–2329.

Lopez, A. and Ra, H. 1980. Organic Compounds in an Industrial Wastewater. Their Transport Into Sediments.

Loraine, G.A. and Pettigrove, M.E. 2006. Seasonal variations in concentrations of pharmaceuticals and personal care products in drinking water and reclaimed wastewater in southern California. Environ. Sci. Technol. 40(3): 687–695.

Lyndall, J., Fuchsman, P., Bock, M., Barber, T., Lauren, D., Leigh, K. et al. 2010. Probabilistic risk evaluation for triclosan in surface water, sediments, and aquatic biota tissues. Integrated Environmental Assessment and Management: Int. J.

Mahapatra, S., Samal, K. and Dash, R.R. 2022. Waste Stabilization Pond (WSP) for wastewater treatment: A review on factors, modelling and cost analysis. J. Environ. Manage 308: 114668.

Maia, J., Cruz, J.M., Sendón, R., Bustos, J., Sanchez, J.J. and Paseiro, P. 2009. Effect of detergents in the release of bisphenol A from polycarbonate baby bottles Int. Food Res. J. 42(10): 1410–1414.

Marchini, S., Tosato, M.L., Norberg-King, T.J., Hammermeister, D.E. and Hoglund, M.D. 1992. Lethal and sublethal toxicity of benzene derivatives to the fathead minnow, using a short-term test. Environ. Toxicol. Chem.: Int. J. 11(2): 187–195.

Maszkowska, J., Stolte, S., Kumirska, J., Łukaszewicz, P., Mioduszewska, K., Puckowski, A. et al. 2014. Beta-blockers in the environment: Part II. Ecotoxicity study. Sci. Total Environ. 493: 1122–1126.

Matongo, S., Birungi, G., Moodley, B. and Ndungu, P. 2015. Pharmaceutical residues in water and sediment of Msunduzi River, kwazulu-natal, South Africa. Chemosphere 134: 133–140.

Matongo, S., Birungi, G., Moodley, B. and Ndungu, P. 2015. Pharmaceutical residues in water and sediment of Msunduzi River, kwazulu-natal, South Africa. Chemosphere 134: 133–140.

McAvoy, D.C., Schatowitz, B., Jacob, M., Hauk, A. and Eckhoff, W.S. 2002. Measurement of triclosan in wastewater treatment systems. Environ. Toxicol. Chem.: An International Journal 21(7): 1323–1329.

Mejía-García, A., Islas-Flores, H., Gómez-Oliván, L.M., SanJuan-Reyes, N., Ortega-Olvera, J.M. and Hernández-Navarro, M.D. 2020. Overview of non-steroidal anti-inflammatory drugs as emerging contaminants. Non-Steroidal Anti-Inflammatory Drugs in Water: Emerging Contaminants and Ecological Impact 41–53.

Merchant Research and Consulting. 2015. Bisphenol A (BPA): 2015 World Market Outlook and Forecast up to 2019. [Strategic Report-PDF], Market Publishers Report Database, ID: B6B1814C5BBEN,

Available at: <https://marketpublishers.com/report/industry/chemicals_petrochemicals/bisphenol_a_world_market_outlook_n_forecast.html>.

Novak, M., Žegura, B., Modic, B., Heath, E. and Filipič, M. 2017. Cytotoxicity and genotoxicity of anticancer drug residues and their mixtures in experimental model with zebrafish liver cells. Sci. Total Environ. 601: 293–300.

Nunes, B., Antunes, S.C., Santos, J., Martins, L. and Castro, B.B. 2014. Toxic potential of paracetamol to freshwater organisms: A headache to environmental regulators? Ecotoxicol. Environ. Saf. 107: 178–185, https://doi.org/10.1016/j. ecoenv.2014.05.027.

Ogunwole, G.A., Saliu, J.K., Osuala, F.I. and Odunjo, F.O. 2021. Chronic levels of ibuprofen induces haematoxic and histopathology damage in the gills, liver, and kidney of the African sharptooth catfish (Clarias gariepinus). Environ. Sci. Pollut. Res. 28: 25603–25613.

Oishi, S. 2002. Effects of propyl paraben on the male reproductive system. Food and Chemical Toxicology 40(12): 1807–1813.

Ojoghoro, J.O., Scrimshaw, M.D. and Sumpter, J.P. 2021. Steroid hormones in the aquatic environment. Sci. Total Environ. 792: 148306.

Peck, A.M. 2006. Analytical methods for the determination of persistent ingredients of personal care products in environmental matrices. Anal. Bioanal. Chem. 386: 907–939.

Peng, X., Ou, W., Wang, C., Wang, Z., Huang, Q., Jin, J. and Tan, J. 2014. Occurrence and ecological potential of pharmaceuticals and personal care products in groundwater and reservoirs in the vicinity of municipal landfills in China. Sci. Total Environ 490: 889–898.

Peng, X., Ou, W., Wang, C., Wang, Z., Huang, Q., Jin, J. et al. 2014. Occurrence and ecological potential of pharmaceuticals and personal care products in groundwater and reservoirs in the vicinity of municipal land fi lls in China. Sci. Total Environ. 490: 889–898.

Peng, Y., Gautam, L. and Hall, S.W. 2019. The detection of drugs of abuse and pharmaceuticals in drinking water using solid-phase extraction and liquid chromatography-mass spectrometry. Chemosphere 223: 438–447.

Quednow, K. and Püttmann, W. 2009. Temporal concentration changes of DEET, TCEP, terbutryn, and nonylphenols in freshwater streams of Hesse, Germany: possible influence of mandatory regulations and voluntary environmental agreements. Environmental Science and Pollution Research, 16: 630–640.

Rani, M., Keshu, Meenu, Sillanpää, M. and Shanker, U. 2022. An updated review on environmental occurrence, scientific assessment and removal of brominated flame retardants by engineered nanomaterials. J. Environ. Manag. 321: 115998. https://doi.org/10.1016/j.jenvman.2022.115998

Rathi, B.S., Kumar, P.S. and Show, P.L. 2021. A review on effective removal of emerging contaminants from aquatic systems: Current trends and scope for further research. J. Hazard. Mater. 409: 124413.

Raut, S.A. and Angus, R.A. 2010. Triclosan has endocrine-disrupting effects in male western mosquito fish, Gambusia affinis. Environmental Toxicology and Chemistry 29(6): 1287–1291.

Reisch, M.S. 2005. Battle tested. Chem. Eng. News, 83: 18–22.

Rodil, R., Quintana, J.B., Concha-Graña, E., López-Mahía, P., Muniategui-Lorenzo, S. and Prada-Rodríguez, D. 2012. Emerging pollutants in sewage, surface and drinking water in Galicia (NW Spain). Chemosphere 86(10): 1040–1049.

Rodil, R., Quintana, J.B., López-Mahía, P., Muniategui-Lorenzo, S. and Prada-Rodríguez, D. 2008. Multiclass determination of sunscreen chemicals in water samples by liquid chromatography–tandem mass spectrometry. Anal. Chem. 80(4): 1307–1315.

Roosens, L., Covaci, A. and Neels, H. 2007. Concentrations of synthetic musk compounds in personal care and sanitation products and human exposure profiles through dermal application. Chemosphere 69(10): 1540–1547.

Routledge, E.J., Parker, J., Odum, J., Ashby, J. and Sumpter, J.P. 1998. Some alkyl hydroxy benzoate preservatives (parabens) are estrogenic. Toxicol. Appl. Pharmacol. 153(1): 12–19.

Samal, K. and Trivedi, S. 2020. A statistical and kinetic approach to develop a Floating Bed for the treatment of wastewater. J. Environ. Chem. Eng. 8(5): 104102.

Samal, K. and Dash, R.R. 2021. Modelling of pollutants removal in Integrated Vermifilter (IVmF) using response surface methodology. Cleaner Engineering and Technology 2: 100060.

Samaras, V.G., Stasinakis, A.S., Mamais, D., Thomaidis, N.S. and Lekkas, T.D. 2013. Fate of selected pharmaceuticals and synthetic endocrine disrupting compounds during wastewater treatment and sludge anaerobic digestion J. Hazard. Mater. 244: 259–267.

Sandstrom, M.W., Kolpin, D.W., Thurman, E.M. and Zaugg, S.D. 2005. Widespread detection of N, N-diethyl-m-toluamide in US Streams: Comparison with concentrations of pesticides, personal care products, and other organic wastewater compounds. Environ. Toxicol. Chem.: Int J. 24(5): 1029–1034.

Schlumpf, M., Cotton, B., Conscience, M., Haller, V., Steinmann, B. and Lichtensteiger, W. 2001. *In vitro* and *in vivo* estrogenicity of UV screens. Environ. Health Perspect. 109(3): 239–244.

Schreurs, R., Lanser, P., Seinen, W. and van der Burg, B. 2002. Estrogenic activity of UV filters determined by an in vitro reporter gene assay and an *in vivo* transgenic zebrafish assay. Arch. Toxicol. 76.

Schwaiger, J., Ferling, H., Mallow, U., Wintermayr, H. and Negele, R.D. 2004. Toxic effects of the non-steroidal anti-inflammatory drug diclofenac: Part I: histopathological alterations and bioaccumulation in rainbow trout. Aquat. Toxicol. 68(2): 141–150.

Sehrawat, A., Phour, M., Kumar, R. and Sindhu, S.S. 2021. Bioremediation of pesticides: an eco-friendly approach for environment sustainability. Microbial Rejuvenation of Polluted Environment: Volume 1: 23–84.

Shahid, M.K., Kashif, A., Fuwad, A. and Choi, Y. 2021. Current advances in treatment technologies for removal of emerging contaminants from water—A critical review. Coord. Chem. Rev. 442: 213993.

Sharma, M., Verma, S. and Sharma, P. 2019. Behavioural and genotoxic effects of paracetamol after subchronic exposure to Cyprinus carpio. J. Entomol. Zool. Stud. 7(3): 22–25.

Soni, M.G., Carabin, I.G. and Burdock, G.A. 2005. Safety assessment of esters of p-hydroxybenzoic acid (parabens). Food Chem. Toxicol. 43(7): 985–1015.

Suter, G.W. 2007. Ecological risk assessment, second ed. CRC Press, Boca Raton, FL.

TCC Consortium, 2002. <http://www.epa.gov/chemrtk/tricloca/c14186cv.pdf>.

Ternes, T.A., Herrmann, N., Bonerz, M., Knacker, T., Siegrist, H. and Joss, A. 2004. A rapid method to measure the solid–water distribution coefficient (Kd) for pharmaceuticals and musk fragrances in sewage sludge. Water Res. 38(19): 4075–4084.

The American Chemical Society. CAS Registry—the gold standard for chemical substance information. Available online: https://www.cas.org/support/documentation/chemical-substances (Accessed on 31 March 2021).

Tran, N.H., Reinhard, M., Khan, E., Chen, H., Nguyen, V. T., Li, Y. et al. 2019. Emerging contaminants in wastewater, stormwater runoff, and surface water: Application as chemical markers for diffuse sources. Sci. Total Environ. 676: 252–267.

Tyumina, E.A., Bazhutin, G.A., Cartagena Gómez, A.D.P. and Ivshina, I.B. 2020. Nonsteroidal anti-inflammatory drugs as emerging contaminants. Microbiology 89: 148–163.

United States Environmental Protection Agency, 1998. Reregistration Eligibility Decision (RED) for DEET. EPA 738-R-98-010. Washington, DC.

United States Environmental Protection Agency 2014. Bisphenol A alternatives in thermal paper final report, [pdf], (last accessed 3 Nov. 2015), available at:<http://www2.epa.gov/sites/production/files/2014-05/documents/bpa_final.pdf>.

Ura, K., Kai, T., Sakata, S., Iguchi, T. and Arizono, K. 2002. Aquatic acute toxicity testing using the nematode Caenorhabditis elegans. J. HEALTH Sci. 48(6): 583–586.

U.S. Department of Defense (DoD), 2011. Emerging Chemical and Material Risks. Chemical and Material Risk Management Program ⟨www.denix.osd.mil/cmrmd/ECMR/RDX/TheBasics.cfm⟩.

US EPA. 2012. Water: Contaminant Candidate List 3. US. Environmental Protection Agency, Washington, DC ⟨http://water.epa.gov/scitech/drinkingwater/dws/ccl/ccl3.cfm⟩.

Veldhoen, N., Skirrow, R.C., Osachoff, H., Wigmore, H., Clapson, D.J., Gunderson, M.P. et al. 2006. The bactericidal agent triclosan modulates thyroid hormone-associated gene expression and disrupts postembryonic anuran development. Aquat. Toxicol. 80(3): 217–227.

Venkidasamy, B., Subramanian, U., Samynathan, R., Rajakumar, G., Shariati, M.A., Chung, I.M. et al. 2021. Organopesticides and fertility: where does the link lead to?. Environ. Sci. Pollut. Res. 28: 6289–6301.

Vulliet, E. and Cren-Olivé, C. 2011. Screening of pharmaceuticals and hormones at the regional scale, in surface and groundwaters intended to human consumption. Environ. Pollut. 159: 2929–2934.

Waleng, N.J. and Nomngongo, P.N. 2022. Occurrence of pharmaceuticals in the environmental waters: African and Asian perspectives. Environmental Chemistry and Ecotoxicology 4: 50–66.

Wilkinson, J.L., Swinden, J., Hooda, P.S., Barker, J. and Barton, S. 2016. Markers of anthropogenic contamination: A validated method for quantification of pharmaceuticals, illicit drug metabolites, perfluorinated compounds, and plasticisers in sewage treatment effluent and rain runoff. Chemosphere 159: 638–646.

Wilson, B.A., Smith, V.H., deNoyelles, F. and Larive, C.K. 2003. Effects of three pharmaceutical and personal care products on natural freshwater algal assemblages. Environ. Sci. Technol. 37(9): 1713–1719.

Wood, T.P., Basson, A.E., Duvenage, C. and Rohwer, E.R. 2016. The chlorination behaviour and environmental fate of the antiretroviral drug nevirapine in South African surface water. Water Res., 104: 349–360.

Wood, T.P., Du Preez, C., Steenkamp, A., Duvenage, C. and Rohwer, E.R. 2017. Database-driven screening of South African surface water and the targeted detection of pharmaceuticals using liquid chromatography-high resolution mass spectrometry. Environ. Pollut. 230: 453–462.

Yamazaki, E., Yamashita, N., Taniyasu, S., Lam, J., Lam, P.K., Moon, H.B. et al. 2015. Bisphenol A and other bisphenol analogues including BPS and BPF in surface water samples from Japan, China, Korea and India. Ecotoxicol. Environ. Saf. 122: 565–572.

Yang, G.C., Yen, C.H. and Wang, C.L. 2014. Monitoring and removal of residual phthalate esters and pharmaceuticals in the drinking water of Kaohsiung City, Taiwan. J. Hazard. Mater. 277: 53–61.

Yang, Y., Lu, L., Zhang, J., Yang, Y., Wu, Y. and Shao, B. 2014. Simultaneous determination of seven bisphenols in environmental water and solid samples by liquid chromatography–electrospray tandem mass spectrometry. J. Chromatogr. A 1328: 26–34.

Yang, Y., Wang, P., Shi, S. and Liu, Y. 2009. Microwave enhanced Fenton-like process for the treatment of high concentration pharmaceutical wastewater. Hazard. Mater. 168(1): 238–245.

Zenker, A., Cicero, M.R., Prestinaci, F., Bottoni, P. and Carere, M. 2014. Bioaccumulation and biomagnification potential of pharmaceuticals with a focus to the aquatic environment. J. Environ. Manage. 133: 378–387.

Zhang, K., Zhao, Y. and Fent, K. 2020. Cardiovascular drugs and lipid regulating agents in surface waters at global scale: Occurrence, ecotoxicity and risk assessment. Sci. Total Environ. 729: 138770.

Zhang, P., Zhou, H., Li, K., Zhao, X. and Liu, Q. 2018b. Occurrence of pharmaceuticals and personal care products, and their associated environmental risks in a large shallow lake in north China. Environ. Geochem. Health. https://doi.org/10.1007/s10653-018-0069-0.

Zhang, Y., Gu, A.Z., Cen, T., Li, X., He, M., Li, D. et al. 2018a. Sub-inhibitory concentrations of heavy metals facilitate the horizontal transfer of plasmid-mediated antibiotic resistance genes in water environment. Environ. Pollut. 237: 74–82.

CHAPTER 11

Legal and Safety Issues of ECs

Manviri Rani,[1],* Rishabh,[2] Keshu,[1,2] Shikha Sharma[2] and Uma Shanker[2],*

1. Introduction

A wide class of compounds known as Emerging Contaminants (ECs) are found in the environment as a result of routine human endeavours. Despite ECs being in the environment is not novel, their attention has been increased as a result of the development of more contaminant-sensitive techniques. which have shown that these pollutants can be found in the earth's biota at concentrations as low as ppt (parts per trillions). ECs are made up of about 3000 different types of compounds and their derivatives, including pharmaceutically active compounds, heavy metals, pesticides, microplastics, fertilizers, personal care products and hormones that usually end up in the ecosystem (Keshu et al., 2022; Gabriel et al., 2015). These kinds of pollutants are categorized as pseudo-persistent pollutants as they are polar, have a short half-life, and can be identified in trace amounts in the ecosystem (Keshu et al., 2021; Rani et al., 2021; Puri et al., 2022). The different sources of ECs are shown in Fig. 11.1. (WWTPs) Urban wastewater treatment plants are thought to be a potential foundation of ECs in the ecosystem because they can not be removed using old conventional treatments like disinfection (Rizzo et al., 2019), filtration (Krzeminski et al., 2019) and activated sludge (Rizzo et al., 2015). According to Wee and Aris (2018), ECs can cause reproductive and endocrine impairments in both humans and wildlife. Antibiotics and acquired Antibiotic Resistance Genes (ARGs), activated by Antibiotic-Resistant Bacteria (ARB), are among the biological and chemical contaminants that are currently prevalent in the environment and constitute a threat to the ecosystems and public health.

[1] Department of Chemistry, Malaviya National Institute of Technology Jaipur, Rajasthan-302017-India.

[2] Department of Chemistry, Dr B R Ambedkar National Institute of Technology, Jalandhar, Punjab, India-144011.

* Corresponding authors: Shankeru@nitj.ac.in; manviri.chy@mnit.ac.in

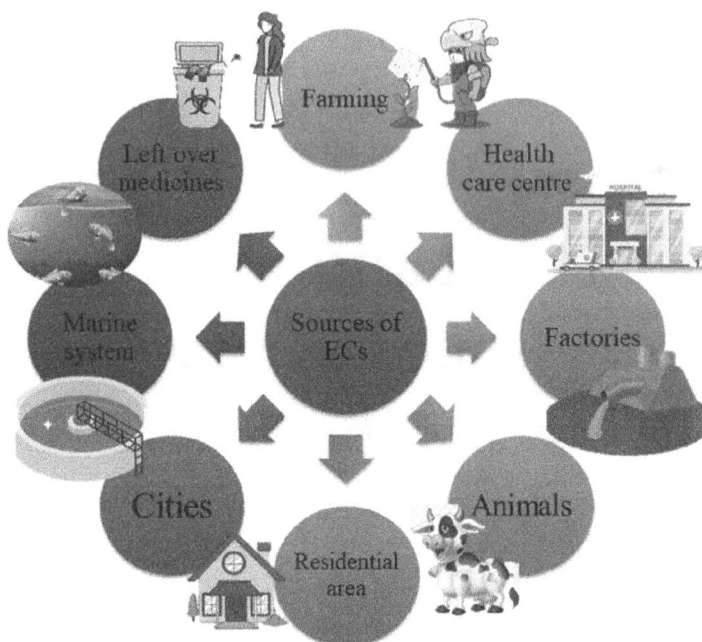

Figure 11.1 Different sources of ECs.

Due to their widespread presence in the ecosystem, drinking water and food chain, ARGs and ARB are considered a serious public health concern by the European Union and other international organizations (Berendonk et al., 2015). Environmental policies usually benefit from international agreements on the management of waste and synthetic chemicals. Following the Basel Convention, many nations implemented their own national regulations for exporting and importing hazardous waste. Many developing countries continue to export hazardous materials like e-waste, textiles, and plastics despite having ratified numerous international conventions because these agreements lack strong regulatory enforcement (UNEP, 2019a). Recently writings by Sanganyado Centre have drawn attention on the difficulties in controlling environmental chemicals. The studies look at the effects on the populace of the region's high health burden in sub-Saharan Africa and provide fresh approaches to move towards competent environmental control, respectively (Sanganyado, 2022). This chapter outlines the current state of knowledge regarding the major global issue of unregulated chemical discharge into the environment, pointing out the flaws in the policies addressing the issue, and makes the case for the need of strict policies to control overall pathway of ECs in the environment. The report also analyzes the management of EC pollution on a global scale, citing the laws and policies of both emerging and industrialized nations that rank among the top 10 producers of chemicals globally (Cefic, 2022). In order to assess the necessity to regulate the new or unidentified ECs in the ecosystem, the chapter summarizes the status of policies in developed and developing nations. This calls for the creation of a special cell and the inclusion of ECs in regular checking programmes of governments. Highlighting the shortcomings of the current policies, which are location-specific and lack comprehensive, enforceable laws and regulations that should be put into place on a worldwide scale.

2. Overview of Emerging Contaminants in the Environment

These refer to the recently identified toxins that are mobile over long distances, dangerous to human health, and have an impact on both terrestrial and aquatic ecosystems even at low concentrations (Ma et al., 2018). These anthropogenic toxins are divided into groups based on how likely it is that they could threaten both human health and the environment (USEPA, 2012). The three most frequent sources of ECs, in addition to other sources, are septic tank seepage, WWTP effluent and landfill leachate (Maria et al., 2018; Roberts et al., 2015). Chemical micropollutant-containing waste discharges from commercial, animal/poultry farms residential, agricultural, and industrial decrease biological activity, which eventually lowers the effectiveness of removing ECs from various environmental sources (Lopez-serna et al.,). The fate of these compounds in aquatic systems, the atmosphere and soil is determined by the ECs' persistence, polarity, environmental compartment characteristics and volatility. Low rates of biological, photodegradation and chemical are all possible. They frequently bioaccumulate in fauna, sediments, and river vegetation. As a result, there is essentially no evidence of their disappearing from the aquatic environment (Verlicchi et al., 2012; Yan et al., 2014). EDCs as shown in Fig. 11.2, which include phenols, phthalates, flame retardants, and many more, that are known to disturb a person's physiological homeostasis (Puri et al., 2023).

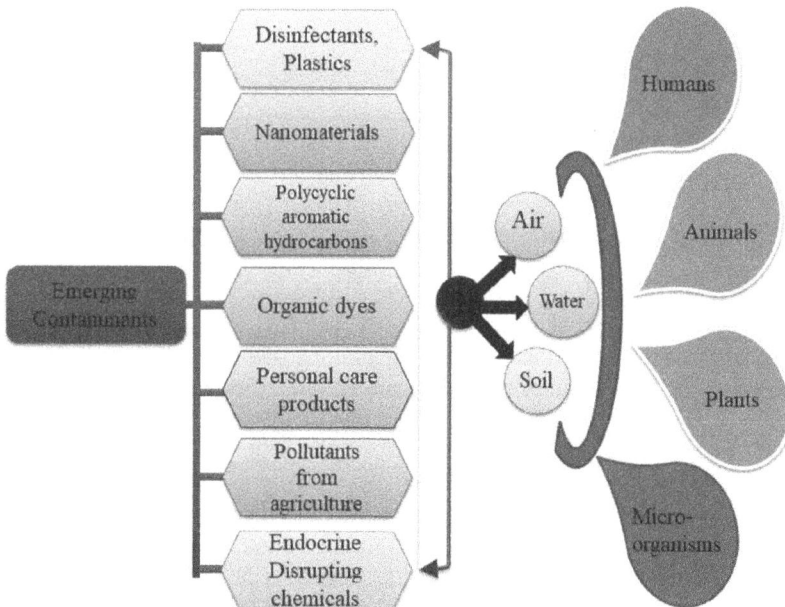

Figure 11.2 Different types of emerging contaminants.

Estrogenic hormones such 17-estradiol (E2) and 17-ethinylestradiol (EE2) were detected in sediment and WWTPs in Spain and California at 48.1 g/kg and 3.9 ng/L, respectively. At the same time, triclosan levels in Indian WWTPs ranged from 145 to 4890 ng/L (Chaturvedi et al., 2021). These substances have reproductive toxicity, are cytotoxic and genotoxic, and are carcinogenic (Xu et al., 2020). Flame retardants, which are used in the chemical and textile industries , are detected in drinking water at levels of ND-2.4 ng/L, ND-4.2 ng/L, and 462-409 ng/L (Yao et al., 2021). They cause toxicity at different stages of a creature's growth. Furthermore, 70% of all agricultural land worldwide is impacted by pesticide use. The three subcategories of pesticides are fungicides, herbicides, and insecticides. The fungicide market reached about US $ 17.58 billion in 2022. Three pesticides that are most frequently used—chlorpyrifos, carbaryl, and terbuthylazine—cause severe growth consequences, respiratory issues, and neurotoxicity in an organism. While 158 pesticide intermediates were tested in a different study by Reemtsma et al., (2013) with concentrations ranging from 0.33 g/L in surface water to 0.62 g/L in groundwater to . The National Primary Drinking Water Regulations (NPDWRs) of the EPA regulate 88 water pollutants, while the National Secondary Drinking Water Regulations (NSDWRs) regulate 15 more pollutants. Due to the presence of pollutants, these recommendations could have considerable aesthetic consequences on taste, aroma, or colour (Salimi et al., 2017). Mixed ECs' global sensitivity analysis infers harmful impacts on the biota, as well as on micro- and macroinvertebrates.

In their study of the combined ECs nonylphenols (NP) and BPA, Martins et al. (2022) highlighted the cascading impacts on the food chain. By altering the physico-chemical characteristics of estuarine water and sediments, the study even highlighted deadly and non-lethal harmful consequences on biota. Table 11.1 listed the different classes of ECs with their concentration detected and effects on human and environment.

3. Status of Policy and Legislation Among Developed Countries

3.1 *Status of Policy and Legislation Among Developed Countries*

3.1.1 *United States (US)*

Pharmaceuticals, Personal Care Products (PPCPs), per- and polyfluoroalkyl substances (PFAS), and other developing pollutants are generally regulated and controlled in the United States through a combination of current environmental laws and focused regulatory efforts. Here is a summary of the laws and policies the US has listed to control emerging contaminants:

The Safe Drinking Water Act (SDWA), originally approved in 1974 and later updated, is the major federal law that controls public drinking water sources in the United States. The SDWA gives the Environmental Protection Agency (EPA) authority to set drinking water standards, such as Maximum Contamination Limits (MCLs) for regulated contaminants. Under the SDWA, the EPA has the jurisdiction to regulate certain emerging contaminants, and it taken made steps in recent years to address PFAS pollution in drinking water (US EPA 2020a). FIFRA (Federal

Table 11.1 Different class of EC's with their concentration detected and effects on human and biota.

S. no.	Class of EC's	Environmental matrice	Concentration detected	Effects of ECs on human and biota	Refs
1.	Personal care products	Drinking water	ND-0.22 µg/L	Induce vitellogenin production in juvenile rainbow trout, reduce plasma testosterone by more than 50% in goldfish, and impact female follicular growth, fertilization, and implantation.	Ebele et al., 2020
2.	Pharmaceuticals	Surface water	2.9–3422 ng/L	Mollusc genotoxicity, neurotoxicity, and oxidative stress, Reduced growth (algae, fish), decreased growth of algae communities Hormone disruption (mammals, including humans).	Galindo-miranda et al., 2019; OECD, 2018
3.	Pharmaceuticals, personal care products, and hormones	Surface water	25.5–2187.5 ng/L	Infertility, hormone-dependent tumour, folliculogenesis, spermatogenesis, and steroidogenesis.	Biswas and Vellanki, 2021; Sangeetha et al., 2021
4.	Microplastics	Sediments	10–520 items/kg	Oxidative stress on both juvenile and adult sea cucumbers and humans leads to cytotoxicity and reproductive damage, blockage of the alimentary canal in fishes, crabs, mussels, oysters, whales, and plankton.	Naidu et al., 2016
5.	Flame retardants	Wastewater	0.09–511.60 µg/L	Inhibition in growth and reproduction and decreased survival rate of crustaceans and zebrafish.	Chen and Ma, 2021
6.	Pesticides	Wastewater	0.085–3.4 µg/L	Deleterious effects on fish gills, Feminization of aquatic organisms, Reproductive and sexual systems of humans are severely affected.	Saidulu et al., 2021
7.	Organohalides	Groundwater	7300.0 µg/L	Male fertility, obesity, and puberty in humans.	Alyasiri et al., 2022; Bertolini et al., 2021
8.	Polyfluoroalkyl substances (PFAS)	Soil	29–14,300 pg/g –	Dysregulation of thyroid hormones and adverse kidney health in humans.	Jha et al., 2021
9.	Nanomaterials	Surface water	0.62 ± 0.12 mg/L	Effects on the liver, gills, and GI tract in Cayprinus carpio.	Banu et al., 2021

Insecticide, Fungicide, and Rodenticide Act): The Federal Insecticide, Fungicide, and Rodenticide Act of 1947, as amended over time, governs pesticide registration, distribution, sale, and use in the United States. The EPA examines and approves pesticide use under FIFRA, including those used to control emerging pollutants such as toxic algal blooms and invading species. (/US EPA, 2021a). The Clean Water Act (CWA), as stated earlier, concerns the regulation and control of water pollution in the United States. While the CWA primarily addresses conventional pollutants, it also includes a regulatory framework for dealing with emerging contaminants that may have an impact on water quality. Under the CWA, the EPA and state regulatory agencies use a number of techniques and tactics to assess and regulate developing pollutants in surface waters. (US EPA, 2020b). The Toxic Substances Control Act (TSCA), adopted in 1976 and extensively updated in 2016, gives the EPA jurisdiction to regulate chemicals in the United States. Recently , the EPA has used its TSCA power to address new pollutants such as PFAS, which are both persistent and bio accumulative. The EPA has taken measures under TSCA to examine and regulate PFAS, including classifying specific PFAS as dangerous compounds and proposing regulatory actions to limit their use (US EPA, 2021b). Individual states in the United States have implemented their own policies and legislation to address emerging contaminants, in addition to federal regulations. To address contamination challenges, several states have created their own water quality standards for specific developing contaminants, established monitoring programmes, and conducted targeted actions. These efforts at the state level can differ, and each state could have its own legislation and procedures to regulating new pollutants. (ASDWA, 2021)

It is crucial to note that controlling emerging pollutants is a continuous process, and that regulatory actions change as new scientific understanding develops.

3.1.2 Europe

In Europe, emerging pollutants are generally regulated and controlled by a combination of European Union (EU) legislation and national rules imposed by different countries. Here is an overview of European policies and legislation for controlling emerging contaminants:

REACH is a comprehensive EU legislation that went into effect in 2007. It stands for Registration, Evaluation, Authorization, and Restriction of Chemicals. It discusses how chemicals are made, imported, and used inside the EU. Manufacturers and importers are required by REACH to register any compounds they generate or import in excess of one tonne annually. Additionally, it provides a framework for assessing and controlling the dangers of chemicals, including new contaminants. (ECA, 2021). The Water Framework Directive (WFD): Adopted in 2000, the Water Framework Directive creates a framework for the protection and sustainable management of water resources in the European Union. It establishes goals for improving the ecological and chemical status of surface waters and groundwater. The WFD mandates member nations to monitor and analyze the quality of bodies of water, including the existence of new pollutants, and to create programmes of pollution-reduction measures (EC, 2020a). Plant Protection Products Regulation (PPPR): The Plant Protection Products Regulation (PPPR) is an EU regulation that establishes the rules for the authorization, placement on the market, and the use of plant protection products (pesticides). It includes rules for assessing pesticide dangers, including their effects on human health and the environment. The PPPR attempts to maintain pesticide sustainability while mitigating hazards, especially those presented by new pollutants (EC, 2020b). Pharmaceutical Law: Specific laws governing the approval, security, and environmental effects of pharmaceutical products are in place in the EU. Pharmaceutical safety and environmental effect are evaluated in part by the European Medicines Agency (EMA) and the European Directorate for the Quality of Medicines and Healthcare (EDQM). There are initiatives underway to address the potential environmental effects of pharmaceutical residues and to promote the creation of more environmentally friendly medications (EMA, 2020). Individual EU member states may also have their own national regulations and guidelines in place to address emerging contaminants. These restrictions could differ from country to country depending on their specific demands, environmental conditions, and developing contamination concerns. To handle new contaminants, member states frequently establish their own environmental quality standards, monitoring programmes, and regulatory measures (EEA, 2020).

It is necessary to highlight that the EU rules stated above are always evolving, and new measures to combat emerging pollutants are being established based on scientific understanding and risk assessments. National policies and efforts taken by each member state also help to control developing pollutants in Europe.

3.1.3 Japan

In Japan, the national legislation and regulations enacted by various government agencies are primarily responsible for the management and control of developing pollutants. Here is an overview of Japan's policies and legislation for controlling emerging contaminants:

Water Pollution Control Law: The Water Pollution Control Law, enacted in 1970 and revised through time, is the major act of law in Japan for dealing with water pollution. Its goal is to keep water clean and prevent contamination. The law establishes regulations for the discharge of pollutants into bodies of water, including emerging contaminants, and mandates water quality monitoring and reporting (JP MOE, 2021). Chemical Substances Control Law (CSCL): The CSCL is a comprehensive law that governs the production, importation, and use of chemicals in Japan. The CSCL requires that all chemical substances, both new and old, go through safety evaluations that take the environment into account. The law provides a framework for the management and regulation of new chemicals and pollutants (JP METI, 2021). Pharmaceutical Affairs Law: In Japan, the Pharmaceutical Affairs Law governs the licencing, creation, sale, and consumption of pharmaceutical goods. It has provisions for evaluating the effectiveness and safety of medications, as well as any potential environmental effects. The law strives to guarantee the safety and quality of medications while keeping in mind any potential environmental repercussions (JP MHLW, 2021a). Food Sanitation Law: The Food Sanitation Law outlines the requirements for Japan's food safety and sanitation. In addition to developing contaminants, it has measures for the control of additives, contaminants, and residues in food items. The law aims to safeguard the general populace's health and the quality of food consumed in Japan (JP MHLW, 2021b). Act on Counter Measures Against

Soil Contamination: The Soil Contamination Counter Measures Act was passed in Japan in order to prevent and address soil contamination. It provides a structure for locating contaminated areas, carrying out investigations and evaluations, and putting in place the proper countermeasures. The law tackles a number of contaminants, including newly developing toxins that could endanger human health and soil quality (JP MOE, 2021).

These are some of the most important laws and regulations in Japan that deal with regulating and controlling emerging pollutants. The government organizations in charge of carrying out these laws collaborate to manage developing pollutants, evaluate the hazards associated with them, and take the required steps to safeguard the environment and the general public's health.

3.1.4 Australia

In Australia, national laws and regulations put into effect by various government organizations serve as the primary means of regulating and controlling new contaminants. Here is a summary of Australia's laws and policies governing the management of emerging contaminants:

The Act of 1999 Concerning Environment Protection and Biodiversity Conservation (EPBC Act): The main section of national legislation in Australia for preserving biodiversity and the environment is the EPBC Act. It controls activities like the release of toxins that could have a big influence on issues of national environmental significance. The Act offers a framework for evaluating and controlling how activities that could lead to the emergence of pollutants may have an impact on the environment (AU AWE, 2021). A uniform framework for the evaluation and management of contaminated sites is provided by the National Environmental Protection Measure (NEPM) for Contaminated Sites in Australia. For the identification, evaluation, and remediation of contaminated sites, especially those impacted by new pollutants, it lays down rules and criteria. The NEPM assists in ensuring that human health and the environment are safeguarded from any possible dangers brought on by emerging pollutants (AU NEPM, 2021). The National Health and Medical Research Council created the Australian Drinking Water Guidelines (ADWG), which provide recommendations for managing drinking water quality in Australia. For numerous pollutants, including emerging contaminants, it lays forth maximum permissible concentrations and health-based recommendations. The standards support water providers and regulators in ensuring the public has access to clean drinking water (AU NHMRC, 2021). The Therapeutic Goods Act of 1989 created a legal framework for the effectiveness, safety, and efficacy of therapeutic goods in Australia, including medications and personal care items. It outlined procedures for the assessment and registration of therapeutic items as well as continuing safety and quality inspections. By assuring the safety of medications and personal care items on the Australian market, the Act helped to limit new pollutants (AU TGA, 2021). Legislation by the Australian States and Territories: In addition to the federal law, each Australian state and territory has its own rules and policies for handling new pollutants. These could include special standards for pollution prevention, site remediation, and water quality. Environmental protection departments and agencies are in charge of implementing these laws in each state and territory (AU ENV GOV, 2021).

These are some of the most important laws and regulations in Australia that deal with regulating and controlling emerging pollutants. The government organizations in charge of carrying out these regulations collaborate to monitor, evaluate, and manage emerging pollutants, taking the required steps to safeguard the environment and the general public's health.

3.1.5 Canada

In Canada, national laws and regulations put into effect by various government organizations serve as the primary means of regulating and controlling developing contaminants. A summary of Canada's laws and policies governing the management of emerging contaminants is provided here:

The main object of federal legislation in Canada that regulates the management and control of harmful substances, including developing pollutants, is the CEPA or Canadian Environmental Protection Act, which was passed in 1999. The Minister of the Environment and Climate Change is empowered by CEPA to evaluate and control the dangers posed by substances of concern. To safeguard both human health and the environment, the Act contains procedures for the assessment, management, and control of developing pollutants (CA GC ENV, 1999). Pest Control Products Act (PCPA): The PCPA governs the use and distribution of pesticides and other pest control products in Canada. These products must be registered and their health and environmental risks assessed. The Act establishes guidelines for their use and includes measures for evaluating and controlling the hazards posed by developing pollutants in order to safeguard both human health and the environment (CA GC PC, 2021). The Food and Pharmaceuticals Act in Canada is in charge of policing the efficacy, effectiveness, and safety of food, pharmaceuticals, and cosmetics. To assure the safety of pharmaceuticals and personal care items for human use, it includes provisions for their examination and approval. The Act tackles new pollutants that could be found in these items and establishes guidelines to safeguard the public's health (CA GC FD, 2021). Canadian Drinking Water Guidelines: These recommendations address how Canada should manage the quality of its drinking water. They provide health-based recommendations for contaminants in drinking water, including emerging contaminants, and were produced by the Federal-Provincial-Territorial Committee on Drinking Water. The recommendations aid in ensuring the availability of clean drinking water throughout the nation (CA DWG, 2021). Regulations from the provinces and territories: In addition to the federal law, each province and territory in Canada has its own regulations and policies for handling new pollutants. Specific standards for site cleanup, pollution management, and water and air quality could be included in these rules. To administer and enforce these rules, each province and territory has its own environmental protection departments and agencies (CA CCME, 2021). These are some of the most important laws and regulations in Canada that deal with regulating and controlling emerging pollutants. The government organizations in charge of carrying out these regulations collaborate to monitor, evaluate, and manage emerging pollutants, taking the required steps to safeguard the environment and the general public's health.

4. Status of Policy and Legislation Among Developing Countries

4.1 Brazil

In Brazil, national laws and regulations put into effect by various government bodies serve as the primary means of regulating and controlling developing contaminants. A summary of Brazil's laws and policies governing the management of emerging contaminants is provided here:

Environmental Crimes Law, Law No. 9.605/1998: Environmental offences, such as those involving the release and contamination of pollutants, are defined by the Environmental Crimes Law, which also specifies prohibitions and punishments. It tries to stop and punish behaviours like the discharge of new pollutants that could harm the ecosystem. A framework for monitoring, policing, and punishing environmental offences is provided by the law (BR PLANALTO 198). Law No. 6.938/1981, National Environmental Policy Act: Brazilian environmental protection and sustainable development are governed by the National Environmental Policy Act, a comprehensive piece of legislation. It outlines rules for the prevention and management of pollution, including new pollutants. The law intends to encourage the preservation of natural resources and guard against harmful effects on the environment (BR PLANALTO 1981). Law No. 9.433/97, National Water Resources Policy: Brazil's administration and conservation of its water resources are governed under the National Water Resources Policy. It lays down concepts and rules for managing, protecting, and planning the water resources. With a view of ensuring the sustainable use and preservation of water resources, the law addresses the prevention and management of water pollution, including new contaminants (BR PLANALTO, 1997). ANVISA (Agency National for Sanitary Vigilance) Resolutions Pharmaceuticals, personal care items, and other substances that can have an impact on public health are subject to registration, control, and monitoring regulations through ANVISA, the National Health Surveillance Agency. The assessment, registration, and control of these products, as well as the management of new pollutants, are governed by many decisions that ANVISA has issued (BR ANVISA, 2021). Legislation at the state and local levels: In addition to the federal law, Brazil's states and municipalities could also have their own rules and regulations for handling new toxins. Specific standards for water quality, pollution prevention, and site rehabilitation could be included in these regulations. To implement and enforce these rules, each state and municipality has its own environmental departments and agencies (BR SEMA, 2021).

These are some of the most important laws and regulations in Brazil that deal with regulating and controlling emerging pollutants. The government organizations in charge of carrying out these regulations collaborate to monitor, evaluate, and manage emerging pollutants, taking the required steps to safeguard the environment and the general public's health.

4.2 China

In China, national laws and regulations put into effect by various government organizations serve as the primary means of regulating and controlling developing toxins. A summary of China's laws and policies governing the management of emerging contaminants is provided here:

Environmental Pollution by Solid Waste: Prevention and Control Law of the People's Republic of China. This law tries to stop and regulate environmental contamination brought on by new toxins and solid waste. It lays down rules for sorting, gathering, moving, storing, and getting rid of solid waste. The law also promotes the use of environmentally friendly technologies and practices and creates penalties for noncompliance (CN NPC, 2017). Water Pollution Prevention and Control Law of the People's Republic of China: The goal of this regulation is to prevent and manage water contamination, including the presence of new contaminants. It lays down rules for managing wastewater discharge, protecting water resources, and preventing and controlling water pollution. The law aims to safeguard the public's health and the quality and security of water supplies (CN NPC, 2008). The People's Republic of China's Environmental Protection Law: In China, environmental protection is governed by this extensive statute. It lays down guidelines for resource conservation, ecosystem protection, and the avoidance and control of pollution. Through actions like planning for pollution prevention, conducting environmental impact analyses, and establishing emission standards, the law addresses the control of emerging contaminants (CN NPC, 2014).

Procedures for the Environmental Management of New Chemical compounds: The Ministry of Ecology and Environment in China has put these procedures into place to control the management of new chemical compounds. Before the production or import, they demand that makers and importers register novel chemical substances and submit them to risk evaluations. The actions are intended to ensure chemical use safety and stop the discharge of emerging pollutants into the environment (CN MEE, 2010). The National Standards for Drinking Water Quality in China set standards and maximum permissible limits for a number of contaminants, including emerging contaminants, in drinking water. The criteria, which are published by the Ministry of Health, encourage the distribution of public drinking water that is both safe and hygienic (CN MOH, 2021).

4.3 India

In India, national laws and regulations put into effect by various government organizations serve as the primary means of regulating and controlling developing contaminants. A summary of India's laws and policies governing the management of emerging contaminants is provided here:

The 1986s Environment (Protection) Act: The main section of legislation in India for environmental protection is the Environment (Protection) Act. It provides a plan for the reduction, prevention, and control of pollution. The monitoring and control of pollution, including emerging pollutants, is the responsibility of the Central Pollution Control Board (CPCB) and State Pollution Control Boards (SPCBs) under this Act.

The Act gives the government the authority to implement the required actions to safeguard and enhance the environment (IN MOEF, 1986).

Act of 1974 on Water (Prevention and Control of Pollution): Water pollution prevention and control are goals of the Water (Prevention and Control of Pollution) Act. It lays forth rules for policing and limiting the release of toxins into water bodies, including newly emerging contaminants. The Act gives the CPCB and SPCBs the authority to regulate effluent standards, monitor water quality, and take action to stop the contamination of water resources (IN MOEF, 1974). The 1940 Drugs and Cosmetics Act governs the production, marketing, and distribution of pharmaceuticals and cosmetics in India. To safeguard the public's health, it has provisions for quality assurance and safety requirements. The Act tackles new pollutants in cosmetics and pharmaceuticals and establishes standards for testing, labelling, and registration (IN CDSCO, 1984). National Ambient Air Quality Standards: To monitor and manage air pollution in India, the Central Pollution Control Board (CPCB) established the National Ambient Air Quality Standards (NAAQS). To safeguard the environment and the general public's health, the NAAQS established allowable limits for different air pollutants, including emerging contaminants. The standards provide recommendations for managing, assessing, and monitoring air quality (IN CPCB, 2021), the 2016 Rules for the Management and Transboundary Movement of Hazardous and Other Wastes: These regulations control how hazardous wastes are handled, managed, and disposed off in India. They lay down guidelines for recognizing, handling, and safe disposal of hazardous wastes, including newly emerging pollutants. The regulations' primary goals are to stop pollution and shield the ecosystem from contaminated hazardous waste (IN MOEF, 2016).

5. Limitations of the Current International Regulations Concerning Emerging Contaminants

Gaps in Policy Intervention: New synthesized drug classes that have the potential to harm people and the environment were added to the Basel Convention. However, as of 2019, the USA had not ratified or acceded to any significant conventions, such as the Minamata Conventions, Rotterdam, Basel, and Stockholm despite being an exporter of hazardous waste and a significant producer. While being participants to multiple international agreements, many poor countries still export hazardous materials like textiles, e-waste, and plastics because their governments have not passed federal legislation that complies with the conventions. This is because international agreements lack enforceable regulatory authority (United Nations Environment Programme, 2019b). Regulatory framework for emerging pollutants is insufficient: Many countries and regional organizations have established regulatory frameworks for maintaining and assessing the safety of produced synthetic chemicals throughout their lifecycle. These regulatory systems contain a registration of chemicals imported, made, and used in that country (Wang et al., 2020). Food additives, cosmetics, medicines, industrial chemicals, and pesticides are just a few examples of the many categories that frequently have several inventories, and various regulatory authorities occasionally keep an eye on these inventories. For instance, the Tobacco Research Board and the Medicines Control Authority in Zimbabwe require registration and certification of all pesticides and pharmaceuticals used for commercial purposes. Environmental behaviour and ecotoxicology: Manufacturer secrecy policies are mostly to blame for the lack of information on synthetic chemicals that end up in drinking water. According to a recent study, 350,000 compounds were included in international and national inventories. However, because they are regarded as proprietary by the producers, 50,000 to 70,000 compound names are still unknown to the public or have insufficient descriptions (Wang et al., 2020). Even though there are a few unofficial recycling facilities, the direct impact of imports to Sub-Saharan Africa on the environmental contamination and garbage that results are neither well-documented nor reported to customers. There is no information on disclosure of information regarding reductions in persistent poverty, environmentally friendly industrialization, and urbanization in the Sub-Saharan Africa region, despite resource organization development and sustainable environmental strategies (Dagestani et al., 2022). The lack of connection between science and policy: To develop environmental rules that safeguard freshwater systems from new contaminants, a substantial amount of applicable and competent scientific evidence is needed. Therefore, there needs to be efficient and fruitful cross-communication between the general public scientists and policymakers. The working groups in Europe offer European regulators with a summary of the scientific data on the ecotoxicology and environmental behaviour of pesticides. The management of synthetic compounds is currently the focus of a few international groups on the emerging contaminant classes, including pesticides, antibiotics, emerging contaminant-containing garbage lead, and persistent organic pollutants (Sanganyado et al., 2020).

6. Hazard Assessment

For PCPs with environment concentration data and sufficient chronic toxicity, a preliminary risk evaluation was carried out using hazard quotients. A hazard quotient is a ratio that can be used to estimate potential negative consequences. It is calculated by dividing the exposure concentration by the toxicological benchmark concentration (Suter, 2007). Hazard quotients above 1 signify possible impacts. Only triclocarban and triclosan have the potential to produce chronic effects based on published reports of toxicity and environmental concentrations, based on hazard quotients > 1. After being exposed for 96 hr to a concentration of 0.12 lg L^{-1} TCS, the growth of a natural algal assemblage from a stream in Kansas, USA, was markedly inhibited (Wilson et al., 2003), producing a hazard quotient of 19. In addition, TCC exhibited a hazard quotient of 10.9 or higher for Americamysis bahia following a 28-d exposure to 0.13 lg L^{-1} (TCC Consortium, 2002). However, given that the environmental concentrations used were the highest recorded (roughly 20 times the median concentrations) and that the most vulnerable species and biological endpoint were also considered , both these hazard quotients could be regarded as worst-case scenarios. As a result, the actual risk is probably substantially lower than what is assumed, which is consistent with early risk evaluations made elsewhere (Costanzo et al., 2007; Lyndall et al., 2010). Hazard quotients could not be determined for UV filters since there are not enough long-term *in vivo* data available. Due to the lack of *in vivo* studies on aquatic organisms, hazard quotients for putative endocrine impacts

Table 11.2 The status of policy and legislation in managing emerging contaminants among different countries globally

S. No.	Countries	Emerging Contaminant	Policy/Legislation	Status/Remarks	Reference
1.	United States	PFAS	EPA's PFAS Action Plan	Ongoing regulatory actions and monitoring efforts	U.S. Environmental Protection Agency (2019). PFAS Action Plan.
		Pharmaceuticals	Safe Drinking Water Act (SDWA)	Regulates drinking water quality and contaminants	U.S. Environmental Protection Agency (2020). Safe Drinking Water Act.
		Microplastics	No specific federal regulations	Some state and local initiatives in place	Law Library of Congress (2020). Regulation of Microplastics in Drinking Water: United States.
		Emerging Chemicals	Toxic Substances Control Act (TSCA)	Regulates chemical substances and testing	U.S. Environmental Protection Agency (2021). Toxic Substances Control Act (TSCA).
2.	Europe	PFAS	EU REACH Regulation	Evaluating and potentially restricting certain PFAS	European Chemicals Agency (2021). REACH Regulation.
		Pharmaceuticals	EU Water Framework Directive (WFD)	Addresses pharmaceuticals as priority substances	European Commission (2000). Water Framework Directive.
		Microplastics	EU Single-Use Plastics Directive	Bans certain single-use plastic products	European Parliament and Council. (2019). Single-Use Plastics Directive.
		Emerging Chemicals	EU REACH Regulation	Regulates chemicals and their use in Europe	European Chemicals Agency (2021). REACH Regulation.
3.	Japan	PFAS	Regulation of Specified Chemical Substances	Monitoring and regulating PFAS compounds	Ministry of Health, Labour, and Welfare (2021). Regulation of Specified Chemical Substances.
		Pharmaceuticals	Pharmaceutical Affairs Act	Regulates pharmaceutical products and safety	Pharmaceuticals and Medical Devices Agency (2021). Pharmaceutical Affairs Act.
		Microplastics	No specific national regulations	Some regional initiatives and research projects	Ministry of the Environment (2021). Measures against Marine Plastic Litter.
		Emerging Chemicals	Chemical Substances Control Law (CSCL)	Regulates chemical substances and their use	Ministry of Economy, Trade and Industry (2021). Chemical Substances Control Law (CSCL).

No.	Country	Category	Regulation/Act	Description	Reference
4.	Australia	PFAS	PFAS National Environmental Management Plan (NEMP)	Framework for managing PFAS contamination	Department of Agriculture, Water, and the Environment (2020). PFAS National Environmental Management Plan.
		Pharmaceuticals	National Industrial Chemicals Notification and Assessment Scheme (NICNAS)	Regulates industrial chemicals and their assessment	Australian Government, Department of Health (2021). National Industrial Chemicals Notification and Assessment Scheme (NICNAS).
		Microplastics	No specific national regulations	Some state and local initiatives and research	Australian Government, Department of Agriculture, Water, and the Environment (2018). National Plastics Plan.
		Emerging Chemicals	Industrial Chemicals Act 2019	Regulates the introduction of new chemicals	Australian Government, Department of Health (2021). Industrial Chemicals Act 2019.
5.	Canada	PFAS	Canada-wide Strategy on Zero Plastic Waste	Includes actions to address PFAS in plastic waste	Government of Canada (2018). Canada-wide Strategy on Zero Plastic Waste.
		Pharmaceuticals	Food and Drugs Act	Regulates pharmaceutical products and safety	Government of Canada (2021). Food and Drugs Act.
		Microplastics	No specific national regulations	Some provincial and municipal initiatives	Environment and Climate Change Canada (2019). Draft Science Assessment of Plastic Pollution.
		Emerging Chemicals	Canadian Environmental Protection Act (CEPA)	Regulates chemical substances and their use	Government of Canada (2021). Canadian Environmental Protection Act (CEPA).
6.	China	PFAS	National Environmental Protection List (NEPL)	Includes monitoring and regulation of PFAS	Ministry of Ecology and Environment of China (2019). National Environmental Protection List (NEPL).
		Pharmaceuticals	Drug Administration Law of the People's Republic of China	Regulates pharmaceutical products and safety	National Medical Products Administration (2019). Drug Administration Law of the People's Republic of China.
		Microplastics	No specific national regulations	Some regional initiatives and research projects	Ministry of Ecology and Environment of China (2020). Control Plan for Plastic Pollution.
		Emerging Chemicals	Law on the Prevention and Control of Environmental Pollution by Solid Wastes	Regulates management and control of solid waste	National People's Congress of the People's Republic of China (2020). Law on the Prevention and Control of Environmental Pollution by Solid Wastes.

Table 11.2 cond....

... *Table 11.2 cond.*

S. No.	Countries	Emerging Contaminant	Policy/Legislation	Status/Remarks	Reference
7.	Brazil	PFAS	No specific national regulations	Some state and local initiatives and research	Ministry of the Environment of Brazil (2020). National Plan for the Adequate Management of PFAS.
		Pharmaceuticals	National Health Surveillance Agency (ANVISA)	Regulates pharmaceutical products and safety	National Health Surveillance Agency (ANVISA).
		Microplastics	No specific national regulations	Some regional initiatives and research projects	Brazilian Association of Public Sanitation and Environmental Engineering (ABES) (2019). National Survey on Microplastics.
		Emerging Chemicals	National Policy on Solid Waste	Regulates management and disposal of solid waste	Ministry of the Environment of Brazil (2010). National Policy on Solid Waste.
8.	India	PFAS	No specific national regulations	Some state-level initiatives and research projects	Central Pollution Control Board (CPCB) (2020). Guidelines for Handling, Treatment, and Disposal of Perfluorinated Compounds.
		Pharmaceuticals	Drugs and Cosmetics Act	Regulates pharmaceutical products and safety	Ministry of Health and Family Welfare (1940). Drugs and Cosmetics Act.
		Microplastics	No specific national regulations	Some research and awareness initiatives	Ministry of Environment, Forest, and Climate Change (2019) Action Plan for Control of Marine Litter.
		Emerging Chemicals	Chemical Accidents (Emergency Planning, Preparedness and Response) Rules	Deals with management of chemical accidents	Ministry of Environment, Forest, and Climate Change (1996). Chemical Accidents (Emergency Planning, Preparedness and Response) Rules.

were also not estimated; nonetheless, research on mammals suggests UV filters are most likely to have endocrine effects. Table 11.2 lists the status of policy and legislation in managing emerging contaminants among different countries globally.

7. Conclusion

ECs are prevalent practically everywhere in the environment, which is unquestionably disturbing. They come from a number of sources and enter the water and sediments. Fish, amphibians, and algae that are not part of the target biota are significantly impacted by ECs and their by-products. Although ECs are only found in trace amounts, their continued and persistent presence in the environment has serious negative effects, which is the main cause for concern. The (GBD) Global Burden of Disease estimates that 12 synthetic compounds caused 1.3 million fatalities in 2016 either directly or indirectly. As a result, during the 5th (UNEA meeting) environmental pollution was recognized as one of the three main global challenges, along with climate change and biodiversity loss (GBD, 2017; UN, 2021). The lack of proof for a chemical that is typically thought of as an environmentally beneficial material characterizes the current state of these substances in conventional ways. Even manufacturers have contributed to this vicious cycle by altering small functional groups while preserving the general biological and physicochemical properties, making the substituted synthetic molecules equally as hazardous as the originals. The domination of chemical-by-chemical techniques and, in certain cases, the relaxation of environmental rules to benefit polluting companies are due to a lack of two-way communication between scientists and politicians. It should be a top priority to guarantee that there are few or no new EC emissions in order to preserve human health and freshwater ecosystems. A model-based study comparable to the one that You et al. (2022) is cited. In order to raise public awareness about ECs, the SIS model can be used to analyze prevention strategies and public opinion guidance. This analysis can then be used to help the government authorities form a risk assessment and an early warning mechanism. High output *in vitro* toxicity testing could make it easier to assess mixture toxicity, replace animals in toxicity testing, and clarify the modes of action of specific substances or groups of substances by exponentially raising the number of chemicals that can be successfully analyzed at a specific time. The ECs registration framework's class-based approach can be tested via enantiomer-based testing since the two biological activities of chiral substances' enantiomers can point to different differences in toxicity and fate. The US EPA's addition of 12 PPCP/EDC compounds to CCL List 3 reflects the developing status of rules and regulations in ECs. The presence of a substance on the CCL 3 list indicates that further research into its occurrence and safety is necessary. Additionally, in order to determine the prevalence of ECs, their possible impacts, and the importance of including these ECs in water quality standards, such as the more recent chemical perfluoro compounds, these international regulations have gathered data to identify the environmental concentrations of ECs. Globally, it is seen that a number of chemicals are present in drinking water and surface waters at concentrations of less than 0.1 and 0.05 g/L, respectively. The WHO has established benchmark values of 0.001 g/L for 17-beta-oestradiol, 0.3 g/L for nonylphenol, and

0.1 g/L for bisphenol A for these substances in surface waters, which are mixed with untreated and treated wastes. Above these levels, the compounds are known to have toxic epigenetic effects, neurotoxic and genotoxic, which have a significant impact on the environment's biota. As a result, there are various issues with the supervisory method that slow down the process, including the lack of detailed information on each prospective EC chemical, the omission of environmental risk assessment from the ecotoxicology analysis, and the continuing addition of new ECs. The need for class-based and effect-based methodologies that can significantly facilitate advancements in biological and chemical analysis, such as the screening and detection of nontarget pollutants with effect-directed analysis, is therefore critical. Creating a long-lasting organization for science-policy interface forums that can be accessed by the general public, academics, and decision-makers, and at regional, national, and worldwide levels is another important step.

Acknowledgments

One of the authors Dr Manviri Rani is grateful for the funding from DST-SERB, New Delhi (Sanction order no. SRG/2019/000114) and TEQIP-III MNIT Jaipur, India. Dr Uma Shanker wishes to thank TEQIP-III, NIT Jalandhar for partial funding. Keshu, Rishabh, and Shikha are thankful to the Ministry of Education, New Delhi, India for providing funding.

References

Agência Nacional de Vigilância Sanitária (ANVISA). 2021. Resoluções da Diretoria Colegiada. Retrieved from https://www.gov.br/anvisa/pt-br/assuntos/medicamentos/resolucoes-da-diretoria-colegiada: Government of Canada. 2021. Canadian Environmental Protection Act, 1999. Retrieved from https://laws-lois.justice.gc.ca/eng/acts/c-15.31/

Alyasiri, Thura, M.H. and AL-Chalabi, S.M. 2022. Environmental and public health impacts of plastic pollution. Br. J. Glob. Ecol. Sustain. Dev. 6: 20–28. https://journalzone.org/index.php/bjgesd/article/view/77.

Association of State Drinking Water Administrators. 2021. Emerging Contaminants. Retrieved from https://www.asdwa.org/emerging-contaminants/

Australian Government, Department of Agriculture, Water and the Environment. 2018. National Plastics Plan.

Australian Government, Department of Health. 2021. National Industrial Chemicals Notification and Assessment Scheme (NICNAS). Retrieved from https://www.nicnas.gov.au/

Australian Government, Department of Health. 2021. Industrial Chemicals Act 2019. Retrieved from https://www.legislation.gov.au/Details/C2021C00137

Banu, A. Najitha., Kudesia, Natasha., Raut, A.M., Pakrudheen, I. and Wahengbam, J. 2021. Toxicity, bioaccumulation, and transformation of silver nanoparticles in aqua biota: a review. Environ. Chem. Lett. 19: 4275–4296. https://doi.org/10.1007/s10311-021-01304-w.

Barboza, L.G.A. and Gimenez, B.C.G. 2015. Microplastics in the marine environment: current trends and future perspectives. Mar. Pollut. Bull. 97(1–2): 5–12.

Berendonk, T.U., Manaia, C.M., Merlin, C., Fatta-Kassinos, D., Cytryn, E., Walsh, F. et al. 2015. Tackling antibiotic resistance: the environmental framework. Nature Reviews Microbiol. 13(5): 310–317.

Bertolini, Martina., Zecchin, Sarah., Beretta, Giovanni Pietro., Nisi, Patrizia De., Ferrari, Laura., Cavalca, L. 2021. Effectiveness of Permeable Reactive Bio-Barriers for Bioremediation of an Organohalide-Polluted Aquifer by Natural-Occurring Microbial Community. Water 13: 1–22. https://doi.org/10.3390/w13172442.

Biswas, P. and Vellanki, B.P. 2021. Occurrence of emerging contaminants in highly anthropogenically influenced river Yamuna in India. Sci. Total Environ. 782: 146741. https://doi.org/10.1016/j. scitotenv.2021.146741.

Brazilian Association of Public Sanitation and Environmental Engineering (ABES). 2019. National survey on microplastics. Retrieved from http://abes-dn.org.br/abes-realiza-pesquisa-sobre-microplasticos/

Canadian Council of Ministers of the Environment. 2021. Provincial and Territorial Environmental Legislation. Retrieved from https://www.ccme.ca/en/resources/provincial_and_territorial_environmental_legislation.html

Cefic. 2022. Facts and Figures of the European Chemical Industry [WWW

Central Pollution Control Board (CPCB). 2020. Guidelines for Handling, Treatment, and Disposal of Perfluorinated Compounds. New Delhi: Ministry of Environment, Forest and Climate Change.

Central Pollution Control Board, Ministry of Environment, Forest and Climate Change, Government of India (IN CPCB). 2021. National Ambient Air Quality Standards. Retrieved from http://www.cpcb. nic.in/national-ambient-air-quality-standards-naaqs/

Chaturvedi, P., Shukla, P., Giri, B.S., Chowdhary, P., Chandra, R. et al. 2021. Prevalence and hazardous impact of pharmaceutical and personal care products and antibiotics in environment: A review on emerging contaminants. Environ. Res. 194: 110664.

Chen, M. and Ma, W. 2021. A review on the occurrence of organophosphate flame retardants in the aquatic environment in China and implications for risk assessment. Sci. Total Environ. 783: 147064. https:// doi.org/10.1016/j.scitotenv.2021.147064.

Costanzo, S.D., Watkinson, A.J., Murby, E.J., Kolpin, D.W. and Sandstrom, M.W. 2007. Is there a risk associated with the insect repellent DEET (N, N-diethyl-m-toluamide) commonly found in aquatic environments?. Sci. The Total Environ. 384(1–3): 214–220.

Dagestani, A.A., Qing, L. and Abou Houran, M. 2022. What Remains Unsolved in Sub-African Environmental Exposure Information Disclosure: A Review. J. Risk. Finan. Manag 15(10): 487.

Department of Agriculture, Water and the Environment, Australian Government. 2021. Environment Protection and Biodiversity Conservation Act 1999. Retrieved from https://www.awe.gov.au/legislation/environment-protection-and-biodiversity-conservation-act-1999

Department of Agriculture, Water and the Environment, Australian Government. 2021. National Environmental Protection Measures (AU NEPM). Retrieved from https://www.environment.gov. au/protection/nepms.

Department of Agriculture, Water and the Environment, Australian Government (AU AWE). 2021. Environment Protection and Biodiversity Conservation Act 1999. Retrieved from https://www.awe. gov.au/legislation/environment-protection-and-biodiversity-conservation-act-1999.

Department of Agriculture, Water and the Environment. 2020. PFAS National Environmental Management Plan. Retrieved from https://www.environment.gov.au/protection/publications/pfas-national-environmental-management-plan.

Department of the Environment and Energy, Australian Government (AU ENV GOV). (2021). Environmental Protection in Australia. Retrieved from https://www.environment.gov.au/about-us/legislation.

Ebele, A.J., Oluseyi, T., Drage, D.S., Harrad, S. and Abdallah, M.A. 2020. Nigeria. Emerg. Contam. 6: 124–132. https://doi.org/10.1016/j.emcon.2020.02.004.

Environment and Climate Change Canada. 2019. Draft Science Assessment of Plastic Pollution. https:// www.canada.ca/en/environment-climate-change/services/canadian-environmental   protection-act-registry/publications/draft-science-assessment-plastic-pollution.html

European Chemicals Agency. 2021. REACH - Registration, Evaluation, Authorization and Restriction of Chemicals. Retrieved from https://echa.europa.eu/regulations/reach

European Chemicals Agency. 2021. REACH Regulation. https://echa.europa.eu/regulations/reach/overview

European Chemicals Agency. 2021. REACH Regulation. Retrieved from https://echa.europa.eu/regulations/reach

European Commission. 2000. Directive 2000/60/EC of the European Parliament and of the Council of 23 October 2000 establishing a framework for Community action in the field of water policy. Official Journal of the European Communities L 327: 1–72.

European Commission. 2020. Plant Protection Products. Retrieved from https://ec.europa.eu/food/plant/pesticides_en

European Commission. 2020. Water Framework Directive. Retrieved from https://ec.europa.eu/environment/water/water-framework/index_en.html

European Environment Agency. 2020. Water policies and measures in Europe. Retrieved from https://www.eea.europa.eu/themes/water/policy-and-measures

European Medicines Agency. 2020. Human regulatory. Retrieved from https://www.ema.europa.eu/en/human-regulatory

European Parliament and Council. 2019. Directive (EU) 2019/904 of the European Parliament and of the Council of 5 June 2019 on the reduction of the impact of certain plastic products on the environment. Official Journal of the European Union, L 155/1.

Gabriel, L., Barboza, A., Carolina, B. and Gimenez, G. 2015. Microplastics in the marine environment : current trends and future perspectives. Mar. Pollut. Bull. https://doi.org/10.1016/j.marpolbul.2015.06.008.

Galindo-miranda, J.M., Guízar-gonzález, C., Becerril-bravo, E.J., Moeller-chávez, G., Leon-becerril, E. and Vallejo-rodríguez, R. 2019. Occurrence of emerging contaminants in environmental surface waters and their analytical methodology—a review. Water Supply 1–14. https://doi.org/10.2166/ws.2019.087

GBD, 2017. Global , regional , and national age-sex specific mortality for 264 causes of death, 1980 – 2016 : a systematic analysis for the Global Burden of Disease Study 2016. Glob. Heal. Metrics 390: 1150–1210. https://doi.org/10.1016/S0140-6736 (17)32152-9.

Government of Canada. 2018. Canada-wide Strategy on Zero Plastic Waste. Retrieved from https://www.canada.ca/en/environment-climate-change/services/managing-reducing-waste/zero-plastic-waste/canada-wide-strategy.html

Government of Canada. 2021. Canadian Drinking Water Guidelines. Retrieved from https://www.canada.ca/en/health-canada/services/publications/healthy-living/guidelines-canadian-drinking-water-quality-summary-table.html

Government of Canada. 2021. Canadian Environmental Protection Act (CEPA). Retrieved from https://laws-lois.justice.gc.ca/eng/acts/c-15.31/

Government of Canada. 2021. Food and Drugs Act. Retrieved from https://laws-lois.justice.gc.ca/eng/acts/f-27/

Government of Canada. 2021. Food and Drugs Act. Retrieved from https://laws-lois.justice.gc.ca/eng/acts/f-27/

Government of Canada. 2021. Pest Control Products Act. Retrieved from https://laws-lois.justice.gc.ca/eng/acts/p-9.01/

Government of Canada. 2021. Canadian Environmental Protection Act (CA GC ENV), 1999. Retrieved from https://laws-lois.justice.gc.ca/eng/acts/c-15.31/.

Government of Canada. 2021. Pest Control Products Act (CA GC PC). Retrieved from https://laws-lois.justice.gc.ca/eng/acts/p-9.01/.

Government of Canada. 2021. Food and Drugs Act (CA GC FD). Retrieved from https://laws-lois.justice.gc.ca/eng/acts/f-27/.

Government of Canada. 2021. Canadian Drinking Water Guidelines (CA DWG). Retrieved from https://www.canada.ca/en/health-canada/services/publications/healthy-living/guidelines-canadian-drinking-water-quality-summary-table.html.

Jha, Gaurav., Kankarla, Vanaja., McLennon, Everald., Pal, Suman., Sihi, Debjani., Dari, Biswanath et al. 2021. Per- and Polyfluoroalkyl Substances (PFAS) in Integrated Crop–Livestock Systems_ Environmental Exposure and Human Health Risks. Internaional J. Environ. Res. Public Heal. 18: 1–20. https://doi.org/10.3390/ijerph182312550.

Keshu, Rani, M. and Shanker, U. 2022. Efficient removal of plastic additives by sunlight active titanium dioxide decorated Cd–Mg ferrite nanocomposite: Green synthesis, kinetics and photoactivity. Chemosphere, 290, 133307. https://doi.org/10.1016/j.chemosphere.2021.133307 .

Keshu, Rani, M., Yadav, J., Chaudhary, S., & Shanker, U. (2021). An updated review on synthetic approaches of green nanomaterials and their application for removal of water pollutants: Current challenges, assessment and future perspectives. J. Environ. Chem. Eng. 9(6): 106763. https://doi.org/10.1016/j.jece.2021.106763

Krzeminski, P., Tomei, M.C., Karaolia, P., Langenhoff, A., Almeida, C.M.R., Felis, E. et al. 2019. Performance of secondary wastewater treatment methods for the removal of contaminants of emerging concern implicated in crop uptake and antibiotic resistance spread: A review. Sci. Total Environ. 648: 1052–1081.

Law Library of Congress. 2020. Regulation of Microplastics in Drinking Water: United States. Retrieved from https://www.loc.gov/law/help/microplastics/microplastics-unitedstates.pdf

Lopez-Avila, V., & Hites, R. A., 1980. Organic compounds in an industrial wastewater. Their transport into sediments. Environ. Sci. Technol. 14(11): 1382–1390.

L´opez-serna, R., Posadas, E., García-encina, P.A. and Mu~noz, R. 2019. Removal of contaminants of emerging concern from urban wastewater in novel algal-bacterial photobioreactors. Sci. Total Environ. 662: 32–40. https://doi.org/10.1016/j.scitotenv.2019.01.206.

Lyndall, J., Fuchsman, P., Bock, M., Barber, T., Lauren, D., Leigh, K. and Capdevielle, M. 2010. Probabilistic risk evaluation for triclosan in surface water, sediments, and aquatic biota tissues. Integr. Environ. Assess. Manag. 6(3): 419–40. doi: 10.1897/IEAM_2009-072.1.

Ma, B., Arnold, W.A. and Hozalski, R.M. 2018. The relative roles of sorption and biodegradation in the removal of contaminants of emerging concern (CECs) in GAC-sand biofilters. Water Res. 146: 67–76.

Maria, A., Starling, C.V.M. and Le, M.D. 2018. Occurrence, control and fate of contaminants

Martins, I., Soares, J., Neuparth, T., Barreiro, A.F., Xavier, C., Antunes, C. et al. 2022. Prioritizing the Effects of Emerging Contaminants on Estuarine Production under Global Warming Scenarios. Toxics 10(2): 46.

Ministry of Ecology and Environment of China. 2019. National Environmental Protection List (NEPL). Retrieved from http://www.mee.gov.cn/gzfw_13107/zcfg/hjyw/201909/t20190906_726353.html

Ministry of Ecology and Environment of China. 2020. Control plan for plastic pollution. Retrieved from http://www.mee.gov.cn/xxgk2018/xxgk/xxgk15/202012/t20201229_819784.html.

Ministry of Ecology and Environment of the People's Republic of China (CN MEE). 2010. Measures for the Environmental Management of New Chemical Substances. Retrieved from http://www.mee.gov.cn/ywgz/fgbz/bz/bzwb/dqhjnwzlbz/201007/t20100702_187805.shtml.

Ministry of Economy, Trade and Industry, Government of Japan. 2021. Chemical Substances Control Law. Retrieved from https://www.meti.go.jp/english/policy/chemical_management/cscl.html\

Ministry of Economy, Trade and Industry. 2021. Chemical Substances Control Law (CSCL). Retrieved from https://www.meti.go.jp/english/policy/chemical_management/index.html.

Ministry of Environment, Forest and Climate Change, Government of India (IN MOEF). 1974. Water (Prevention and Control of Pollution) Act, 1974. Retrieved from http://www.moef.gov.in/sites/default/files/Water%28P%29Act,%201974_0.pdf.

Ministry of Environment, Forest and Climate Change, Government of India (IN MOEF). 1986. Environment (Protection) Act, 1986. Retrieved from http://www.moef.gov.in/sites/default/files/E%28P%29Act,%201986.pdf.

Ministry of Environment, Forest and Climate Change, Government of India (IN MOEF). 2016. Hazardous and Other Wastes (Management and Transboundary Movement) Rules, 2016. Retrieved from http://www.moef.gov.in/sites/default/files/So%201442%28E%29_14.06.2016_0.pdf.

Ministry of Environment, Forest and Climate Change. 1996. Chemical Accidents (Emergency Planning, Preparedness and Response) Rules.

Ministry of Environment, Forest and Climate Change. 2019. Action Plan for Control of Marine Litter. New Delhi: Government of India.

Ministry of Health and Family Welfare, Government of India. 1940. Drugs and Cosmetics Act, 1940. Retrieved from https://cdsco.gov.in/opencms/opencms/system/modules/CDSCO.WEB/elements/common_download.jsp?num_id_pk=MTU4Mw==

Ministry of Health and Family Welfare. 1940. Drugs and Cosmetics Act. Retrieved from http://legislative.gov.in/actsofparliamentfromtheyear/drugs-and-cosmetics-act-1940.

Ministry of Health of the People's Republic of China (CN MOH). 2021. National Standards for Drinking Water Quality. Retrieved from http://www.moh.gov.cn/zwgkzt/s9491/202103/98c930eaa43d41b0b22c8b01f694fb3d.shtml.

Ministry of Health, Labour and Welfare, Government of Japan (JP MHLW).(2021). Pharmaceutical Affairs Law. Retrieved from https://www.mhlw.go.jp/english/policy/health-medical/pharmaceuticals/index. html.

Ministry of Health, Labour and Welfare, Government of Japan. 2021. Pharmaceutical Affairs Law. Retrieved from https://www.mhlw.go.jp/english/policy/health-medical/pharmaceuticals/index.html

Ministry of Health, Labour and Welfare, Government of Japan. 2021. Food Sanitation Law. Retrieved from https://www.mhlw.go.jp/english/policy/health-medical/food/index.html

Ministry of Health, Labour and Welfare. 2021. Regulation of Specified Chemical Substances.

Ministry of the Environment of Brazil. 2010. National Policy on Solid Waste.

Ministry of the Environment of Brazil. 2020. National Plan for the Adequate Management of PFAS. Retrieved from https://www.gov.br/mma/pt-br/assuntos/agenda-ambiental-urbana/residuos-solidos/plano-nacional-de-gerenciamento-de-pfas

Ministry of the Environment, Government of Japan. 2021. Soil Contamination Countermeasures Act. Retrieved from https://www.env.go.jp/en/law/soil/

Ministry of the Environment, Government of Japan. 2021. Water Pollution Control. Retrieved from https://www.env.go.jp/en/water/water_pollution/

Ministry of the Environment. 2021. Measures against Marine Plastic Litter. Retrieved from https://www.env.go.jp/en/focus/plasticlitter/

Naidu, R., Jit, J., Kennedy, B. and Arias, V. 2016. National guidance for contaminants of emerging concern in Australia chemosphere emerging contaminant uncertainties and policy: The chicken or the egg conundrum. Chemosphere 154: 385–390. https://doi.org/10.1016/j.chemosphere.2016.03.110.

National Health and Medical Research Council, Australian Government. 2021. Australian Drinking Water Guidelines. Retrieved from https://www.nhmrc.gov.au/health-advice/water-quality/australian-drinking-water-guidelines

National Health Surveillance Agency. 2021. About ANVISA. Retrieved from http://portal.anvisa.gov.br/about-anvisa.

National Health and Medical Research Council, Australian Government (AU NHMRC). (2021). Australian Drinking Water Guidelines. Retrieved from https://www.nhmrc.gov.au/health-advice/water-quality/australian-drinking-water-guidelines.

National Medical Products Administration. 2019. Drug Administration Law of the People's Republic of China. http://www.nmpa.gov.cn/WS04/CL2138/334040.html.

National People's Congress of the People's Republic of China (CN NPC). 2008. Law of the People's Republic of China on the Prevention and Control of Water Pollution. Retrieved from http://www.npc.gov.cn/npc/c30834/200908/06fe3bfc3c0541fcbda17405522a4110.shtml.

National People's Congress of the People's Republic of China (CN NPC). 2017. Law of the People's Republic of China on the Prevention and Control of Environmental Pollution by Solid Waste. Retrieved from http://www.npc.gov.cn/npc/c30834/201705/97ed1da704e346a5a7a8d99c868fbc32.shtml.

National People's Congress of the People's Republic of China (CN NPC). 2014. Environmental Protection Law of the People's Republic of China. Retrieved from http://www.npc.gov.cn/npc/c30834/201412/d5914f41c2c9474ab9f7416e4adbcbe0.shtml.

National People's Congress of the People's Republic of China. 2020. Law on the Prevention and Control of Environmental Pollution by Solid Wastes. Retrieved from http://www.npc.gov.cn/npc/c30834/202010/2b36da1a71c1464b8b231f0fa0a546de.shtml.

OECD, 2018. Managing Contaminants of Emerging Concern in Surface Waters : Scientific Developments and Cost-Effective Policy Responses, 5 February 2018 Summary Note.

of emerging concern in environmental compartments in Brazil. J. Hazard Mater. 372: 17–36.

Pharmaceuticals and Medical Devices Agency. 2021. Pharmaceutical Affairs Act. Retrieved from https://www.pmda.go.jp/files/000238221.pdf

Presidência da República, Casa Civil. 1981. Lei No. 6.938/1981. Retrieved from http://www.planalto.gov.br/ccivil_03/leis/L6938.htm.

Presidência da República, Casa Civil. (BR PLANALTO) 1997. Lei No. 9.433/1997. Retrieved from http://www.planalto.gov.br/ccivil_03/leis/L9433.htm.

Presidência da República, Casa Civil. 1998. Lei No. 9.605/1998. Retrieved from http://www.planalto.gov.br/ccivil_03/leis/L9605.htm.

Puri, M., Gandhi, K. and Kumar, M.S. 2023. The occurrence, fate, toxicity, and biodegradation of phthalate esters: An overview. Water Environ. Res. 95(1): e10832.

Rani, M., Keshu, Meenu, Sillanpää, M. and Shanker, U. 2022. An updated review on environmental occurrence, scientific assessment and removal of brominated flame retardants by engineered nanomaterials. Journal of Environmental Management 321: 115998. https://doi.org/10.1016/j.jenvman.2022.115998

Rani, M., Yadav, J. and Shanker, U. 2022. Environmental, legal, health, and safety issues of green nanomaterials. In Green Functionalized Nanomaterials for Environmental Applications (pp. 567–594). Elsevier.

Reemtsma, T., Alder, L. and Banasiak, U. 2013. Emerging pesticide metabolites in groundwater and surface water as determined by the application of a multimethod for 150 pesticide metabolites. Water Res. 47(15): 5535–5545.

Rizzo, L., Fiorentino, A., Grassi, M., Attanasio, D. and Guida, M., 2015. Advanced treatment of urban wastewater by sand filtration and graphene adsorption for wastewater reuse: Effect on a mixture of pharmaceuticals and toxicity. J. Environ. Chem. Eng. 3(1): 122–128.

Rizzo, L., Malato, S., Antakyali, D., Beretsou, V.G., Đolić, M.B., Gernjak, W. et al. 2019. Consolidated vs new advanced treatment methods for the removal of contaminants of emerging concern from urban wastewater. Sci. Total Environ. 655: 986–1008.

Roberts, J., Kumar, A., Du, J., Hepplewhite, C., Ellis, D.J., Christy, A.G. and Beavis, S.G. 2015. Pharmaceuticals and personal care products (PPCPs) in Australia's largest inland sewage treatment plant , and its contribution to a major Australian river during high and low fl ow. Sci. Total Environ. https://doi.org/10.1016/j.scitotenv.2015.03.145.

Roberts, J., Kumar, A., Du, J., Hepplewhite, C., Ellis, D.J., Christy, A.G. et al. 2016. Pharmaceuticals and personal care products (PPCPs) in Australia's largest inland sewage treatment plant, and its contribution to a major Australian river during high and low flow. Sci. Tot. Environ. 541: 1625–1637.

Saidulu, D., Gupta, B., Kumar, A. and Sarathi, P. 2021. A review on occurrences, eco-toxic effects , and remediation of emerging contaminants from wastewter: Special emphasis on biological treatment based hybrid systems. J. Environ. Chem. Eng. 9: 105282. https://doi.org/10.1016/j.jece.2021.105282.

Salimi, M., Esrafili, A., Gholami, M., Jonidi Jafari, A., Rezaei Kalantary, R., Farzadkia, M. et al. 2017. Contaminants of emerging concern: a review of new approach in AOP technologies. Environ. Monit. Asses. 189: 1–22.

Sanganyado, E. 2022. Policies and regulations for the emerging pollutants in freshwater ecosystems: Challenges and opportunities. Emerg. Freshwater Pollut. 361–372.

Sanganyado, E., Lu, Z. and Liu, W. 2020. Application of enantiomeric fractions in environmental forensics: Uncertainties and inconsistencies. Environ. Res. 184: 109354.

Sangeetha, S., Vimalkumar, K. and Loganathan, B.G. 2021. Environmental Contamination and Human Exposure to Select Endocrine-Disrupting Chemicals : A Review. Sustain. Chem. 2: 343–380. https://doi.org/10.3390/suschem2020020

Secretaria de Estado do Meio Ambiente (BR SEMA). 2021. Legislação Ambiental. Retrieved from http://www.sema.ce.gov.br/legislacao/legislacao-ambiental.

Suter, G.W. 2007. Ecological Risk Assessment, second ed. CRC Press, Boca Raton, FL.

Tas, J.W., Balk, F., Ford, R.A. and van der Plassche, E.J. 1997. Environmental risk assessment of musk ketone and musk xylene in the Netherlands in accordance with the EU-TGD. Chemosphere 35: 2973–3002.

Terasaki, M., Makino, M. and Tatarazako, N. 2009. Acute toxicity of parabens and their chlorinated by-products with Daphnia magna and Vibrio fischeri bioassays. J. Appl. Toxicol. 29:242–247.

Therapeutic Goods Administration, Australian Government (AU TGA). (2021). Therapeutic Goods Act 1989. Retrieved from https://www.tga.gov.au/about-us/legislation.

U.S. Environmental Protection Agency. 2019. PFAS Action Plan. https://www.epa.gov/pfas

U.S. Environmental Protection Agency. 2021. Toxic Substances Control Act (TSCA). Retrieved from https://www.epa.gov/laws-regulations/summary-toxic-substances-control-act

UN. 2021. For People and Planet: the United Nations Environment Programme Strategy for 2022–2025 to Tackle Climate Change, Loss of Nature and Pollution. Fifth Session of the UN Environment Assembly (UNEA-5). Nairobi, Kenya [WWW Document]. UNEP. URL. https://www.unep.org/environmentassembly/unea5.

United Nations Environment Programme, 2019b. Global Chemicals Outlook II - from Legacies to Innovative Solutions: Implementing the 2030 Agenda for Sustainable Development.

United States Environmental Protection Agency. 2020. Safe Drinking Water Act. Retrieved from https://www.epa.gov/sdwa

United States Environmental Protection Agency. 2020. Safe Drinking Water Act. Retrieved from https://www.epa.gov/sdwa

United States Environmental Protection Agency. 2020. The Clean Water Act. Retrieved from https://www.epa.gov/cwa

United States Environmental Protection Agency. 2021. Federal Insecticide, Fungicide, and Rodenticide Act (FIFRA). Retrieved from https://www.epa.gov/laws-regulations/summary-federal-insecticide-fungicide-and-rodenticide-act

United States Environmental Protection Agency. 2021. Toxic Substances Control Act (TSCA). Retrieved from https://www.epa.gov/tsca

USEPA, 2012. Water: Contaminant Candidate List 3. US. Environmental Protection Agency, Washington, DC.

Verlicchi, P., Al Aukidy, M. and Zambello, E. 2012. Occurrence of pharmaceutical compounds in urban wastewater: removal, mass load and environmental risk after a secondary treatment—a review. Sci. Tot. Environ. 429: 123–155.

Wang, Z., Walker, G.W., Muir, D.C. and Nagatani-Yoshida, K. 2020. Toward a global understanding of chemical pollution: a first comprehensive analysis of national and regional chemical inventories. Environ. Sci. TechnoL. 54(5): 2575–2584.

Wee, S.Y. and Aris, A.Z. 2018. Occurrence and public-perceived risk of endocrine disrupting compounds in drinking water. *In*: Npj Clean Water, vol. 2. https://doi.org/10.1038/s41545-018-0029-3.

Wilson, B.A., Smith, V.H., deNoyelles, F. and Larive, C.K. 2003. Effects of three pharmaceutical and personal care products on natural freshwater algal assemblages. Environ. Sci. Technol. 37(9): 1713–1719.

Xu, L., Wang, X., Sun, Y., Gong, H., Guo, M., Zhang, X., Meng, L. and Gan, L., 2020. Ultrasonics - sonochemistry Mechanistic study on the combination of ultrasound and peroxymonosulfate for the decomposition of endocrine disrupting compounds. Ultrason. Sonochem. 60: 104749 https://doi.org/10.1016/j.ultsonch.2019.104749.

Yan, Z., Yang, X., Lu, G., Liu, J., Xie, Z. and Wu, D. 2014. Potential environmental implications of emerging organic contaminants in Taihu Lake, China: comparison of two ecotoxicological assessment approaches. Sci. Tot. Environ. 470: 171–179.

Yao, C., Yang, H. and Li, Y. 2021. A review on organophosphate flame retardants in the environment: Occurrence, accumulation, metabolism and toxicity. Sci. Tot. Environ. 795: 148837.

You, G., Gan, S., Guo, H. and Dagestani, A.A. 2022. Public opinion spread and guidance strategy under COVID-19: a SIS model analysis. Axioms 11: 1–19. https://doi.org/10.3390/axioms11060296.

CHAPTER 12

COVID-19 and Plastic Pollution
Health, Environment

Meenu,[1] Manviri Rani[1],* and Uma Shanker[2]

1. Introduction

The COVID-19 infection has been spread globally, with tremendously increasing cases every day, which was initially detected in China (Sahin et al., 2020; Han et al., 2021). This outbreak has glinted a public health emergency on high alert and was declared a pandemic by international health organizations (WHO 2019). As of May, 2023, about 687 million confirmed cases have been detected, and above 6.8 million died worldwide because of this virus (WHO, 2023). The origin of infection is still controversial; however, it resembles mammal infection caused by bat-SARC-CoV-2. Numerous new variants of COVID-19 infection (coronavirus) appeared in the fall of 2021. To avoid spreading virus infection, health authorities advised regular use of hand wash, gloves, face masks, and Personal protective equipment (PPE) kits by the public, frontline workers, and healthcare workers, as shown in Fig. 1 and Fig. 2 (WHO, 2019).

Along with, the antiviral, antimicrobial drugs, and COVID-19 specific vaccination drugs as a precaution to fight against this coronavirus (Paital et al., 2020; Kumar et al., 2020). To halt the pandemic, most nations have imposed a lockdown to safeguard social distancing and decrease the pressure on medical societies (Bahukhandi et al., 2020). Social distancing shall remain the norm for many months,

[1] Department of Chemistry, Malaviya National Institute of Technology Jaipur, Jaipur, Rajasthan, India-302017.
[2] Department of Chemistry, Dr. B R Ambedkar National Institute of Technology Jalandhar, Jalandhar, Punjab, India-144011
* Corresponding author: manviri.chy@mnit.ac.in

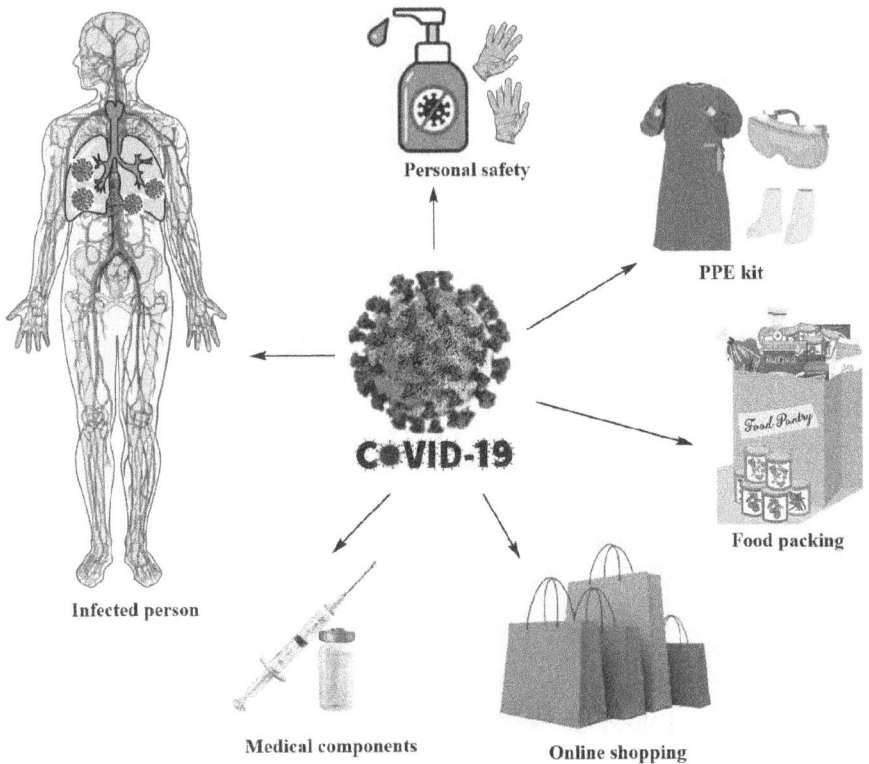

Figure 12.1 Plastic waste components during the Covid-19 pandemic.

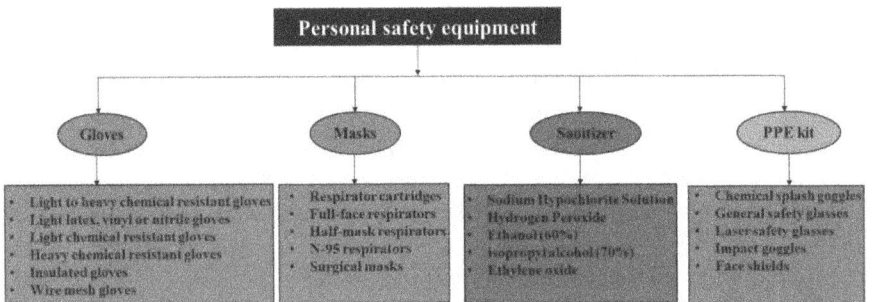

Figure 12.2 Classification of personal safety equipment used during Covid-19 pandemic.

even after lockdowns, until herd immunity is built or/and the availability of COVID-19-specific vaccination and antiviral drug (Paital et al., 2020; Kumar et al., 2020).

Consequently, restrictions in anthropogenic activities due to COVID-19 triggered a load of plastic debris by frequent use of plastics disposal and packing (online shopping, food, and medical items), which direct or indirect affect the health, economic, and environmental perspectives before and after this pandemic (Drury et al., 2021; Liu et al., 2020; Abbasi et al., 2020; da Costa et al., 2021; Dharmaraj et al., 2021). Nevertheless, the pandemic causes a negative paradigm shift in the world's gross domestic product (GDP) as almost all sectors, including business, agriculture, and tourism get shut down because of lockdown and social distancing norms for safety reasons (Bahukhandi et al., 2020; UNDESA 2020). Export and import are also breakdowns that directly influence the food and medical chain supply. Due to Improper management or deficiency of medical supplies, the world population suffers even more health emergencies. Lack of working staff, poor recycling facilities, absence of advanced technologies, and improper disposal of plastic waste led to plastic pollution (Tripathi et al., 2020). Landfill and improper disposal of plastic waste generate toxic chemicals that interfere with the natural habitat and cause adverse effects on biota (Silva et al., 2021; Haque et al., 2020). The level of drugs, plastics, especially PPE kits and personal care products debris to COVID-19 protection increased many folds (Silva et al., 2021; Gorrasi et al., 2021; De-la-Torre et al., 2021). Along with the release of residential wastewater effluents, floods (Rafey et al., 2021) and mismanagement of WWTP affect the quality and demand of drinking water for good hygiene (Silva et al., 2021; Ejima et al., 2021). Also, Increased Contribution of microfibers and hazardous chemical compounds released from disposable facemasks (Selvaranjan et al., 2021; Padgelwar et al., 2021). The virus may be present on the surface of the masks and enter the water bodies, threatening those in contact with it (Gryseels et al., 2021). Thus, medical waste is further a suspect for transmission of infection. The open-burning landfill produces toxic chemicals and gases that mix with wastewater effluent and end up in oceans through water bodies like lakes and rivers. By weathering moisture contact, this plastic debris further ferments into small sizes called micro and nano plastics (size < 5 nm). Increasing the level of microplastic in marine causes the synanthropic predators (e.g., crows), altering assemblage structure in wildlife species (Gilby et al., 2021) (e.g., small mammals, reptiles, crustaceans) and interaction with PPE (Hiemstra et al., 2021) (ingestion and entanglement). In the end, increased transmission of virus (including SARS-CoV-2) infection (Padgelwar et al., 2021) and microplastic pollution through sea animals as food cycles (Nabi et al., 2021; Hu et al., 2019) (susceptibility of virus). This chapter aims to provide a comprehensive overview of plastic pollution and its consequence on health, the environment, and the economy, along with a glance at the COVID-19 origin, classification of various variants, detection, and treatment. The chapter also summarizes the data about plastic waste pollution, current management steps, recycling challenges, and environmental concerns by various organizations and countries to deal with the overload of plastic waste due to this health crisis.

2. Background of the COVID-19 Pandemic

The increasing death rate due to virus infection (later named COVID-19) worldwide was declared a pandemic by WHO on March 11, 2020 (WHO 2020). The clinical features of this infection were similar to that of viral pneumonia (Wu et al., 2020; Holshue et al., 2021). After analyzing the respiratory sample, a group of researchers from the PRC Centre for Disease Control announced that pneumonia was caused by a novel coronavirus, far ahead known as Novel Coronavirus (Wang et al., 2020; Wang et al., 2021). This novel coronavirus has identical symptoms to severe acute

Table 12.1 Stability of Covid-19 on different environmental surfaces.

Sr. No.	Materials	Contagious or stability Time (Hours) of virus
1.	Tissue paper	0.5
2.	Aerosols	3
3.	Copper	4
4.	Wood	6
5.	Cardboard	24
6.	Banknotes	48
7.	Glass	48
8.	Sewage	72
9.	Stainless steel	72
10.	Plastics	72
11.	Solid faeces	96
12.	Mask (inner layer)	96
13.	Mask (outer layer)	168

respiratory syndrome (SARS), later termed SARS-CoV-2 (Wang et al., 2020; Cheng et al., 2020). The classification of this virus into two subgroups or subfamilies, such as Torovirinae and Coronavirinae, and further subdivided into four genera, viz. Alphacoronavirus, Betacoronavirus, Gammacoronavirus, and Deltacoronavirus (Alanagreh et al., 2020; Jindal et al., 2020) detailed discussion in Table 12.2. The total number of cases of a different variant of COVID-19 infection worldwide is discussed in Table 12.3. COVID-19 comes into the category of β-coronavirus declared as an HG3 pathogen (Hazardous Group 3). The virus is non-segmented and enveloped with positive-sense RNA (Conceicao et al., 2020; Shi et al., 2020). The origin of this virus considers like other mammal innovation diseases such as Tuberculosis (T.B.) and Human Immunodeficiency Virus (HIV). Reports suggested that this β-coronavirus was closely related to two bat-derived severe acute respiratory syndrome (SARS)-like coronaviruses, bat-SL-CoVZXC21 and bat-SL-CoVZC45 (Chirumbolo 2020; Li et al., 2021). The study reveals that genome sequences of COVID-19, RaTG13 is more related to Bat CoV RaTG13 (a bat coronavirus previously detected in Rhinolophus affine from Yunnan Province, China) (Peng et al., 2020; Kumar et al., 2021). However, some other reports proposed the virus's origin from anteaters/pangolins (Gong et al., 2020). Therefore, the origin of COVID-19 remains debated and needs more critical analyses to confirm its origin.

2.1 Mode of transmission of coronavirus

The main routes of COVID-19 transmission were as: a) from animal to human; b) human-to-human and in particular: I) respiratory transmission; II) oral-fecal transmission; III) asymptomatic patients' transmission; IV) healthcare worker (HCW)-patients transmission; V) mother-to-child, in the uterus transmission; VI) children-adult transmission; c) air, surface-human transmission described in Fig. 12.3. Initially, it was reported that the genomic sequence of COVID-19 infection was closely related to mammal infection (SARS-derived virus family), and this virus

Table 12.2 Selected characteristics and identification of Covid-19 Variants with their classification.

Variant types	ᵃWHO Label	Spike Protein Substitutions	First Identified	Attributes	References
Variants of Interest	Eta (B.1.525)	A67V, 69del, 70del, 144del, E484K, D614G, Q677H, F888L	United Kingdom and Nigeria– December 2020	Potential reduction in neutralization by some Emergency Use Authorization (EUA) monoclonal antibody treatments. Potential reduction in neutralization by convalescent and post-vaccination sera	Shen et al., 2020
	Iota (B.1.526)	L5F, (D80G*), T95I, (Y144-*), (F157S*), D253G, (L452R*), (S477N*), E484K, D614G, A701V, (T859N*), (D950H*), (Q957R*)	United States (New York)– November 2020	Reduced susceptibility to the combination of bamlanivimab and etesevimab monoclonal antibody treatment; however, the clinical implications of this are not known. Alternative monoclonal antibody treatments are available. Reduced neutralization by convalescent and post-vaccination sera	Annavajhala et al., 2021
	Kappa (B.1.617.1)	(T95I), G142D, E154K, L452R, E484Q, D614G, P681R, Q1071H	India–December 2020	Potential reduction in neutralization by some EUA monoclonal antibody treatments. Potential reduction in neutralization by post vaccination sera	Sonabend et al., 2021
	None (B.1.617.3)	T19R, G142D, L452R, E484Q, D614G, P681R, D950N	India–October 2020	Potential reduction in neutralization by some EUA monoclonal antibody treatments. Potential reduction in neutralization by post-vaccination sera	Pascarella et al., 2021
Variant of Concern	Alpha (B.1.1.7)	69del, 70del, 144del, (E484K*), (S494P*), N501Y, A570D, D614G, P681H, T716I, S982A, D1118H (K1191N*)	United Kingdom	~ 50% increased transmission. Potential increased severity based on hospitalizations and case fatality rates. No impact on susceptibility to EUA monoclonal antibody treatments. Minimal impact on neutralization by convalescent and post-vaccination sera	Sapkal et al., 2021

Table 12.2 contd. ....

... *Table 12.2 contd.*

Variant types	ᵃ WHO Label	Spike Protein Substitutions	First Identified	Attributes	References
	Beta (B.1.351, B.1.351.2, B.1.351.3)	D80A, D215G, 241del, 242del, 243del, K417N, E484K, N501Y, D614G, A701V	South Africa	~ 50% increased transmission. Significantly reduced susceptibility to the combination of bamlanivimab and etesevimab monoclonal antibody treatment, but other EUA monoclonal antibody treatments are available. Reduced neutralization by convalescent and post-vaccination sera	Nath and Aditya 2021; Celik et al., 2021
	Delta (B.1.617.2, AY.1, AY.2, AY.3, AY.4, AY.5, AY.6, AY.7, AY.8, AY.9, AY.10, AY.11, AY.12)	T19R, (V70F*), T95I, G142D, E156-, F157, R158G, (A222V*), (W258L*), (K417N*), L452R, T478K, D614G, P681R, D950N	India	Increased transmissibility Potential reduction in neutralization by some EUA monoclonal antibody treatments. Potential reduction in neutralization by post-vaccination sera	Celik et al., 2021
	Gamma (P.1, P.1.1, P.1.2)	L18F, T20N, P26S, D138Y, R190S, K417T, E484K, N501Y, D614G, H655Y, T1027I	Japan/Brazil	Significantly reduced susceptibility to the combination of bamlanivimab and etesevimab monoclonal antibody treatment. but other EUA monoclonal antibody treatments are available. Reduced neutralization by convalescent and post-vaccination sera	Chan et al., 2021
Variant of High Consequence	None	None	None	Currently, there are no SARS-CoV-2 variants that rise to the level of high consequence	Vasireddy et al., 2021

Note: ᵃFact Sheet For Health Care Providers Emergency Use Authorization (Eua) Of Bamlanivimab And Etesevimab 02092021 (fda.gov) external icon: *Detected in some sequences but not all

Table 12.3 Status of Covid-19 Vaccines within the WHO EUL/PQ evaluation process.

S. No.	Manufacturer/WHO EUL holder	Name of Vaccine	*NRA of Record	Platform	Status of assessment
1.	BioNTech Manufacturing GmbH	BNT162b2/COMIRNATY Tozinameran (INN)	EMA, USFDA	Nucleoside modified mNRA	Finalized
2.	Astra Zeneca, AB	AZD1222 Vaxzevria	EMA	Recombinant ChAdOx1 adenoviral vector encoding the Spike protein antigen of the SARS-CoV-2	Finalized
3.		AZD1222 Vaxzevria	MFDS KOREA	Recombinant ChAdOx1 adenoviral vector encoding the Spike protein antigen of the SARS-CoV-2	Finalized
4.		AZD1222 Vaxzevria	Japan MHLW/PMDA	Recombinant ChAdOx1 adenoviral vector encoding the Spike protein antigen of the SARS-CoV-2	Finalized
5.		AZD1222 Vaxzevria	Australia TGA	Recombinant ChAdOx1 adenoviral vector encoding the Spike protein antigen of the SARS-CoV-2	Finalized
6.	Serum Institute of India Pvt. Ltd	Covishield (ChAdOx1_nCoV-19)	DCGI	Recombinant ChAdOx1 adenoviral vector encoding the Spike protein antigen of the SARS-CoV-2	Finalized
7.	Janssen–Cilag International NV	Ad26.COV2.S	EMA	Recombinant, replication-incompetent adenovirus type 26 (Ad26) vectored vaccine encoding the (SARS-CoV-2) Spike (S) protein	Finalized
8.	Moderna Biotech	mRNA-1273	EMA, USFDA	mNRA-based vaccine encapsulated in lipid nanoparticle (LNP)	Finalized
9.	Beijing Institute of Biological Products Co., Ltd. (BIBP)	SARS-CoV-2 Vaccine (Vero Cell), Inactivated (InCoV)	NMPA	Inactivated, produced in Vero cells	Finalized
10.	Sinovac Life Sciences Co., Ltd.	COVID-19 Vaccine (Vero Cell), Inactivated/ CoronavacTM	NMPA	Inactivated, produced in Vero cells	Finalized

Table 12.3 contd. ...

... *Table 12.3 contd.*

S. No.	Manufacturer/WHO EUL holder	Name of Vaccine	*NRA of Record	Platform	Status of assessment
11.	Gamaleya Research Institute of Epidemiology and Microbiology	Sputnik V	Russian NRA	Human Adenovirus Vector-based Covid-19 vaccine	Finalized
12.	Bharat Biotech, India	SARS-CoV-2 Vaccine, Inactivated (Vero Cell)/COVAXIN	DCGI	Whole-Virion Inactivated Vero Cell	Finalized
13.	Sinopharm/WIBP2	Inactivated SARS-CoV-2 Vaccine (Vero Cell)	NMPA	Inactivated, produced in Vero cells	Ongoing
14.	CanSino Biologics	Ad5-nCoV	NMPA	Recombinant Novel Coronavirus Vaccine (Adenovirus Type 5 Vector)	Ongoing
15.	Novavax	NVX-CoV2373/Covovax	EMA	Recombinant nanoparticle prefusion spike protein formulated with Matrix-M™ adjuvant	Ongoing
16.	Sanofi	CoV2 preS dTM-AS03 vaccine	EMA	Recombinant, adjuvanted	-
17.	Serum Institute of India Pvt. Ltd.	NVX-CoV2373/Covovax	DCGI	Recombinant nanoparticle prefusion spike protein formulated with Matrix-M™ adjuvant	-
18.	Clover Biopharmaceuticals	SCB-2019	NMPA	Novel recombinant SARS-CoV-2 Spike (S)-Trimer fusion protein	-
19.	CureVac	Zorecimeran (INN) concentrate and solvent for dispersion for injection; Company code: CVnCoV/CV07050101	EMA	mRNA-based vaccine encapsulated in lipid nanoparticle (LNP)	-
20.	Vector State Research Centre of Virology and Biotechnology	EpiVacCorona	Russian NRA	Peptide antigen	-
21.	Zhifei Longcom, China	Recombinant Novel Coronavirus Vaccine (CHO Cell)	NMPA	Recombinant protein subunit	-
22.	IMBCAMS, China	SARS-CoV-2 Vaccine, Inactivated (Vero Cell)	NMPA	Inactivated	-
23.	BioCubaFarma–Cuba	Soberana 01, Soberana 02 Soberana Plus Abdala	CECMED	SARS-CoV-2 spike protein conjugated chemically to meningococcal B or tetanus toxoid or Aluminum	-

* National regulatory authority (NRA)

transferred from animal to human via food consumption of infected mammals (Lu et al., 2020; Wan et al., 2020; Cai et al., 2020). However, various reports suggested that coronavirus is powerfully pathogenic and transmissible from human to human through respiratory fomites (Conly et al., 2020; Rana et al., 2020; Karia et al., 2020). Furthermore, an increasing body of evidence suggests that human-to-human transmission may occur during the COVID-19 asymptomatic incubation period, estimated to last from 2 to 10 days (Wei et al., 2021). Human-to-human transmission is either by respiratory droplets (Yadav and Saxena 2020) or by indirect contact with the touched surface (metallic, plastic, and other) from the infected person (Vella et al., 2020) (Table 12.4). The retaining time of the COVID-19 virus on different surfaces is discussed in Table 12.1. Ding and Liang (2020) showed that COVID-19 could be found in feces. Thus stool samples can contain the virus even when it is not detectable in the respiratory tract; therefore, the fecal-oral transmission route is possible. The occurrence of gastrointestinal symptoms, in particular diarrhea, in COVID-19-positive subjects can contribute to the spreading of the virus (Xiao et al., 2020).

Several investigations were suggested (Gao et al., 2021; Zhou et al., 2020). Due to the weakening of immunity, the virus can reproduce and transmit the virus in asymptomatic patients (Oran and Topol, 2020). According to the 6th Guide of WHO, an asymptomatic affected role in spreading COVID-19 infection (WHO 2019). According to Abrol et al. (2020), another critical route of spreading COVID-19 is hospital-associated transmission, mainly through healthcare workers (HCWs) (Abrol et al., 2020). This study revealed that 17.3% of HCWs were positive for COVID-19.

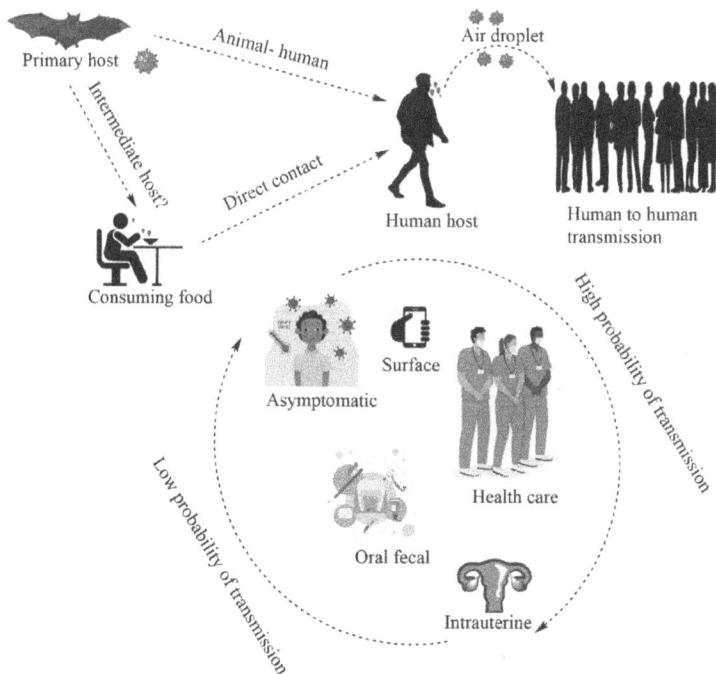

Figure 12.3 Transmission pathways of COVID-19 infection.

Table 12.4 Practices for infectious waste separation, storage, transportation, and disposal in Asian countries.

Country	Practices for COVID-19 Waste Generated from Healthcare Facilities	COVID-19 Waste Treatment and Disposal
India	Use dedicated trolleys and collection bins in COVID-19 isolation wards, laboratories and test centers. Used masks are discarded and collected in separate 'yellow colour-coded plastic bags' (suitable for biomedical waste collection) labelled 'COVID-19 waste.' Disinfect inner and outer surfaces of containers, trolleys and bins with 1% NaClO solution daily. Depute dedicated sanitation workers for biomedical and general solid waste collection and timely transfer to temporary storage. Use vehicles with GPS and barcoding systems for containers containing HCW for waste tracking. Label vehicles with 'Biohazard' sign.	Common biomedical waste treatment facility (CBWTF). Disposal permitted by deep burial only in rural or remote areas without CBTWF facilities. Large volume of yellow colour-coded (incinerable) COVID-19 waste beyond the capacity of existing CBWTFs and BMW incinerators, necessitates permitting HW incinerators' usage at existing treatment, storage and disposal facilities (TSDFs) or captive industrial incinerators if any exist in the state/union territory. In such cases, ensure separate arrangement for handling and waste feeding.
China	Wuhan improved the process by closing the municipal solid waste disposal site and enhancing disinfection and sterilisation of waste-related facilities. Medical waste collection is performed with strict technical guidelines to avoid virus transmission. Medical waste is collected from individual containers placed at medical institutions and public areas. Medical waste is transported and disposed through specially designed registered vehicles operated by professional workers at scheduled times. Healthcare workers are equipped with protective equipment, including masks, gloves, goggles, protective clothing and disinfectants.	Temporary incinerator installation suggested for waste management. Municipal solid waste incinerators to co-process medical waste in rotary kiln. Hazardous waste is thoroughly incinerated in high temperature flue gas and slag residue after 60 min of high temperature (850ºC) incineration
Bangladesh	Use colour-coded bins (red: sharp waste, yellow: infectious/pathological waste and black: non-hazardous waste). Store the bins on premises. They are regularly collected by covered vehicles for transportation to treatment sites.	Incineration
Indonesia	Identify the classification and communication (labels, symbols) means. Designate COVID-19 infectious waste bins. Conduct internal sterilisation and disinfection before bags are tied. Label bags 'Danger, do not open.' Disinfect bags before collection. Schedule regular waste transportation by cleaning services on weekdays.	Mostly incineration, disinfecting at source and transporting to the disposal site, open burning (if no incinerator) or hazardous waste landfill.

Japan	Separate and store infectious, non-infectious and general wastes and sharp objects from other infectious wastes with proper containers. Seal easy-to-use and durable containers. Transport by a designated cart to avoid scattering and spilling wastes within facilities. Use short storage periods. Access storage rooms, if you are an authorised person. Apply clear labelling on infectious waste containers at storage rooms. Incineration, melting, steam sterilisation (autoclave), dry sterilisation and disinfection followed by shredding and disposal to sanitary landfills.	Incineration, melting, steam sterilisation (autoclave), dry sterilisation and disinfection followed by shredding and disposal to sanitary landfills.
Malaysia	Do not separate COVID-19 waste from other infectious waste. Equip cold rooms in bigger healthcare facilities. Collect daily or three times a week depending on the quantity. Transport only by a special lorry licensed to transport hazardous waste.	Mostly incineration
Nepal	Designate waste storage in health facilities. Use specific trollies for transportation within hospitals. Use specific vehicles for transportation from healthcare facilities to waste management service providers (WMSPs).	Mostly burned, small-scale incineration or dumped in backyards, municipal landfills
Thailand	Separate into sharp and non-sharp COVID-19 waste. Disinfect and use double bags. Designate specific storage areas. Send waste from community healthcare facilities to district healthcare facilities once a week. Temperature-controlled storage available at the district level. Transport by licensed WMSPs (Requires temperature-controlled vehicles). Treat within 48 h after transportation. Disinfect vehicles and bin daily with NaClO.	Incineration, autoclave, WMSP, sanitary landfill.

The mortality rate among residents, taking into account deaths of patients resulting positive for COVID-19 or with pseudo-influenza symptoms, was 3.1% but went up to 6.8 in the Region of Lombardy (Abrol et al., 2020). The healthcare professionals were working in a high-risk of COVID-19 infection for self or candidate of spreading the virus to non-COVID-19 patients55. Furthermore, it was confirmed that transmission of infection from mother to child and child to adult person (Yang et al., 2020; Kasraeian et al., 2020). In summary, infectious persons (human, child, and healthcare staff) and infectious inanimate surfaces (glass, metal, and plastic) become the source of virus transmission, consequently community transfer (Kampf, 2020).

2.2 Tools and Techniques for Covid-19 Detection

COVID-19 infection has a wide range of clinical manifestations varying from asymptomatic to symptomatic, including pneumonia, respiratory symptoms, shortness of breath, fever, heart failure, cough, renal failure, dyspnea, severe acute respiratory syndrome, and in severe cases, death (Yao et al., 2020). However, the leading cause of death linked to COVID-19 infection because of respiratory failure, followed by renal failure, septic shock, hemorrhage, and heart failure (Fig. 12.6). Several diagnostic techniques in clinical and public health applications were accustomed to detecting the COVID-19 virus or infection. Initially, screening or diagnosis of COVID-19 infection, by oropharyngeal and nasopharyngeal swab, sputum, bronchial aspirate, bronchoalveolar lavage fluid, blood, later on, CT-Scan or X-ray are usually recommended (Reginelli et al., 2021; Karam et al., 2021; Cui and Zhou, 2021). Accuracy and sensitivity of tests must be kept the main objective whenever a clinical method is developed for the diagnostic test of the virus through quick and accurate results. Virus infection is either confirmed by direct or indirect tests instantly by measuring the viral RNA or virus antibodies in a host, respectively. The screening of specimens from bronchoalveolar lavage (BAL) fluid or lower respiratory tract by using nucleic acid amplification tests (NAATs) like RT-PCR; however, obtaining specimens from the lower respiratory is not always 100 % probability (Shammus et al., 2020; Axell-House et al., 2020; Bilal et al., 2020).

Along with clinical or laboratory screening, diagnostic imaging can also accompany assessing the COVID-19 infection in the lower respiratory tract or other anatomical sites. Diagnostic imaging techniques consist of computed tomography (C.T.) scan, chest radiography or chest X-ray (CXR), resonance imaging (MRI), positron emission tomography-CT (PET/CT), and ultrasound, magnetic (Zimmerman et al., 2021; Gitman et al., 2021; Oishee et al., 2021). Some of these laboratory testing techniques are discussed in this chapter (Fig. 12.4):

(A) Nucleic acid tests

Nucleic acid tests are based on amplification techniques such as the polymerase chain reaction of nucleic acid (targeted virus consists of several nucleic acids). Such techniques amplify the small amount of infected nucleic acid (genomic material) of patient samples into many folds of the DNA chain. Further, numerous methods have been developed to detect these amplified DNA chains. In this direction, real-time reverse transcription-quantitative

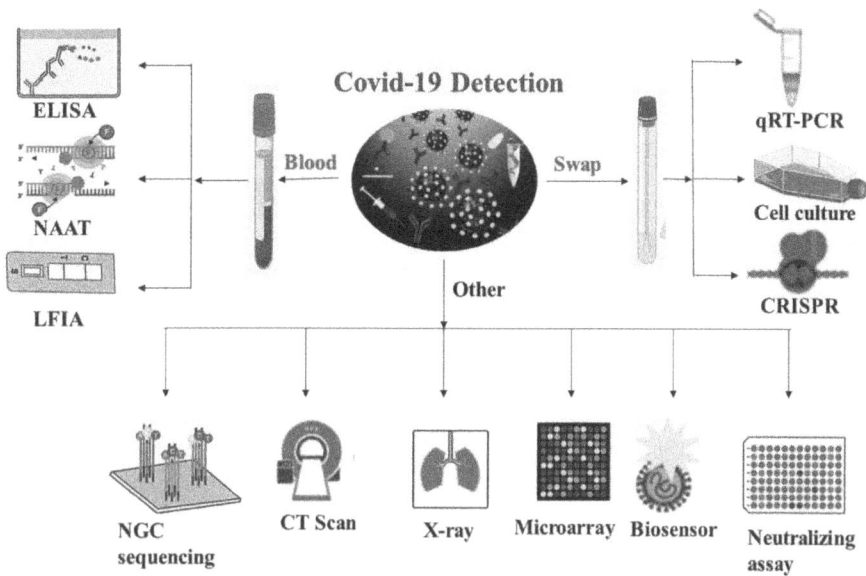

Figure 12.4 Various techniques for the detection of COVID-19.

polymerase chain reaction (RT-qPCR) is recommended for both symptomatic and asymptomatic COVID-19 cases (Russo et al., 2020). Real-time reverse transcription-quantitative polymerase chain reaction (RT-qPCR), transcription-loop-mediated isothermal amplification, and antigen-based tests (Fig. 12.5). RT-qPCR has all international diagnostic protocols with an accuracy and sensitivity range of 98.8%–100% for asymptomatic and symptomatic COVID-19 infection; however, it is a relatively costly diagnosis tool for public health (Russo et al., 2020; Van Walle et al., 2021). In this race for detection of COVID-19 infection, several other techniques play their role discussed in Figs. 4 and 5.

(b) Antigen tests

Antigen tests detect COVID-19 infection by observing the antibodies protein unit in infected persons (Yakoh et al., 2021). These tests' sensitivity and specificity are less than nucleic acid tests as their detection mode is very rapid. Cause of less accuracy, a negative report of a person is not a good agreement; they may be infected, and for this reason, WHO does not recommend this for diagnosis of COVID-19 infection (Carpenter et al., 2021). Currently, the researcher is working on the accuracy level improvement of antigen tests.

(c) Serological (antibody) tests

Serological tests usually detect the COVID-19 virus within or after the infection symptom. These might give false positives because of their rapidness or low specificity and sometimes false positives (Bastos et al., 2021). Nevertheless, varying amounts and variety of antibodies generated by individuals with COVID-19 infected persons limit the detection of these testes.

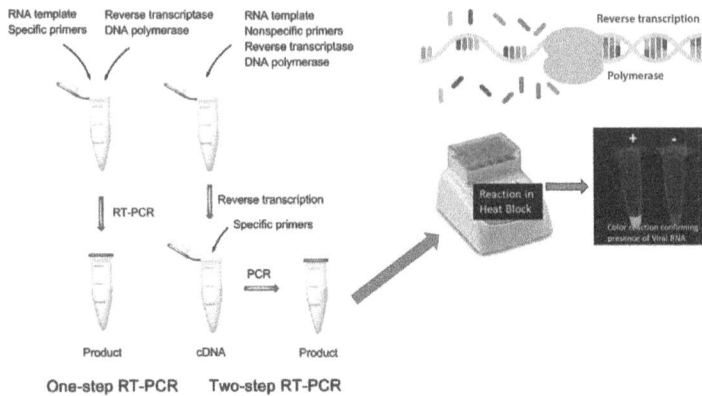

Figure 12.5 Real-time *reverse transcriptase*-polymerase chain reaction (RT-PCR) mode of detection.

2.3 Precaution and Vaccination

The primary precaution to break the transmission of COVID-19 infection, regular hand washing, and wearing face masks are advised by WHO. WHO recommended various disinfection with different compositions of chemicals, such as 62–71% ethanol, 0.1% sodium hypochlorite, 0.5% hydrogen peroxide, and 1:100 of 5% sodium hypochlorite for disinfecting small surfaces (Kampf 2020). The regular use of PPEs kits, face masks, gloves, hand sanitizer, and other essential immunity booster pharmaceuticals or food appliances are recommended by health advisory to restrict the community transfer of COVID-19 infection (Wu et al., 2020; Wu et al., 2020; Dadhich 2021). Several efforts have been focused on developing vaccines against COVID-19 to deter the pandemic, and a majority of vaccines have been advanced using the S-protein of the COVID-19 virus (Chen 2020; Chung et al., 2021). Up to July 2, 2020, about 158 vaccines have developed, of which 135 are in the trial phase or exploratory stage at clinical levels across the world (Kaur and Gupta 2020). Some vaccines such as Ad5-nCoV (CanSino Biologicals), mRNA-1273 (Moderna), ChAdOx1 (University of Oxford), Pathogen-specific, APC (Shinzen Geno-Immune Medical Institute), LV-SMENP-DC, and INO-4800 (Inovio, Inc.), have arrived the clinical trials. All vaccines produced either by live attenuation or inactivation of viruses such as virus-like particles (VLP), DNA, nanoparticles, RNA, viral vector (replicating and non-replicating), and protein sub-unit has their pros and cons (Rather et al., 2020). Furthermore, to boost immunogenicity and help in developing vaccines, numerous technologies like MF-59 (Novartis), AS03 (GSK), and CpG 1018 (Dynavax) have been grasped by scientists (Damodharan et al., 2020).

3. Health and Economic Crisis during the COVID-19 Pandemic

There are economic and health losses worldwide due to this epidemic, and reports suggest the mismanagement of lockdowns of food and medical supplies is the main reason for these crises (Shammi et al., 2020). Meanwhile, this infection causes many health issues such as blood pressure, skin irritation, loss of taste, heart problems, and mental health issues, even the recovery of this viral infection (Fig. 12.6) as lockdown is indeed imposed to restrict the transfer of this shitty virus, however many businesses and employment show the negative impact of this action. The world economy stands by exchanging goods among the countries (Export or Import). The world economy shrinks due to the first lockdown imposed in China, the foremost world exporter related to plastics, metals, car parts, textiles, medicine, electronics, and other goods. Many other sectors such as the food industry, agriculture, and tourism are directly or indirectly related to the world economy becoming victims of this pandemic.

Consequently, food and basic needs products prices are touching the height of the sky. According to the International Monetary Fund statics, in 2020, the worst global economic loos or experience negative per capita GDP growth since the 1930s because of the COVID-19 pandemic in 170 nations around the world (Jackson 2021). India has a vast economic sector in business, tourism, agriculture, and other government sectors (Fig. 12.7). Due to the pandemic, about 50 million employees lost their jobs. The lack of working staff in each sector experienced a negative effect on their production and growth, consequently a decrease in gross economic growth worldwide. Another report reveals a 25% fall in the world tourism sector, and about 12–14% of jobs are put at risk due to lockdown to break the transmission of COVID-19 infection (Anwari et al., 2021). The impact of this pandemic is almost identical in all the world's largest economies. For instance, China, Germany, the US, France, Japan, Italy, and Britain are listed in this outbreak's top ten utmost affected economies. The economic decline indirectly influences the health commodities and supplies of basic medicals and food appliances. Another factor is an improper contribution or mismanagement of supplies to halt the pandemic. Shortage of medical facilities such as PPE kits, oxygen cylinders, and other basic treatment supplies for this epidemic reveals the significant loss of health. However, many health issues were observed in the person after the COVID-19 recovery, called post-COVID-19 impact. Several

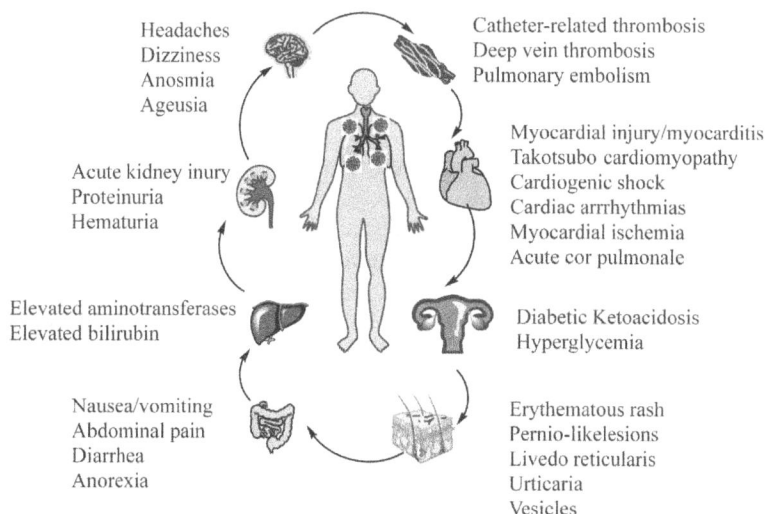

Headaches
Dizziness
Anosmia
Ageusia

Catheter-related thrombosis
Deep vein thrombosis
Pulmonary embolism

Acute kidney inury
Proteinuria
Hematuria

Myocardial injury/myocarditis
Takotsubo cardiomyopathy
Cardiogenic shock
Cardiac arrrhythmias
Myocardial ischemia
Acute cor pulmonale

Elevated aminotransferases
Elevated bilirubin

Diabetic Ketoacidosis
Hyperglycemia

Nausea/vomiting
Abdominal pain
Diarrhea
Anorexia

Erythematous rash
Pernio-likelesions
Livedo reticularis
Urticaria
Vesicles

Figure 12.6 Sever effect of COVID-19 on the human body.

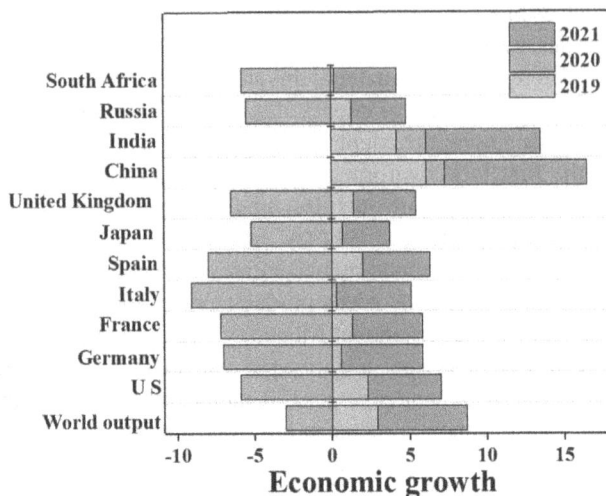

Figure 12.7 Latest world economic outlook growth projections (Real GDP, Annual percent changes).

case studies reported the heart and skin problems observed in many people who once suffered from COVID-19. Briefly, PPE can directly affect organisms via, for instance, ingestion, entanglement, and nest materials in the case of birds; or can indirectly affect organisms through the release of hazardous contaminants, e.g., metals, plasticizers, surfactants (Silva et al., 2020). Several species such as seagulls (Larus sp.), peregrine falcons (Falco peregrinus), crows (Corvus corone), white storks (Ciconia ciconia), foxes (Vulpes vulpes), cats (Felis catus), and dogs (Canis lupus familiaris) seemed to interact with PPE macro-litter (Hiemstra et al., 2020). Some interactions result in chronic effects, such as restricting feeding to the point of starvation, facilitating predation, exhausting the animal, causing suffocation, infections, severe wounds, and amputations (Silva et al., 2020). In aggravated cases, it can even result in an organism's death. PPE waste in the bird's nest structure can also alter thermal and drainage properties, influencing reproductive success Hiemstra, In addition, as polymeric and lipophilic materials, PPE litter can adsorb environmental contaminants and pathogens, including SARS-type (Hiemstra et al., 2020; Workie et al., 2020). Thus, frequent ingestion of PPE litter can decrease organisms' health due to physical effects and potential chemical body burdens, as observed in Seagulls (Vella et al., 2020; Vanapalli et al., 2021). Although the fraction of hazardous chemicals (e.g., hydrophobic organics) sorbed by plastic debris seems to be low compared to natural particles, particularly in aquatic environments (Endo and Koelmans, 2016); their role as vectors increase when their higher mobility is considered, contributing to long-range environmental transport and affect remote locations (Workie et al., 2020). Thus, Critical studies are obligatory to maintain current supply chains for basic needs (Food and medical) by all types of healthcare commodities, government, and non-government organizations.

4. Plastic Waste Generation (Food, packing, and medical waste)

There is a tremendous amount of waste generation, especially plastic-related waste, due to COVID-19 infection worldwide. This pandemic reverses the ban policy for single-use plastic as there is a sudden increase in plastic-related waste in the form of food packing and biomedical equipment. PPE, such as syringes, masks, respirators, and gloves, is in high demand. Due to the lockdown and fear of consignations of COVID-19 infection, food packing, and online shopping are sources of plastic waste (Benson et al., 2021). Presently, face masks are not endorsed for reuse due to the high risk of virus infection. As a result, the production and waste of PPE touched the crisis level (Torres et al., 2021). To avoid infection and for safety purposes, individuals have to wear a face mask that further inclines the percentage of waste. For safety issues, cafes and restaurants evaded reusable or personal containers as an alternative to using single-use packaging supplies.

Meanwhile, the lockdown has pulled the packing application tremendously because of work from home or restricted travel, increased online shopping, and higher food consumption which contributed to household waste extensively (Sharma et al., 2020). The data retrieved from the Mobile app for food delivery revealed a 149 % and 187% increase in delivery by 46,000 users from January to December and December and March 2020, respectively, in Brazil (Rappi, Food, and Uber Eats (de Albuquerque et al., 2021). According to the WHO estimates, approximately 89 million face masks were demanded to respond to COVID-19 per month. Some reports suggested the inclination to gloves and face mask demands will increase to 0.5 billion and 1 billion, respectively, during phase 2, when social gathering is restricted by many countries (Burnett and Sergi, 2020). Numerous administrations such as Operation Thames (London), Oceans Asia (Hong Kong), and Mer Propre (France) have articulated concern about the chaotic disposal of face masks and PPE that ended up in water sources commanded to plastic waste pollution. Biomedical waste (BMW) is generated from hospitals and quarantine areas, for instance, dining boxes, food, bags, and infusion bottles used by COVID-19 patients and generally disposed of by nurses for recycling (Das et al., 2021). The various report investigates the regular disposal of face masks, and only 24% is recycled. Other is left as landfill, which poses a very dreadful impact on the environment. About 10 million masks are dumped every month. According to World Wide Fund for Nature (WWF), one mask weighing 4 g and only 1% of improper disposal of masks result in 40,000 kg of plastic into the environment, which is a significant concern (Wu et al., 2020; Dadhich, 2021). India is manufacturing approximately 550 tons of biomedical waste (BMW) annually, of which only 198 tons are treated as Common Bio-Medical Waste Treatment Facilities (CBMWTFs), 225 tons as captive incinerators, and other is untreated left (Thind et al., 2021). During this pandemic, Wuhan (China) generated 240 tons of additional biomedical waste each day (Wu et al., 2020; Di Maria et al., 2020). This unprecedented demand for face masks increases the production and plastic waste on the Earth. For instance, in February 2020, the daily production of face masks in China increased by 14.8 million (Al-Omran et al., 2021). According to the Japanese Ministry of Economy, trade, and Industry (METI), about 600 million order of face masks was reported monthly in 2020 (METI, 2020). The number of Covid-19 infected cases rises across the world to 226 million, with over 4 million deaths as of September 15, 2021 (Worldometer, 2021). According to WHO 2020, 500 million testing for COVID-19 confirmation generates more than 15,000 tons of scum waste until August 2020. Most diagnostic tests are manufactured with polypropylene as raw material and incinerated after use to avoid the risk of infection. However, they release poisonous chemicals during the process, carriages air and water pollution (Celis et al., 2021).

Additionally, this waste can be a candidate to spread COVID-19 infection if not properly managed (Al-Omran et al., 2021). On average, 3–4 kg of medical waste was dumped by COVID-19 patients in each hospital during this pandemic in China (ADB 2020). According to the literature, 30–50% of medical waste increased in China, France, and the Netherlands, which is around six times higher than before this outbreak. Only China produced 45 to 247 tons of biomedical waste during this pandemic (Parashar and Hait, 2020). This outbreak increased the production, consumption, and waste generation of PPE and plastic packing globally due to plastic waste disposal problems suffered in developing and developed countries (Wang et al., 2019).

5. Environment Concern about Plastic Waste

The environment was scratched day by day due to continuously added plastic waste during this outbreak. The mismanagement, lack of public awareness, and government policy led to the generation of the waste landfill by a large number of populations globally (Wang et al., 2019). Medical equipment such as face masks, gloves, PPE kits, and other products are frequently used for the sake of safety against virus infection (Kampf, 2020). However, this medical equipment's made of polymeric materials and releases toxic chemicals improperly disposed of into the environment (Thind et al., 2021). A report by Oceans Asia, an association of research and advisory committee on marine pollution, shows the occurrence of various types of face masks with different colors in an ocean. Correspondingly, face masks collected from the sideways of drainage and highway in various regions show the landfill or improper disposal of waste. This pandemic has increased the challenge of plastic pollution and the advent of PPEs kits as environmental litter for aquatic and terrestrial bodies (Aragaw, 2020). For instance, drinking bottles, fast food containers, and plastic packaging materials are the foremost causes of microplastic pollution in the ocean and land (Fadare and Okoffo 2020). The improper management of this plastic waste leads to either landfill or directly disposed into water bodies, later by stormwater (Thind et al., 2021), wind, and drainage systems; these residues enter into rivers or oceans as discussed in Fig. 12.8.

Consequently, they can fragment /degrade or break down into smaller size/ pieces of particles under 5 mm, known as microplastics under environmental conditions. The various investigation concluded that plastic particles indigested by aquatic animals that constitute a foremost part of the food chain and are meant for

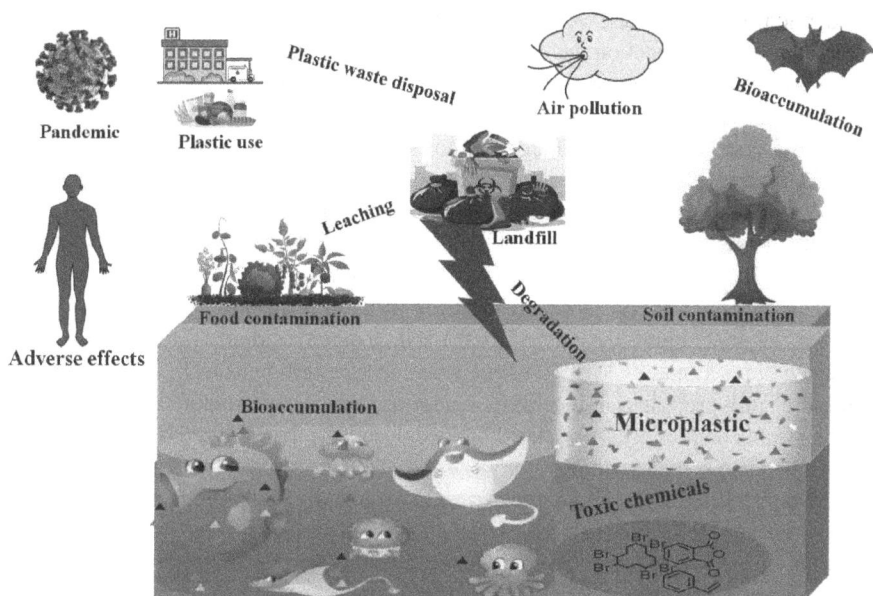

Figure 12.8 Environment concern about microplastic generated by plastic waste during Covid-19.

human consumption threaten the environment and raise a concern about global food safety (Fadare and Okoffo, 2020). The presence of plastic particles in the atmosphere consequences climate change and disturbs the ecosystem due to the emission of harmful gases and carbon, which has a greater risk to the global food chain. Another insinuation of extensively disposed face masks in the open land led to the candidate of spreading virus infection, as plastic particles are recognized to propagate invasive pathogens or microbes (Jamieson et al., 2020). The two river outlets in Jakarta Bay, Indonesia, contain 5% more plastic waste than in the last five years (Cordova et al., 2021). The collected plastic waste from rivers contains 16% weight of various PPEs used during this outbreak. In water bodies, seabirds and animals suffer adverse effects or even die because of plastic waste ingestion by believing it is food (Workie et al., 2020). Municipal solid waste (MSW) used high solids anaerobic digestion (HSAD) to recycle face masks and other PPE equipment in their waste management facilities. The observations conclude that HSAD of plastic waste produced a considerable amount of methane, severely affecting biota (de Albuquerque et al., 2021). The production of maximum and total cumulative methane was reduced up to 12–29% and 18%, compared to the control digester (without face masks), which reveals the additional load of methane on the environment during face mask recycling.

Furthermore, the face masks may have fungicidal, bactericidal, and antiviral properties (traces of Cu and Ag nanoparticles) to combat pathogens, leading to emerging contaminants by these nanoparticles. However, complete degradation of plastic waste is not possible; recycling by incineration or other processes often leaves residues of micro and nano-sized plastic particles, causing a dreadful effect on both ecosystems and the atmosphere. Furthermore, incineration emits poisonous gases (Celis et al., 2021). The composition of biomedical waste during the COVID-19 pandemic produces a massive extent of plastics/microplastics and causes a direct or indirect negative impact on the ecosystem (Das et al., 2021).

There is scarce data about the potential degradation and toxicological effect of PPE kits. Buried (e.g., landfilled) or in water compartments, littered PPE will undergo fragmentation and biodegradation due to physicochemical and biochemical processes while releasing a myriad of micro and nano plastics and leachable hazardous chemicals such as plasticizers, metals, organophosphate esters, as recently reported in laboratory conditions (Wang et al., 2020; Ammendolia and Walker 2021; Auta et al., 2017). The only available data suggests that disposable face masks (as individual layers or as a blend) can easily decompose in natural topsoils (Celis et al., 2021) (75% of the water holding capacity, 25°C), with a mean residence time of 2 to 3 days, releasing approximately 3 to 5% of the total masque carbon as CO_2. In addition, the release of polypropylene microfibres resulting from the mechanical fragmentation of disposable face masks (1 g/kg soil) decreased the reproduction and growth of springtails (Folsomia candida) by 48% and 92%, respectively (Kwak, 2021). In Earthworms (Eisenia andrei), acute exposure to such microfibres decreased esterases activity by 62% (enzymes actively involved in the resistance of several contaminants such as insecticides and spermatogenesis (vital for earthworms reproduction) (Chowdhury et al., 2020). Therefore, COVID-19 triggered a load of plastic waste from production to waste management, persistence in the environment, and adverse effects on different biological systems are the major threats to environmental compartments on a short- and long-term basis.

6. Challenges in Plastic Waste Management During the COVID-19 Pandemic

During this pandemic, all the countries suffer from recycling facilities due to the absence of employees at work (risk of contamination), lack of management, and government policies. To the literature, low-income nations (World Bank classification) recycle only 20% of total waste. Another is dumped openly over the land is a hygienic point of view; infectious PPE waste is incinerated at high temperatures, releasing toxic gases (Chowdhury et al., 2020; Auta et al., 2017). In contrast, high-income countries recycle about 51% of total waste and have better management facilities. A survey on beverage or plastic container recycling reveals that there is 45% less than in the past year in the USA (Ammendolia and Walker 2021). For sustainability, recycling plastic is a must, but some researchers argued that the emission of global CO_2 and global waste during recycling harm the environment (Kareem et al., 2021). Furthermore, developing lands face a higher risk of infection through waste than developed countries with more protected, and greener approaches to waste management. Because of poor management in developing countries, the waste separation step is mainly avoided, and waste is directly disposed of without any safety gear by healthcare workers or rag pickers (Vanapalli et al., 2021). This hazardous waste landfill further spreads the virus and complicates management tracing (Fig. 12.9). According to WHO classification, medical waste is categorized as pathological waste such as body parts and human tissues; metallic waste consists

Covid -19

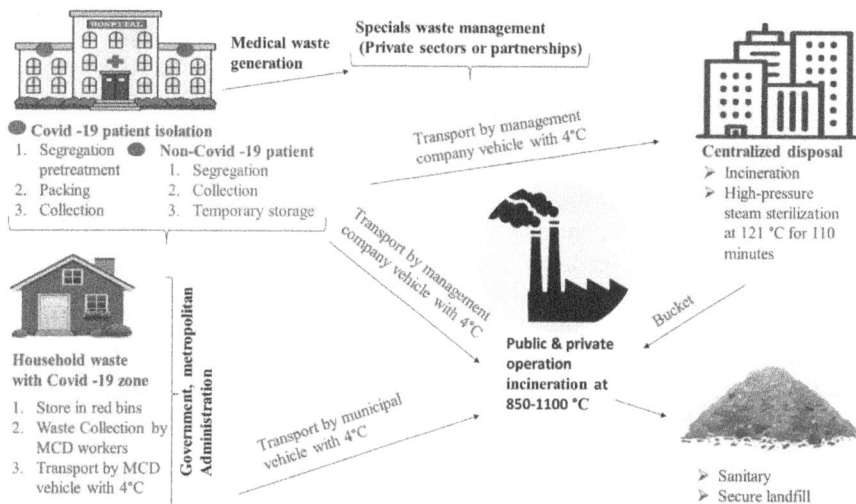

Figure 12.9 Scheme of biomedical and domestic waste management during Covid-19 outbreak.

of needles and surgical blades; chemical waste includes sterilants, disinfectants, and mercury; pharmaceutical waste includes contaminated and expired drugs, and plastic waste consists PPEs (WHO, 2020). Most countries recycle about 10–15% of plastic waste compared to last year as only one-fourth of recycling units are working due to a lack of workers and mismanagement of facilities. For an instant, the Netherlands suffers from the demand for recycling products related to plastic and textile due to this outbreak (Ibn-Mohammed et al., 2020). International Solid Waste Association (ISWA) has enclosed three objectives to handle plastic waste throughout the COVID-19 outbreak: (1) The waste management applies should not be negotiated any place across the globe, and the safety of health workers not suffer; they must be well fortified with protective equipment's or gear (2) The waste management or recycling action requirements to be repeated to cross-contamination or transmission of infection. (3) infectious or biomedical waste should be safely disposed of, and it does not create infection or secondary pollutants (Vella et al., 2020). Waste generators (infectious patients or citizens) and waste management staff should be aware of basic guidelines (temporary) for the disposal of COVID-19 infectious waste. For safe and easy recycling, municipal waste service providers are advised to use mechanical-manual handling systems instead of manual handling. For Citizens or quarantined person waste, it is advised to keep waste in different bags to avoid combinations (Silva et al., 2021; Silva et al., 2020). Infectious waste is the first disinfectant disposed of in a red bin or can be stowed in paper bags compared to plastic bags as the residence time of virus on plastic is more than on paper.

7. Global Policies for Plastic Production and Utilization

In the early 2000s, the initiative was taken by the United Nations' war on plastic' of plastic pollution, specifically in the field of the ocean. Plastic pollution is now gaining momentum as it is involved in every government's agenda worldwide, as pandemics generate unexpected plastic waste (Karasik et al., 2020). This movement is also sped up by public awareness and engagement of NGOs throughout the globe. One of the reasons for the suspension of plastic waste recycling or management is the government's lack of proper policies and their implementation on production companies and users for their (Pedra and Gonçalves, 2020). There is a sudden pause in implementing single-use plastic policies or taxes because of this pandemic (ARPBA, 2019). By taking benefit of the pandemic, these groups of plastic producers emphasized the single use of plastic to combat this global health crisis in their address to the U.S. (Mekonnen and Aragaw, 2021; Kulkarni and Anantharama, 2020). Department of Health and Human Services (HSS) by the Plastics Industry Association (PIA). Since 2000, nearly 28 global policies have been documented to address plastic pollution globally. Some of them are discussed here:

(1) 2010: Decision Adopted by the Conference of the Parties to the Convention on Biological Diversity at its Tenth Meeting (UNEP/CBD/COP/DEC/X/2) "The Strategic Plan for Biodiversity 2011–2020 and the Aichi Biodiversity Targets".

(2) 2011: The Honolulu Strategy—A Global Framework for Prevention and Management of Marine Debris.

(3) 2011: Resolution MEPC. 201(62) Amendments to the Annex of the Protocol of 1978 Relating to the International Convention for the Prevention of Pollution from Ships, 1973 "Revised MARPOL Annex V."

(4) 2012: UNGA (United Nations General Assembly) Resolution A/Res/66/288 "The Future We Want."

(5) 2014: UNEP/CMS/Resolution 11.30 Eleventh Meeting of the Conference of the Parties to the Convention on Migratory Species—Management of Marine Debris.

(6) 2014: UNEA (United Nations Environmental Assembly)/Resolution 1/6 "Marine Plastic Debris and Microplastics."

(7) 2015: UNGA Resolution A/Res/70/1 "Transforming Our World: The 2030 Agenda for Sustainable Development".

(8) 2016: CBD/COP/DEC/XIII/10 "Decision Adopted by the Conference of the Parties to the Convention on Biological Diversity—Addressing Impacts of Marine Debris and Anthropogenic Underwater Noise on Marine and Coastal Biodiversity." (9) 2016: UNEA Resolution 2/11 "Marine Plastic Litter and Microplastics" (10)2017: Thirteenth meeting of the Conference of the Parties to the Basel Convention—BC-13/11: Technical assistance; Work Programme 2018–2019.

(11) 2018: UNEA Resolution 3/7 "Marine Litter and Microplastics."

(12) 2019: UNEA Resolution 4/6 "Marine Plastic Litter and Microplastics."

(13) 2019: UNEA Resolution 4/9 "Addressing Single-Use Plastic Products Pollution."
(14) 2019: BC-14/13 Fourteenth Meeting of the Conference of the Parties to the Basel Convention—Further actions to address plastic waste under the Basel Convention.

8. Conclusions and Future Perspective

According to the current scenario, the pandemic is expected to surmount beyond 2025; thus, generating a long-term plan for plastic waste management is necessary. Because of the COVID-19 pandemic, used face masks, PPE kits, and other biomedical items waste tremendously litter the environment, which is a source of plastic pollution. Instead of single-use masks, reusable ones can be sterilized and used. To enhance the recycling facilities and ensure the safety of healthcare workers or waste management, staff should be trained to manage infection. This pandemic shows both health and economic crises and needs a proper policy or regulation for plastic waste management for producers and consumers. Additionally, there is an urgent need for an eco-friendly replacement of waste management authorities to strengthen and sustain the solution of plastic waste pollution for environmental protection. Furthermore, awareness of COVID-19 infectious waste, especially plastic amounts to citizens, healthcare workers, and municipal staff, makes waste management easy and safer. Who knows? Plastic pollution may be the next world pandemic.

Acknowledgment

One author, Dr. Manviri Rani, is grateful to DST-SERB, New Delhi (Sanction order no. SRG/2019/000114) and TEQIP-III, MNIT Jaipur, India, for the financial assistance. Dr. Uma Shanker wishes to thank TEQIP-III, NIT Jalandhar, for his financial support. Ms. Meenu is thankful to the Ministry of Human Resources and Development (MHRD) in New Delhi for a research fellowship.

References

Abbasi, 2020. J. Jama 323: 1881–1883.
Abrol, A., Abrol, S., Mahajan, S., Abrol, R.K. and Banga, G. 2020. Journal homepage: http://www. ijcmas. com 9.
Alanagreh, L.A., Alzoughool, F. and Atoum, M. 2020. Pathogens 9: 331.
Al-Omran, K., Khan, E., Ali, N. and Bilal, M. 2021. Sci. Total Environ. 149642.
Ammendolia, J. and Walker, T.R. 2021. Sci. Total Environ. 149957.
Anwari, N., Ahmed, M.T., Islam, M.R., Hadiuzzaman, M. and Amin, S. 2021. Transp. Res. Interdiscip. Perspect. 9: 100334.
Aragaw, T.A. 2020. Mar. Pollut. Bull. 159: 111517.
Auta, H.S., Emenike, C.U. and Fauziah, S.H. 2017. Environ. Int. 102: 165–176.
Axell-House, D.B., Lavingia, R., Rafferty, M., Clark, E., Amirian, E.S. and Chiao, E.Y. 2020. J. Infect. 81: 681–697.
Bahukhandi, K., Agarwal, S. and Singhal, S. 2020. Int. J. Environ. Anal. Chem. 1–15.
Bastos, M.L., Tavaziva, G., Abidi, S.K., Campbell, J.R., Haraoui, L.P., Johnston, J.C. et al. 2020. bmj 370.
Benson, N.U., Bassey, D.E. and Palanisami, T. 2021. Heliyon 7: e06343.

Bilal, H., Bilquees, M.S., Tabassum, N. and Akhter, Y. 2020. Epidemiol. Infect. 246.

Burnett, M.L. and Sergi, C.M. 2020. Disaster Med. Public Health Prep. 14: e47–e50.

Cai, J., Sun, W., Huang, J., Gamber, M., Wu, J. and He, G. 2020. Emerg. Infect. Dis. 26: 1343.

Carpenter, C.R., Mudd, P.A., West, C.P., Wilber, E. and Wilber, S.T. 2020. Acad. Emerg. Med. 27: 653–670.

Celis, J.E., Espejo, W., Paredes-Osses, E., Contreras, S.A., Chiang, G. and Bahamonde, P. 2021. Sci. Total Environ. 760: 144167.

Chen, W. 2020. Hum Vaccin Immunother 16: 2604–2608.

Cheng, V.C., Wong, S.C., Chen, J.H., Yip, C.C., Chuang, V.W., Tsang, O.T. et al. 2020. Control. Hosp. Epidemiol. 41: 493–498.

Chirumbolo, S. 2020. J. Med. Virol.

Chowdhury, G.W., Koldewey, H.J., Duncan, E., Napper, I.E., Niloy, M.N.H., Nelms, S.E. et al. 2020. Sci. Total Environ. 143285.

Chung, J.Y., Thone, M.N. and Kwon, Y.J. 2021. Adv. Drug Deliv. Rev. 170: 1–25.

Conceicao, C., Thakur, N., Human, S., Kelly, J.T., Logan, L., Bialy, D. et al. 2020. PLoS Biology 18: 3001016.

Conly, J., Seto, W.H., Pittet, D., Holmes, A., Chu, M. and Hunter, P.R. 2020. Antimicrob. Resist. Infect. Control. 9: 1–7.

Cordova, M.R., Nurhati, I.S., Riani, E. and Iswari, M.Y. 2021. Chemosphere 268: 129360.

Cui, F. and Zhou, H.S. 2021. Biosens. Bioelectron. 165: 112349.

da Costa, J.P. 2021. Sci. Total Environ. 145806.

Dadhich, H. 2021. Jo ARB 8: 21–29.

Damodharan, K., Arumugam, G.S., Ganesan, S., Doble, M. and Thennarasu, S. 2021. RSC Adv. 11: 20006–20035.

Das, A.K., Islam, N., Billah, M. and Sarker, A. 2021. Sci. Total Environ. 146220.

de Albuquerque, F.P., Dhadwal, M., Dastyar, W., Azizi, S.M.M., Karidio, I., Zaman, H. et al. 2021. CSCEE 3: 100082.

De-la-Torre, G.E. and Aragaw, T.A. 2021. Mar. Pollut. Bull. 163: 111879.

Dharmaraj, S., Ashokkumar, V., Pandiyan, R., Munawaroh, H.S.H., Chew, K.W., Chen, W.H. et al. 2021. Chemosphere 275: 130092.

Di Maria, F., Beccaloni, E., Bonadonna, L., Cini, C., Confalonieri, E., La Rosa, G. et al. 2020. Sci. Total Environ. 743: 140803.

Ding, S. and Liang, T.J. 2020. Gastroenterology 159: 53–61.

Drury, J., Carter, H., Ntontis, E. and Guven, S.T. 2021. BJ Psych open 7.

Ejima, K., Kim, K.S., Ludema, C., Bento, A.I., Iwanami, S., Fujita, Y. et al. 2021. Epidemics 35: 100454.

Endo, S. and Koelmans, A.A. 2016. Springer, Cham. 185–204.

Fadare, O.O. and Okoffo, E.D. 2020. Sci. Total Environ. 737: 140279.

Gao, Z., Xu, Y., Sun, C., Wang, X., Guo, Y., Qiu, S. et al. 2021. J. Microbiol. Immunol. Infect. 54: 12–16.

Gilby, B.L., Henderson, C.J., Olds, A.D., Ballantyne, J.A., Bingham, E.L., Elliott, B.B. et al. 2021. Biol. Conserv. 253: 108926.

Gitman, M.R., Shaban, M.V., Paniz-Mondolfi, A.E. and Sordillo, E.M. 2021. Diagnostics 11: 1270.

Gong, Y., Guan, L., Jin, Z., Chen, S., Xiang, G. and Gao, B. 2020. J. Med. Virol. 92: 2551–2555.

Gorrasi, G., Sorrentino, A. and Lichtfouse, E. 2021. Environ. Chem. Lett. 19: 1–4.

Gryseels, S., De Bruyn, L., Gyselings, R., Calvignac-Spencer, S., Leendertz, F.H. and Leirs, H. (2021) Mammal Rev. 51, 272–292.

Han, X., Xu, Y., Fan, L., Huang, Y., Xu, M. and Gao, S. 2021. PNAS 118.

Haque, M.S., Uddin, S., Sayem, S.M. and Mohib, K.M. 2020. J. Environ. Chem. Eng. 104660.

Hiemstra, A.F., Rambonnet, L., Gravendeel, B. and Schilthuizen, M. 2021. Anim. Biol. 71: 215–231.

Holshue, M.L., DeBolt, C., Lindquist, S., Lofy, K.H., Wiesman, J., Bruce, H. et al. 2021. Engl. J. Med. 382: 929–936.

Hu, D., Shen, M., Zhang, Y., Li, H. and Zeng, G. 2019. ESPR 26: 19997–20002.

Ibn-Mohammed, T., Mustapha, K.B., Godsell, J.M., Adamu, Z., Babatunde, K.A., Akintade, D.D. et al. 2020. Resour. Conserv. Recycl. 105169.

Jackson, J.K. 2021. CRS.

Jamieson, A.J., Brooks, L.S.R., Reid, W.D., Piertney, S.B., Narayanaswamy, B.E. and Linley, T.D.R. 2019. Soc. Open Sci. 6: 180667.

Jindal, M.K. and Sar, S.K. 2020. Int. J. Environ. Anal. Chem. 1–12.

Kampf, G. 2020. IPIP 2, 100044.

Karam, M., Althuwaikh, S., Alazemi, M., Abul, A., Hayre, A., Alsaif, A. et al. 2021. JRSM Open 12: 20542704211011837.

Karasik, R., Vegh, T., Diana, Z., Bering, J., Caldas, J., Pickle, A. et al. 2020. NI X. 20: 20–105.

Kareem, K.Y., Adelodun, B., Tiamiyu, A.O., Ajibade, F.O., Ibrahim, R.G., Odey, G. et al. 2021. Detection and Analysis of SARS Coronavirus: Advanced Biosensors for Pandemic Viruses and Related Pathogens 219–242.

Karia, R., Gupta, I., Khandait, H., Yadav, A. and Yadav, A.S.N. 2020. Compr. Clin. Med. 1–4.

Kasraeian, M., Zare, M., Vafaei, H., Asadi, N., Faraji, A., Bazrafshan, K. et al. 2020. J Matern. Fetal. Neonatal. Med. 1–8.

Kaur, S.P. and Gupta, V. 2020. 198114.

Kulkarni, B.N. and Anantharama, V. 2020. Sci. Total Environ. 743: 140693.

Kumar, A., Jain, V., Deovanshi, A., Lepcha, A., Das, C., Bauddh, K. et al. 2021. Environ. Sustain. 1–8.

Kumar, V., Singh, S.B. and Singh, S.J. 2020. Environ. Chem. Eng. 8: 104144.

Kwak, J.I. and An, Y.J. 2021. J. Hazard. Mater. 416: 126169.

Li, M., Yang, Y., Liu, Y., Zheng, M., Sun, D., Li, H. et al. 2021. Asian J. Tradit. Med. 16: 32–52.

Liu, C., Zhou, Q., Li, Y., Garner, L.V., Watkins, S.P., Carter, L.J. et al. 2020. ACS Cent. Sci. 6: 315–331.

Lu, R., Zhao, X., Li, J., Niu, P., Yang, B., Wu, H. et al. 2020. The Lancet. 395: 565–574.

Mekonnen, B.A. and Aragaw, T.A. 2021. COVID-19 117–140.

Nabi, G. and Khan, S. 2021. Environ. Res. 188: 109732.

Oishee, M.J., Ali, T., Jahan, N., Khandker, S.S., Haq, M.A., Khondoker, M.U. et al. 2021. Infect. Drug. Resist. 14: 1049.

Oran, D.P. and Topol, E.J. 2020. Ann. Intern. Med. 173: 362–367.

Padgelwar, S., Nandan, A. and Mishra, A.K. 2021. J. Environ. Anal. Chem. 101: 1894–1906.

Paital, B., Das, K. and Parida, S.K. 2020. Sci. Total Environ. 728: 138914.

Parashar, N. and Hait, S. 2020. Sci. Total Environ. 144274.

Pedra, A.S.A. and Gonçalves, L.C.S. 2020. Rev. Int. Sociol. 11.

Peng, X., Xu, X., Li, Y., Cheng, L., Zhou, X. and Ren, B. 2020. Int. J. Oral Sci. 12: 1–6.

Rafey, A. and Siddiqui, F.Z. 2021. J. Environ. Anal. Chem. 1–17.

Rana, S.S. 2020. J. Dig. Endosc. 11: 27–30.

Rather, R.A., Islam, T., Rehman, I.U. and Pandey, D. 2020. Asian J. Med. Sci. 3–21.

Reginelli, A., Grassi, R., Feragalli, B., Belfiore, M.P., Montanelli, A., Patelli, G. et al. 2021. Biology 10: 89.

Ren, Y., Feng, C., Rasubala, L., Malmstrom, H. and Eliav, E.J. 2020. Dent. 101: 103434.

Russo, A., Minichini, C., Starace, M., Astorri, R., Calò, F. and Coppola, N. 2020. Infect. Drug Resist. 13: 2657.

Sahin, A.R., Erdogan, A., Agaoglu, P.M., Dineri, Y., Cakirci, A.Y., Senel, M.E. et al. 2020. EJMO 4: 1–7

Selvaranjan, K., Navaratnam, S., Rajeev, P. and Ravintherakumaran, N. 2021. Environ. Challenge. 100039.

Shammi, M., Bodrud-Doza, M., Islam, A.R.M.T. and Rahman, M.M. 2020. Heliyon 6: e04063.

Shammus, R., Mahbub, S., Rauf, M.A. and Harky, A. 2020. Acta Bio Medica: Atenei Parmensis 91: e2020019.

Sharma, H.B., Vanapalli, K.R., Cheela, V.S., Ranjan, V.P., Jaglan, A.K., Dubey, B. et al. 2020. Resour. Conserv. Recycl. 162: 105052.

Shi, Y., Wang, G., Cai, X.P., Deng, J.W., Zheng, L., Zhu, H.H. et al. 2020. Zhejiang Univ. Sci. B 21: 343–360.

Silva, A.L.P., Prata, J.C., Walker, T.R., Campos, D., Duarte, A.C., Soares, A.M et al. 2020. Sci. Total Environ. 742: 140565.

Silva, A.L.P., Prata, J.C., Walker, T.R., Duarte, A.C., Ouyang, W., Barceló, D. et al. 2021. Chem. Eng. Sci. 405: 126683.

Thind, P.S., Sareen, A., Singh, D.D., Singh, S. and John, S. 2021. Environ. Pollut. 276: 116621.

Torres, F.G. and De-la-Torre, G.E. 2021. Sci. Total Environ. 147628.

Tripathi, A., Tyagi, V.K., Vivekanand, V., Bose, P. and Suthar, S. 2020. CSCEE 2, 100060.

United Nations Department of Economic and Social Affairs (UNDESA), 2020.https://www.un.org/development/desa/dpad/wpcontent/uploads/sites/45/publication/WESP2020_MYU_Report.pdf.

Van Walle, I., Leitmeyer, K. and Broberg, E.K. 2021. Eurosurveillance 26: 2001675.

Vanapalli, K.R., Sharma, H.B., Ranjan, V.P., Samal, B., Bhattacharya, J., Dubey, B.K. et al. 2021. Sci. Total Environ. 750: 141514.

Vella, F., Senia, P., Ceccarelli, M., Vitale, E., Maltezou, H., Taibi, R. et al. 2020. Health 1.

Vella, F., Senia, P., Ceccarelli, M., Vitale, E., Maltezou, H., Taibi, R. et al. 2020. Health 1.

Wan, Y., Yan, S., Zhang, Y., An, S., Yang, K., Xu, H. et al. 2020. Preprints 2020020289.

Wang, L., Wang, Y., Ye, D. and Liu, Q. 2021. Int. J. Antimicrob. Agents. 55: 105948.

Wang, M.H., He, Y. and Sen, B. 2019. Environ. Pollut. 248: 898–905.

Wang, Y., Kang, H., Liu, X. and Tong, Z. 2020. J Med Virol. 92: 1401–1403.

Wei, Y., Wei, L., Liu, Y., Huang, L., Shen, S., Zhang, R. et al. 2021. Infection 1–11.

Workie, E., Mackolil, J., Nyika, J. and Ramadas, S. (2020) Current Research in Environ. Sustain. 100014

Wu, H.L., Huang, J., Zhang, C.J., He, Z. and Ming, W.K.E 2020. Clinical Medicine 21: 100329.

Wu, Y., Xu, X., Yang, L., Liu, C. and Yang, C. 2020. Brain Behav. Immun. 87: 55.

Wu, Y.C., Chen, C.S. and Chan, Y.J. 2020. J. Chin. Med. Assoc. 83: 217.

Xiao, Y., Huang, S., Yan, L., Wang, H., Wang, F., Zhou, T. et al. 2020. Int. Emerg. Nurs. 52: 100912.

Yadav, T. and Saxena, S.K. 2020. Coronavirus Disease 2019 (COVID-19) 33.

Yakoh, A., Pimpitak, U., Rengpipat, S., Hirankarn, N., Chailapakul, O. and Chaiyo, S. 2021. Biosens. Bioelectron. 176: 112912.

Yang, H., Wang, C. and Poon, L.C. 2020. UOG 55: 435–437.

Yao, X., Ye, F., Zhang, M., Cui, C., Huang, B., Niu, P. et al. 2020. Clin. Infect. Dis. 71: 732–739.

Zhou, R., Li, F., Chen, F., Liu, H., Zheng, J., Lei, C. et al. 2020. IJID 96: 288–290.

Zimmerman, P.A., King, C.L., Ghannoum, M., Bonomo, R.A. and Procop, G.W. 2021. Pathog. Immun. 6: 135.

Index

About the Editors

Dr. Uma Shanker

Dr. Uma Shanker is an Associate Professor in the Department of Chemistry, Dr. B R Ambedkar National Institute of Technology Jalandhar, Punjab, India. Dr. Shanker has twelve years of experience in teaching and research in green nanomaterials fabrication and their applications. He has published 80 research and review papers in peer-reviewed international journals with > 3847 citations, h-index 35, and i-10 index of 61 as per Google Scholar. Dr. Shanker has edited several scientific monographs and handbooks published with Springer (01 major reference work), Elsevier (02 books), and CRC Press (01 book). To date, he has contributed to over 55 book chapters in books of international publishers and has delivered over 20 invited talks in India and abroad. Dr. Shanker has been featured among the top 2% of scientists around the globe as per the report of Stanford University, USA and Elsevier consecutively in 2022 and 2023. Dr. Shanker received the best teacher award in 2021 by the Chairman BOG, Dr. B R Ambedkar NIT Jalandhar, India. He has also received a highly cited author award from the Royal Society of Chemistry in 2017. Dr. Shanker has completed many R&D projects from DST, New Delhi and World bank. His teaching interests are Nanotechnology, Green Chemistry, Materials Chemistry, and Instrumental Methods of Analysis.

Dr. Manviri Rani

Dr. Manviri Rani (Born in Modinagar, India) is an Assistant Professor at Department of Chemistry, Malaviya National Institute of Technology Jaipur, Rajasthan, India. Dr. Rani obtained her PhD degree in Analytical Chemistry from Department of Chemistry, Indian Institute of Technology, Roorkee, Uttarakhand, India. She was a post-doctoral scientist from August 2012 to February 2015 at Korea Institute of Ocean Science and Technology (KIOST), South Korea. Dr. Rani has published several research/review papers (> 80) in journals of international repute with more than 4596 citations and h-index 35; i-10: 61 as per Google Scholar. She has presented scientific papers in several conferences (> 50) at international and national level. Dr. Rani has edited many books of Elsevier, Springer and CRC Press, Taylor & Francis publisher and also contributed > 55 chapters in the books of international publishers. Her research interests include, Green and sustainable nanotechnology, environmental management, environmental analysis, analytical method development for persistent organic pollutants, analytical chemistry, environmental remediation, application of nanotechnology and advanced materials for various industries. Dr. Manviri Rani has been featured twice among top 2% (2022 and 2023) of the scientists around the globe as per the report of Stanford University, USA and Elsevier. She has completed many R&D projects from DST, New Delhi and World bank.

For Product Safety Concerns and Information please contact our EU
representative GPSR@taylorandfrancis.com
Taylor & Francis Verlag GmbH, Kaufingerstraße 24, 80331 München, Germany

www.ingramcontent.com/pod-product-compliance
Lightning Source LLC
Chambersburg PA
CBHW060331220326
41598CB00023B/2674

9 7 8 1 0 3 2 3 7 1 8 8 7